International Perspectives on Gender and Mathematics Education

A volume in
International Perspectives on Mathematics Education—Cognition, Equity, & Society
Bharath Sriraman and Lyn English, *Series Editors*

International Perspectives on Mathematics Education— Cognition, Equity, & Society

Bharath Sriraman and Lyn English, *Series Editors*

Challenging Perspectives on Mathematics Classroom Communication
Edited by Anna Chronaki and Iben Maj Christiansen

Mathematical Representation at the Interface of Body and Culture
Edited by Wolff-Michael Roth

Mathematics Education within the Postmodern
Edited by Margaret Walshaw

Researching Mathematics Classrooms: A Critical Examination of Methodology
Edited by Simon Goodchild and Lyn English

Unpacking Pedagogy: New Perspectives for Mathematics
Edited by Margaret Walshaw

Which Way Social Justice in Mathematics Education
Edited by Leone Burton

International Perspectives on Gender and Mathematics Education

Edited by

Helen J. Forgasz
Monash University

Joanne Rossi Becker
San Jose State University

Kyeong-Hwa Lee
Seoul National University

Olof Bjorg Steinthorsdottir
University of North Carolina–Chapel Hill

INFORMATION AGE PUBLISHING, INC.
Charlotte, NC • www.infoagepub.com

Library of Congress Cataloging-in-Publication Data

International perspectives on gender and mathematics education / edited by
Helen J. Forgasz ... [et al.].
    p. cm. – (International perspectives on mathematics education,
cognition, equity, & society)
 Includes bibliographical references.
 ISBN 978-1-61735-041-2 (pbk.) – ISBN 978-1-61735-042-9 (hardcover) –
ISBN 978-1-61735-043-6 (e-book)
1. Mathematics–Study and teaching (Elementary)–Psychological aspects. 2.
Mathematical ability–Sex differences. 3. Gender identity. I. Forgasz,
Helen J.
 QA27.5.I58 2010
 510.71–dc22

                                                                2010012640

CONTENTS

v

SECTION II
NATIONAL FOCUS

SECTION III
HIGH ACHIEVERS

SECTION IV

TERTIARY STUDENTS

CHAPTER 1

INTERNATIONAL PERSPECTIVES ON GENDER AND MATHEMATICS EDUCATION

An Overview

Joanne Rossi Becker
San José State University, USA

Helen Forgasz
Monash University, Australia

Olof Bjorg Steinthorsdottir
University of North Carolina, Chapel Hill, USA

Kyeong-Hwa Lee
Seoul National University, South Korea

INTRODUCTION

Why a book on gender issues in mathematics in the 21st century? Several factors have influenced the undertaking of this project by the editors. First,

International Perspectives on Gender and Mathematics Education, pages 1–11
Copyright © 2010 by Information Age Publishing
All rights of reproduction in any form reserved.

an international volume focusing on gender and mathematics has not appeared since publication of papers emerging from the 1996 International Congress on Mathematical Education (Keitel, 1998). Surely it was time for an updated look at this critical area of mathematics education. Second, we have had lively discussion and working groups on gender issues at conferences of the International Group for the Psychology of Mathematics Education [PME] for the past four years, sessions at which stimulating and ground-breaking research has been discussed by participants from many different countries. Some publication seemed essential to share this new knowledge emerging from a wider variety of countries and from different cultural perspectives. Third, some western countries such as Australia and the USA have experienced in recent years a focus on the "boy problem," with an underlying assumption that issues of females and mathematics have been solved and are no longer worthy of interest. Thus it seemed timely to look more closely at the issue of gender and mathematics internationally.

When the idea for this volume first emerged, invitations were issued to those regularly attending the working and discussion groups at PME. Potential authors were charged to focus on gender issues in mathematics and were given wide scope to hone in on the issues that were central to their own research efforts, or were in receipt or in need of close attention in their own national or regional contexts. As can be seen from the list of authors and chapter titles, international representation has been achieved. We were thrilled when the series editors of *International Perspectives on Mathematics Education–Cognition, Equity & Society* published by Information Age Publishing agreed with us that the collection of chapters that is now found in these pages was worthy of publication.

The resulting collection of chapters fell into four main groupings which make up the four sections of the book:

- History, policy, and non-school factors;
- National focus;
- High achievers; and
- Tertiary students.

In this opening chapter we present an overview of the main issues that emerged.

DEDICATION

Some years ago, the late Leone Burton edited a seminal book entitled "Gender and mathematics: An international perspective" (Burton, 1990). Not long after that she became the founding editor of the book series,

International Perspectives on Mathematics Education–Cognition, Equity & Society. This book, a volume in that same series, re-visits the field of gender and mathematics education across the globe. It is dedicated to the memory of Leone Burton.

SECTION I: HISTORY, POLICY AND NON-SCHOOL FACTORS

Section I of this volume consists of five chapters that provide a historical perspective on issues of gender, illuminate how policy impacts issues of gender equity in mathematics education, and focus on out-of-school factors of parental influence and programs to support young women in mathematics.

Teri Perl's chapter is a reprinting of an article first published in 1979, entitled *The Ladies Diary or Woman's Almanack, 1704–1841.* As Perl reminds us in a new preface to the paper, it is important to reconsider the history of women's involvement in mathematics to remind us that even in the 1700s, women, given educational opportunities, were interested in and enjoyed mathematics. Although it began as a compilation of recipes and household tips, as time passed the *Ladies Diary* became devoted to mathematics in the form of problems and puzzles. While all but one editor and most of the contributors to the *Ladies Diary* were men, women contributed especially in the areas of algebra, geometry, and indeterminate analysis. It is heartening to know that mathematics has been a discipline of interest to women for such a long time. Perl traces the notion, that mathematics is not a discipline for women, to the social roles that women held in those times, primarily centered on the family and household. As mathematics developed as a tool in the realm of work and became less accessible to amateurs, the discipline became the province of men who were involved in the world of work in which mathematics was used.

The chapter by Melfried Olson, Judith Olson, Claire Okazaki and Thuy La entitled *Conversations of parents and children working on mathematics,* provides an in-depth look at how parents of both sexes do mathematics with boys and girls. This chapter reports results from a video study of 32 dyads consisting of both parents and both a male and female child. As the dyads worked on three mathematics tasks, one each in geometry, algebra and number, dialogues were analyzed for differences in the level of cognitively demanding language used with each child. Interesting differences were found, too complex to detail here, but the findings are provocative for those who work with young children or design materials for parent materials and programs.

Lynda Wiest's chapter discusses results of working with young women in out-of-school programs that focus on mathematics. Entitled *Out-of-school-time programs as mathematics support for females,* this chapter reports on results from programs in the USA but the critical characteristics of such programs should be of interest to those concerned with equity in other countries, as these can be adapted for local cultural contexts. Significant features of out of school programs include: clearly articulated program mission, goals, and objectives, with high expectations for all participants; strong, focused academic curriculum with enrichment opportunities; concerted attention to attitudes and beliefs; single-sex setting; female role models and mentors; student-centered instruction; and career and academic information. These and other important characteristics are discussed in detail in this chapter. The chapter ends with a case study of a mathematics and technology intervention program for middle school students developed by the author in Nevada to embody essential features of robust out of school experiences for young women. The five-day residential program includes problem solving, geometry, spatial skills, data analysis and probability, algebra, and use of powerful technological tools such as *Geometer's Sketchpad.* Quantitative and qualitative data over a six-year period indicate that the program improved participants' attitudes while increasing their knowledge, skills, and future participation in mathematics and technology.

Fiona Walls' chapter, *Freedom to choose? Girls, mathematics and the gendered construction of mathematical identity,* reports on a unique longitudinal study of young people from New Zealand who were followed from age seven to 18. This chapter affords a glimpse of how young people develop as mathematical selves and how that is impacted by gender. This qualitative study included interviews with the children, their teachers, and their parents, observations of the children in mathematics class, and examination of textbooks, assessments, and other artifacts related to their schooling. Walls posits a link among children's interests outside of school when young, the schooling they experience, their engagement with mathematics over time, and their eventual vocational aspirations. Schooling and societal experiences reflect male norms and values, such as competition in mathematics class, and persistently undervalue those of females. The self-exclusion of females from the realm of mathematics, an exclusion pervasive throughout the world, reflects the strong connections among masculinity, mathematics and social power.

Colleen Vale examines the construct of gender mainstreaming, a strategy for achieving gender equality adopted by the United Nations at the Fourth World Conference on Women in 1995. In her chapter, *Gender mainstreaming: Maintaining attention on gender equality,* Vale uses Australia as a case study of how attention paid to gender equality in general and in mathematics education in particular has changed over time. By analyzing Australian educational policies from the new feminist movement of the 1970s through the recent, feminist backlash, Vale faults the government in its ability and even commitment to achieve equality for females in mathematics. Since research has highlighted how equal access and equal treatment are not sufficient to guarantee equal outcomes, Vale emphasizes the need to target the diverse mathematical learning needs of young women by using different and appropriate pedagogical techniques. As policies have lost focus on gender inclusive teaching, achievement differences have re-emerged on TIMSS and other assessments in Australia, so that now Australia trails the developed world in gender equity at the secondary level. Vale's chapter underscores the critical need for constant vigilance, monitoring, research and activism to keep gender equality salient in educational policy and practices.

SECTION II: NATIONAL AND INTERNATIONAL FOCUS

Six chapters comprise Section II. Frequently drawing on international research to situate and compare findings, several authors were concerned with trying to understand gender differences in the mathematics achievements of school-level students or drawing attention to identified gender differences deserving of further research within their particular national contexts. Large scale international data comparisons are the focus of other contributions in this section.

There are two chapters from Mexico in which the authors report on a range of settings and circumstances in which gender differences in mathematics learning outcomes were evident in that country.

In their chapter entitled *Studies in Mexico on gender and mathematics*, Sonia Ursini, Martha Ramirez, Claudia Rodriguez, Maria Trigueros, and Dolores Lozano review and synthesize findings from several different research studies. In Mexico, they claim, gender has received increasing research attention, with two main research approaches adopted: descriptive analyses and a focus on socio-cultural dimensions. They conclude that the field of mathematics is distinctly a male domain in Mexico as a result of the dominance of an androcentric culture. Irrespective of level of education, they assert that this permeates girls' and boys', and their teachers', perceptions of the field.

Gender differences when working with algebraic variables: A study with Mexican secondary school students is the title of the chapter by Carolina Rubí Real Ortega and Sonia Ursini. The authors were interested in examining girls' and boys' understanding of the concept, variable, and their capabilities in interpreting, symbolizing, and manipulating variables in various algebraic contexts. A questionnaire was used with grade 9 students. The results were analyzed statistically and some of the students' responses to individual items examined closely. All had difficulties with understanding the concept of variable and both boys and girls tended to interpret variables incorrectly. Gender differences were also evident. Compared to boys, girls left more questions unanswered and had greater difficulty moving between the different ways of using variables. The gender differences identified were considered to stem from a range of factors associated with the mathematics involved but, as in the other chapter from Mexico, the authors also consider the entrenched gendered culture in Mexico to be a major contributor.

Pamela Paek's chapter, *Factors contributing to gender differences in mathematics performance of United States high school students*, focuses on students' problem-solving processes. The participating students were from one school in California and in grade 11 or 12. The students completed a timed pen-and-paper test. As well, in an untimed context, they answered items using an online computer system, IMMEX (Integrated Multi-Media Exercises), which captures what transpires between the presentation of a question and a student's final answer. Paek used SAT (Scholastic Aptitude Test) items because of their centrality in determining students' study pathways following high school in the US. This innovative comparative approach enabled some interesting findings to emerge. The males scored higher than the females on the traditional pen-and-paper, timed, tests. However, in the IMMEX environment, while the females took a greater number of steps to solve problems, they attempted more items, took longer to solve each question, and answered more items correctly than did the males. While unable to generalize beyond her sample, Paek's work suggests that timed, high stakes testing settings may be under-predicting what girls are capable of mathematically and that girls' different approaches to problem-solving are not accommodated in these testing regimes.

Xin Ma draws on several large scale data bases—regional and international, contemporary and from the past—to identify patterns of gender difference in mathematics achievements in his chapter entitled *Gender differences in mathematics achievement: Evidence from regional and international assessments*. In discussing some of the findings, Ma focuses on the countries with the largest gender differences. The following regional assessment regimes—LAB (Latin America), PASEC (Francophone Africa), and SACMEQ (Southern and Eastern Africa)—as well as the two international assessment programs over various years: PISA (Programme for International Student Assessment) and TIMSS (Trends in International Mathematics and Science Study) were examined. Ma's analyses reveal that there is no longer a complete dominance of male superiority in mathematics achievements either regionally or internationally. Girls, he claims, are catching up with boys. Ma identifies the limitations and the advantages of what can be learned at national and international levels by following trends in gender differences over time and encourages nations to participate in large scale assessment programs as a means of monitoring the attainment of gender equity.

Focussing on the *"playschool"* years (18 months to 5 years of age) in Iceland, Olof Steinthorsdottir, Kimberley Dadisman, Dylan Robertson, and Kristjana Steinthorsdottir, explored whether boys' and girls' achievements and mathematical problem-solving skills differed in the final year of playschool. The sample sizes were relatively small. The children participated in pre- and post-tests comprising individual interviews in which the teacher read aloud mathematical questions and the children thought aloud while solving them. Scores improved but no gender differences were found. The authors consider the findings important in that they suggest that it is not until after the playschool years that any gender differences emerge, and that the improvements in the test scores demonstrate that valuable mathematics learning does take place in these very early years of learning.

In their chapter entitled *Mathematics teacher education and gender effects*, Sigrid Blömeke & Gabriele Kaiser examined only the German pre-service mathematics teacher data from the *Mathematics teaching in the 21ˢᵗ century* study in which six countries participated. The study was designed to determine levels of professional competence with respect to mathematics knowledge (various dimensions), knowledge of mathematics pedagogy, and general pedagogical knowledge. The German data presented in the chapter were examined by gender. The sample comprised lower secondary pre-service teachers from two programs: primary and lower secondary, and lower and upper secondary. There were more females than males in the primary and lower secondary programs, and more males than females in the secondary only programs. The content knowledge and pedagogical knowledge tested at the beginning and at the end of the teacher education programs revealed that the males knew more mathematics than the females at both points, but the females gained more knowledge than the males during the teacher education programs. While no appreciable gender difference was found in mathematics pedagogical knowledge at the beginning of the programs, males appeared to gain more knowledge during the programs and had higher post-test scores than the females. When the type of teacher education program was taken into consideration, the gender difference was either insignificant or of low practical relevance. In explaining the variations in the findings, the authors suggest that gender differences are essentially the result of students' educational choices.

SECTION III: HIGH ACHIEVERS

Gender issues among high achieving students are discussed in the three chapters comprising this section of the book. Generally more males than females are found amongst the very highest achievers in mathematics, and participation rates in the most challenging mathematics courses offered are also dominated by male enrolments. Authors from three different countries—South Korea, Australia, and the USA—have looked at different issues associated with high achievement in mathematics.

Discovering the potential of gifted females in mathematics is the title of the chapter by Kyeong-Hwa Lee, Eun-Jung Lee, Seoung-Hey Paik, and HeiSook Lee. The authors provide an overview of Korean society and describe the mathematics learning context and poor career participation rates of females in fields dependent on mathematics. Girls in a gifted program within the *Women into Science and Engineering* [WISE] project in Korea were the focus of the study described. The girls enjoyed group work. Despite their identified talent, they were found to be challenged by some problem types, particularly in some content areas. They lacked confidence and believed they needed lots of practice to be as good as boys. Their families considered the futures of their sons more important than their daughters' and underestimated the girls' mathematical competence. The authors suggested that interventions focusing on problem solving were needed to encourage girls into mathematics.

Gilah Leder and Helen Forgasz present findings from two large sources of mathematics achievement data in Australia in their chapter, *Gender and high achievers in mathematics: Who and what counts?* The data reveal that boys outnumber and outperform girls at the highest levels of mathematics achievement. One data source comprised the publicly available results in the high stakes grade 12 Victorian Certificate of Education over a number of years. At the very highest levels of achievement in each of the three mathematics courses offered, the proportions of boys receiving the very highest grades was found to be disproportionate to their relative enrollment numbers in each subject. The examination results in one of these subjects with two parallel versions running simultaneously—one in which students must use graphics calculators and the other in which the more sophisticated Computer Algebra Systems (CAS) calculator is mandated—lead the authors to suggest that girls may be disadvantaged by the CAS. The Australian Mathematics Competition (AMC) was the second data source. Not only did males greatly outnumber females in winning medals in the competition, but of the those responding to a survey to gather more information about them subsequent to their wins, 15% of the males, but none of the females, described themselves as mathematicians. The AMC medallist data did, however, reveal that females are not only capable of high achievement in mathematics but that it can have positive impacts on their lives.

In their chapter, *What are high achieving young women's perceptions of mathematics over time?*, Amanda Lambertus, Susan Bracken, and Sarah Berenson report findings from a series of 19 semi-structured interviews with a group of nine females over a ten year period. They were interested in the young womens' perceptions of the role of mathematics in their education and future careers over time. The participants were first interviewed while attending a summer camp that was part of the *Girls on Track* program. The analysis of the interviews over time revealed that mathematics had been viewed as a tool with uses that differed for individuals and at different times. Mathematical knowledge was seen as essential for success in other science-related areas and was valued as a means to improve problem solving and reasoning skills. Those successful at mathematics perceived it as a contributor to their success in their chosen careers.

SECTION IV: TERTIARY STUDENTS

This section of the book is comprised of three chapters that deal with a different population of women, albeit all beyond high school. Historically, much of the impetus for research on gender and mathematics derived from the under-representation of women in careers that depend on mathematics, those that have come to be called STEM careers in the USA, standing for science, technology, engineering and mathematics. The first place one might think to search for reasons for this under-representation is indeed at the collegiate level and beyond. Identification of critical variables affecting the choices of college-aged women to study or not study mathematics intensive careers has, as much of the rest of this volume exemplifies, led to considerable research in the pre-college years when decisions are made that affect choice of college major or even college itself. But, with the percentage of women majoring in mathematics increasing to near parity in many places, the question of why so few in STEM careers remains of interest. The three chapters in this section focus on different post-secondary groups of students: undergraduates taking remedial mathematics; undergraduate mathematics majors; and women pursing doctorates in mathematics or mathematics education.

The chapter by Jennifer Hall, *The influence of high school and university experiences on women's pursuit of undergraduate mathematics degrees in Canada*, begins this section by examining data from a western country in which only about one-fourth to one-third of degrees in mathematical sciences are earned by women, although women represent 60% of undergraduate students. Hall's work centers on upper division female undergraduate mathematics majors, and aims to identify the supports and challenges they face as they pursue a degree in mathematics. Six women participated in in-depth interviews that focused on family, peers, personal characteristics, and the formal educational system. While parents and other family members had a positive impact on the women, echoing earlier research, peers seemed indifferent to the subject. And while love of mathematics and determination were personal characteristics that described the sample, a significant subset found lack of confidence in their abilities in mathematics as a hindrance. Informants generally experienced enriched activities in mathematics in pre-college work, which served to pique their interest in the subject. College level mathematics was characterized by male dominance in student body and instructors. Interestingly, all members of the group preferred applied mathematics to a more theoretical orientation, stressing the real-life usefulness of the subject; this orientation was evident in their career choices. And the sense of being "othered" as female mathematics majors seemed to lead to identification outside that small group within the university, even to the exclusion of other female mathematics majors.

The chapter by Ansie Harding, Leigh Wood, and Michelle Muchatuta, with Barbara Edwards, Lucia Falzon, Sibba Gudlaugsdottir, Belinda Huntley, Jillian Knowles, Barbara Miller-Reilly, and Tobia Steyn, has an intriguing title: *Try and catch the wind: Women who do doctorates at a mature stage in their lives.* This chapter provides insights into the influences, obstacles and opportunities facing mature women who pursue a doctorate in mathematics or mathematics education. Women from several different countries: Malta, South Africa, the United States, Iceland, New Zealand, and Australia, provide rich life stories that describe the family commitments that precluded a PhD earlier in life. The women's stories detail the strong intellectual appeal of asking and answering a burning question and the personal and professional growth that can result from achieving the terminal degree. Given the various socio-cultural, political and personal factors that affect women's pursuit of PhDs, particularly the caretaking of families, it is likely that these women's stories will continue to reflect those of women around the world. Strategies to make the transition back to academia easier to manage and navigate would ensure the continuing contribution of mature women to mathematics education.

Jillian Knowles' chapter, *Recognizing gender in mathematics relationships: A relational counseling approach helps teachers and students overcome damaging perceptions,* focuses on students who seek help in beginning college mathematics courses from learning centers. Knowles found that gender stereotyping in students' past mathematical histories affected their ability to progress in a mathematics course and to take advantage of tutoring-counseling available through the learning center. When tutor-counselors were able to confront such gender stereotyping, students were able to overcome them and make progress mathematically. Knowles used an unusual relational psychoanalytic approach to investigate students' mathematical self-identities and their interpersonal mathematics relationships. The results of Knowles' work has implications for anyone who works with young people on mathematical learning, including teachers, college professors, tutors, and even parents.

CONCLUDING WORDS

The work of the authors of the chapters included in this volume presents a contemporary snapshot of the continuing scholarly research being undertaken in the field of gender and mathematics around the world. In countries with shorter histories of research in the field, prior knowledge has been used to effect in the methods employed to conduct the studies and in the interpretation of findings. New knowledge is also being generated. In nations with well-established histories of scholarship in the field, close monitoring of the directions of gender differences and the examination of trends over time continue. New and exciting dimensions have also been investigated. While there is some evidence that the extent of male dominance in participation and achievements may not be as great as in the past, nonetheless it cannot yet be said that gender equity has been achieved in mathematics education. Work in the field must continue to ensure that future generations of young women will benefit from the opportunities afforded through engagement with mathematics.

REFERENCES

Burton, L. (Ed.) (1990). *Gender and mathematics: An international perspective.* London: Cassell.

Keitel, C. (1998). *Social justice and mathematics education. Gender, class, ethnicity and the politics of schooling.* Berlin: IOWME and Freie Universität Berlin.

SECTION I

HISTORY, POLICY, AND NON-SCHOOL FACTORS

CHAPTER 2

THE *LADIES' DIARY* OR *WOMAN'S ALMANACK,* 1704–1841

Teri Perl
San Francisco State University

2010 UPDATE

I initially learned of the existence of the *Ladies' Diary* through an offhand remark by the daughter of Grace Chisholm Young, one of the women mathematicians I was researching for my book *Math Equals.* Kind of a gift— though it had been thrown out as a sarcastic offhand remark: "Well! If you're interested in (implied: *crazy stuff like*) women and mathematics, you should have a look at the *Ladies' Diary*!"

That look raised many questions as well as the incentive to do research in the elegant British Museum Library that houses copies of the *Ladies' Diary* from its birth to its demise. Who was the readership of the magazine? How did its existence tally with the longstanding stereotype of women not being interested in mathematics. And...mathematics for entertainment? Why, indeed, did the *Ladies' Diary,* initially comprised of recipes, household tips, and so forth, become, *by popular demand,* and as time passed, more and more devoted to mathematical problems and puzzles?

It was at the height of the women's movement, during the 1970s, that I was researching the *Ladies' Diary.* Elizabeth Fennema at the University of

International Perspectives on Gender and Mathematics Education, pages 15–32
Copyright © 2010 by Information Age Publishing

Wisconsin was studying the relationships between girls and mathematics. My Stanford dissertation research involved the analysis of a longitudinal data stream to sort out why girls were less likely than boys to elect upper division mathematics courses when the subject became elective in high school. "Perception of usefulness" was the variable that most explained the sex differences here. Expanding Your Horizons was an organization formed at that time to encourage young girls to keep taking math in school, in order to maximize their future career options.

Researching the *Ladies' Diary* in the 1970s was important as a way to help overcome the then current attitude that math was hard for girls—implying that it was easy for boys. Learning about the existence of the *Ladies' Diary* was important then because it showed that, even so long ago, given the proper education and the opportunity, women were certainly capable of understanding and enjoying mathematics. So why were these sex differences in attitudes towards mathematics still persistent in the 1970s?

A lot has changed in the last quarter century. Today, women are encouraged to pursue careers in many fields, including those fields that require significant knowledge of mathematics. Women have been demonstrably successful in many such fields. It becomes more and more apparent that as women receive educations that are equal to that of men, their attitudes toward mathematics become much the same as those of men.

The story of the *Ladies' Diary* as well as biographies of women mathematicians can be interesting reading on their own. Perhaps even more important, these stories can help women understand that mathematics, a subject so basic to developing skills in a variety of professions, has been an area of positive interest to many women for a very long time.

THE LADIES' DIARY OR WOMAN'S ALMANACK, 1704–1841

(reproduced, with permission, from:
Historia Mathematica, 6, 1979, 36–53)
Teri Perl
San Francisco State University

The existence of the Ladies' Diary or the Woman's Almanack, an 18th century English magazine devoted largely to problems and puzzles in mathematics, indicates that stereotypes about the inability of women to understand and enjoy mathematics were less strongly believed in the 18th century than they are today. The beginning of the Ladies' Diary coincides with the popularization of mathematics and the growth of mathematical literacy. However, as mathematical literacy spread in response to developing technology's requirements for more mathematically sophisticated workers, women, not part of this need, were left behind. This effect is reflected in the decline in the number of women contributors over the life of the publication.

The *Ladies' Diary*[1] was one of the first popular magazines to appear in England at the beginning of the 18th century. Published once a year from 1704 to 1841, its initial format was an almanac containing articles of general interest to women as well as the usual calendar and astronomical observations (White, 1970, pp. 23–29). After a few years its emphasis changed, and by 1707 the recipes, sketches of notable women, articles on health and education , and so on, had been replaced by enigmas, queries, and mathematical questions, "the answers to the principal of which . . . are rewarded with a certain number of copies of the work" (Bickley, 1889, p. 113). Since the names of contributors were included with their contributions, it is possible (within certain limitations described below) to estimate the number of women who contributed to the magazine over the years.

The *Ladies' Diary* was small—only about 10 by 16 centimeters in size (see Figure 2.1).[2] Described on the cover of the 1738 issue as "Containing many Delightful and Entertaining Particulars, Peculiarly Adapted for the Use and Diversion of the Fair-Sex," The magazine promised the women who read

Figure 2.1

it that the cultivation of their minds would increase their attractiveness. "Wit join'd to BEAUTY... lead more Captive than the Conqu'ring Sword" was the message. That a woman's magazine should have appeal as a source of information and instruction was not unusual. In fact, a commonly professed goal of early 18th century editors was the improvement of the minds of male and female readers alike (White, 1970).

As an almanac the *Ladies' Diary* was unusual only in its choice of theme. (Almanacs of the period typically focused on themes—for example, medicine or gardening.) The choice of mathematics as a subject, with women as an audience, seems to have been a sound one for its first editor and publisher.

> John Tipper was reimbursed handsomely for the editorial attention he bestowed upon them (women). In a century when almanacs flourished and failed by the score, when with each new year England was deluged with them, the *Ladies' Diary* proved itself one of the longest lived and one of the most remunerative. (Meyer, 1955, p. 65)

Mathematical puzzles were clearly popular with the Diary's readers. Tipper writes in the 1718 issue of the *Ladies' Diary*, "foreigners would be amaz'd when I show them no less than 4 or 500 several letters from so many several woman, with solutions geometrical, arithmetical, algebraical, astronomical, and philosophical" (Meyer, 1955, p. 62).

With its change of format, from topics of general interest to enigmas and mathematics, the *Ladies' Diary* became one of the widely read 18th century magazines devoted to the popularization of science and mathematics; these were addressed mainly to readers with no specialized training in the subjects[3] (Hogan, 1976, p. 404). The *Ladies' Diary* differed from these others primarily in the language used in some of the problems—language that reminds the reader that the problems were addressed to women. Figures 2.2 and 2.3 contain examples of problems from the *Ladies'*

Questions proposed in 1795, and answered in 1796.

1. QUESTION 984, *by Miss* Nancy Mason, *of Clapham*.

Dear ladies fair, I pray declare,
In Dia's page next year,
When first it was I 'gan to pass
My time upon this sphere.
My age so clear ; the first o'th' year,
In years, in months, and days,
With ease you'll find, by what's subjoin'd*,
Exact the same displays.

$$*xy + z = 238$$
$$xz + y = 158$$
$$x + y + z = 39$$

Where $x =$ the years, $y =$ the months, and $z =$ the days of my age, the first of January, 1795.

Figure 2.2

Diary; the first was proposed by a woman and is typical of a type that often appeared in the magazine. These problems were in verse, a requirement for both contributed questions and answers in the early years. However,

---

XIII.   QUESTION 359, *by* Honorius.

Miss's apron grown short, she is full of complaint,
And to merit your pity she looks like a saint!
On the floor falls her tea; then her screams you may hear,
And fainting she sinks in a fit on the chair.
Mamma for the doctor immediately sends,
Who, in honour to miss, in his chariot attends;
He examines her pulse, and appearing so wise,
Descants on the languishing looks of her eyes—
But alas! neither spirits, nor letting Miss blood,
Specifics, nor preaching, are found to do good:
For a surgeon came in, who the cause did declare,
And the doctor's finesse, and his art made appear.
   Mamma now was told 'Miss's hoop was too small,
Therein lay her grievance, disorder, and all;
The question was ask'd—Polly sighing reply'd,
A French hoop will cure me, and so will a bride.
A hoop of the fashion to cure her disease,
Extends from her centre quite round to her knees:
In the right and left wing a French placket * is made,
To her elbows advancing, and forms a parade.

   * *Opens and shuts, forms a pair of bellows, and rises and falls by the means of strings or bowlings.*

   Miss Polly to church now, or play can repair,
And wherever she goes is admir'd for her air!
At the sight of a beau, how her heart beats alarms!
While the winds swell her pride and her legs tell their charms:
Her hidden perfections she knows will invite,
Or ensnare the beholder, should chance give them sight.
   By the pow'r of her hoop Polly steps into fame,
By out-priding the rest she conceals her own shame;
In the country she reigns o'er the 'squire and the clown,
O'er the lords and the fops she's triumphant in town.
Her hoop is the secret—and if you would know
What it holds with her petticoat, seek from below †.

   † *Form of the hoop is the lower frustum of an ellipsoid, with its vertex next the head.*

   Transverse ⎱ *diams.* ⎰ 42 inches ⎱ *above,* ⎰ 48 inches ⎱ *below.*
   Conjugate ⎰   ⎱ 26   ⎰   ⎱ 29   ⎰
            Altitude of the frustum 12 *inches.*

   From the lower part of the hoop's circumference to the bottom of the petticoat, the form is an elliptic cylinder, by the petticoat hanging nearly perpendicular from thence: the altitude of which elliptical cylinder is 18 inches: Quere the content of the whole concavity in wine gallons?

Figure 2.3

Thomas Leybourn expressed his displeasure at the generally poor quality of the verse, complaining that it was not "favorable to the development of Mathematical genius" (Leybourn, 1817, I, viii). It is not yet clear whether the rhyme format was used because of its appeal to 18th century readers in general or to women readers in particular.

Thomas Leybourn describes the *Ladies' Diary* as "a collection of Practical Exercises in almost every branch of Mathematical learning" (Leybourn, 1817, I, vi). In the *Mathematical Questions Proposed in the Ladies' Diary, 1704–1816*, Leybourn includes an index classifying the mathematical problems which had appeared in the magazine into 25 categories (Leybourn, 1817, IV, 437–440). The list of these categories reads like a typical math–science curriculum for the young 18th century gentleman. It includes algebra, astronomy, dynamics, fluxions, geometry, harmonics, hydrostatics, isoperimetry, navigation, optics, pneumatics, spherics, and statics. Leybourn describes the level of difficulty of the problems in the *Ladies' Diary* as considerable: "The geometrical part has always been conducted in a superior style; the Problems proposed have tended to awaken curiosity, and the Solutions to convey instruction, in a much better manner than is always to be found in more splendid publications" (Leybourn, 1817, I, viii).

The *Ladies Diary*, Leybourn concludes, "may be considered as exhibiting a view of the state of Mathematics in this country, and, on this account, must be valuable, as illustrative of the progress of the science in England, during the last century" (Leybourn, 1817, I, vi–vii).

Although we know something about the size of the circulation of the *Ladies' Diary* (Meyer, 1955 refers to continued brisk sales of six or seven thousand in 1718) and that it was published over a long period of time, very little is known about its readership. It was "designed on purpose for the diversion and use of the Fair sex," writes Tipper (1704, p. 59); yet we do not even know if this decrease reflects a change in the proportion of men and women readers. Nor is it clear what class of women read the magazine. Events listed in the almanac section, such as dates of Oxford and Cambridge terms, suggest that the readership of the *Diary* was middle and upper class. However, Cynthia White contends that the audience for the new ladies' magazines, of which the *Ladies' Diary* was one example, was not restricted to the upper classes: "This expanding readership was made up of new recruits from the commercial classes as well as a substantial number of domestic servants whose conditions of work gave them both the facilities for reading and access to reading matter" (White, 1970, p. 24). John Tipper seems to agree. In an early issue he projects a broad audience for the *Ladies' Diary*: "for Ladies, there would be information concerning 'essences, perfumes and unguents,' for Waiting-women and Servants, 'excellent directions in cooking pastry, confectionary'; for Mothers, families, and for Virgins, 'directions for love and marriage'" (White, 1970, p. 25). Whether or not this broad audience materialized is not known.

Another question arising about the women who read the *Ladies' Diary* concerns their ability to read and their possession of some degree of mathematical literacy. Both would have been requirements for patrons of such a magazine. By the 19th century, reading, at least at some minimal level, was a skill that cut across many classes in England, and was available to both men and women. The Charity School movement of the 17th and 18th centuries, designed to increase access to the Bible, had extended the ability to read even to the very poor (Bayne-Powell, 1939).

Concerning mathematical literacy, formal mathematics education in the 18th century was poor for all people, men as well as women. At the beginning of the second half of the 17th century the education given in English schools was based almost entirely on Latin grammar; little, if any, mathematics was taught, while simple arithmetic, which was considered suitable only for clerks, was entirely neglected (in the grammar schools). John Flamsteed (1629–1719), the first Astronomer Royal, left school at the age of 14 because of ill health. He wrote that his father taught him arithmetic, including the doctrine of fractions and the golden rule of three. John Newton remarked in 1677 that he had "never heard of any grammar school in England in which mathematics was taught" (Taylor, 1954, p. ix). Mathematics was not considered a proper academic subject and was therefore not part of the education of a young gentleman. "Mathematicks at that time, with us," Dr. Wallis recalled, "were scarce looked upon as Academical Studies, but rather Mechanical; as the business of Traders, Merchants, Seaman, Carpenters, Surveyors of Lands, or the like, and perhaps some Almanack Makers in London.... For the Study of Mathematics was at that time more cultivated in London than in the Universities" (Taylor, 1954, p. 41). Thus, not having access to formal mathematics education in the 18th century should not have been more of a handicap to women than to men.

Several factors, however, were encouraging the growth of mathematical literacy among men. With the general expansion of trade in England during the Elizabethan period, the study of practical mathematics had spread among England's new skilled workers (Ross, 1975). The growth of English navigation was of primary importance to the expansion of mathematical knowledge since it generated patronage and support for many English mathematicians. The development of new navigational instruments required of their users a mathematical sophistication far beyond that of most naval personnel. Gresham College, which had been founded in London at the end of the 16th century to educate craftsmen and mariners, now became a center for navigational studies. By the beginning of the 18th century other schools were serving the same function: "[S]chools such as Woolwich Military Academy, and Watts Academy, outside of the Oxford-Cambridge

University structure, were training men in mathematics, and producing as well a steady flow of textbooks on applied mathematics for their students and courses" (Taylor, 1966, p. 16).

Further impetus for these changes in education came from the religious upheavals of the latter 17th century, during which puritan scholars were expelled from the universities. As these men formed new institutions (often referred to as the dissenting academies) to replace the ones from which they had been excluded, they introduced crucial changes that reflected their religious beliefs. They believed that with the imminent approach of the millennium, man would be restored to the dominion over nature which had been lost in the fall of Adam. To facilitate this process, or indeed to show that it was already in progress, the Puritans led the way in diverting pedagogy from classical studies toward science and technology. They "encouraged projects that linked 'scientific' analysis with practical innovation: in practical mathematics and surveying.... The sciences with which the Puritan-dominated culture were concerned were practical mathematics, technology of trades, chemistry, and husbandry" (Frank, 1977, pp. 385–386). Thus, under these pressures for reform, mathematics education (for boys) was encouraged and expanded. The teachers in this movement were mainly the mathematical practitioners who by the 17th century had become knowledgeable mathematicians.

The end of the 17th century also marked the appearance of a group of English nobility and gentry whom Ross calls virtuosi (Ross, 1975). These men use the new leisure and the wealth created by the commercial expansion to enjoy learning for its own sake. Together with the members of the Royal Society (founded in 1662), this class of leisured gentlemen enjoyed mathematics as a pastime. In all probability it was the mathematical practitioners, their students, and the virtuosi who made up the majority of readers for the popular mathematics magazines of the 18th century.

It is of interest to consider how this expansion of mathematical literacy affected women. Toward the end of the 17th century there was a strong movement calling for a general improvement in the education of women: "writers a different as Addison and Defoe, Miss Astell and Lady Masham, the poet Shadwell and Lady Chudleigh, Mrs. Makin and Lady Mary Wortley Montagu, Josiah Child and Mrs. Woolley... unite in condemning the slight and superficial education of girls" (Thompson, 1974, pp. 201–202). However, the objectives of these reformers were quite limited: It was not their intention that women be educated to a new role in society. Lady Mary Wortley Montagu, herself a well-known literary figure, wrote in 1715, "I am not arguing for equality of the two sexes. I do not doubt that God and Nature have thrown us into an inferior rank, we are a lower part of creation, we owe obedience and submission to the superior sex" (Stanton, 1957, p. 259). But the reformers argued that education improved men's minds and that educated men would prefer educated wives. Thus, the primary purpose of educating women was to satisfy this preference, "to polish your Souls, that you may glorify God, and answer the end of your Creation, to be help meets to your Husbands" (Thompson, 1974, p. 200). There is no evidence to suggest that mathematics was considered less accessible to women than to men and thus to be excluded from such reforms. And so as men's education in mathematics improved, one may expect that women's access to mathematical knowledge improved as well. Thus, the evidence suggests that despite the limited formal mathematics education available to them, there were women, by the beginning of the 18th century, who were capable of reading and contributing to the *Ladies' Diary*.

The growth of the printing industry and the availability of cheaper books were also factors that gave women greater access to mathematics. Mathematics textbooks used at the colleges training and practical mathematicians were no longer written primarily in Latin. With more such books available in English, both men and women without formal training could have taught themselves enough mathematics to be able to enjoy the diaries at some level. It is very likely the case that among those who had learned some advanced mathematics, more women than men were self-taught or informally taught. However, to have acquired the training, books, and so forth needed to become mathematically literate, the 18th century woman must have been completely dependent on the men with whom she was associated at home. It is probably in homes sympathetic both to mathematics and to education for women that one may look for the women who were the patrons of the *Ladies' Diary*.[4]

More is known about the men and women who contributed to the ladies' Diary and other such magazines than of their readers. For example, we know that the *Ladies' Diary* was founded by John Tipper, who is listed in the *Dictionary of National Biography* as an almanac maker (Lee, 1903, V56, 408). Taylor describes Tipper as a schoolmaster whose specialty was applied mathematics, including navigation who "began to publish an Almanack (*Ladies' Diary*) in 1704, and introduced into it miscellaneous papers on mathematics" (Taylor, 1954, pp. 295–296). With one exception, all of the editors of the *Ladies' Diary* were men. For a brief period following the death in 1743 of the second editor, Henry Beighton, his widow, Caelia Beighton, edited the magazine and continued in that capacity until Robert Heath became editor in 1745.

Thomas Leybourn calls the *Ladies' Diary* "the result of the joint labour of almost all the Mathematicians of eminence, that have appeared in England in the course of the last century" (Leybourn, 1817, p. vi). These men appear to have been mainly self-taught as mathematicians. William Emerson (1701–1782), an instrument-maker who declined membership in the Royal Society because he did not want to pay dues, and John Landen (1719–1790), a surveyor, were two frequent contributors to the *Diaries*. Thomas Simpson (1710–1761), one of the last editors of the *Diary*, had originally been trained and employed as a weaver. Charles Hutton (1735–1823), also an editor, worked for a while as a "hewer" in a coal mine. These men and many of the contributors to the Ladies' Diary were associated with institutions such as Woolwich Military Academy, Watts Academy, and the Royal Military College at Sandhurst, colleges where young men were trained in practical mathematics and natural science. (Hutton, whose *Mathematical and Philosophical Dictionary* was published in 1795, and Simpson were professors at Woolwich Military Academy; Leybourn taught at Sandhurst.) They produced a steady flow of textbooks on applied mathematics and were responsible for the invention of many new instruments and techniques (Taylor, 1966, p. 16).

The main source of information about contributors to the *Ladies' Diary* has been Leybourn's index "Of the Names of the Persons who have Proposed and Answered the Questions" (Leybourn, 1817, iv, 415–436). Most of the contributors to the *Diary* appear to have been men. Even allowing that women sometimes published under male pseudonyms, there is no reason to expect this would have been a significant number in a magazine for women. During the 18th century, it was not uncommon for men as well as women to adopt aliases when writing. This can be found in literary works, in personal letters, and in contributions to magazines. Some of the aliases used by contributors were suggested by figures in ancient Greece or Rome; some had literary or astronomical associations, while others were merely chosen for fun (Timothy Doodle). Noting this practice, Leybourn lists, among others, some aliases used by Simpson, who had been an editor (1745–1760) of the *Diary* and one of its major contributors (Leybourn, 1817, I, ix). The true identities of those whom he knew to have contributed under aliases are parenthetically included in his index. One of the more prolific "women" contributors, Ann Nichols, was an alias used by a Mr. Wales. What that means, we have no idea.[5]

Out of the 813 contributors listed in Leybourn's index, only 32 were women (3.5%). Leybourn mentions no male aliases used by women. Therefore, the only names we count as those of women contributors are those names which are clearly female. Since it is likely that some women used aliases, we may assume this is a minimal estimate. A tabulation of women's contributions over Leybourn's 25 subject classification of the *Diaries'* problems shows that women's contributions tended to be more numerous in algebra, geometry, and indeterminate analysis.[6] This would support the hypothesis that women were largely self-taught: Books in algebra and geometry were the earliest to appear in English and are most likely to have been accessible (Taylor, 1954, pp. 311–341).

As yet we know little or nothing about the women contributions other than their names, and for the later contributors, their class, education, marital status, and any possible connections between them. A good guess probably is that some of these women were associated with the mathematical practitioners. Preliminary investigation in this direction appears promising[7] and suggests that careful research should yield more information.

An analysis of the dates of the contributions made by women shows an interesting cluster over time: More than half of these contributions were made within the first two decades of the magazine's publication. Seventeen women contributed between 1710 and 1725, while only fifteen women contributors are listed from 1738 to 1815. At first we conjectured that this decline may have been caused by changes in editorial policies. However, this proved not to be the case. A comparison of the dates of editorships having the longest intervals during which no women contributors are listed (1725–1748 and 1768–1782) shows that neither interval coincides with the tenure of any single editor or change of editorship.[9] In fact, the longest interval occurred during the later years of the editorship of Henry Beighton, who very likely supported the efforts of women contributors—his wife, Caelia Beighton, knew enough mathematics to have taken over the editorship for a time after his death.

The decline in the number of women contributors seems to reflect the beginning of the current stereotypes about women and mathematics. Although we cannot be entirely certain of the causes of this decline, a pattern emerges. At the beginning of the 18th century, women were not considered less capable of learning mathematics than men. In fact, late 17th and early 18th century evidence suggests that many influential people considered women intellectually capable and educable. Ambrose Phillips, publisher of an early women's journal, the *Free-Thinker*, showed his high regard for the female intellect in the quality and scope of his magazine (White, 1970). Taylor writes about a late 17th century calculating machine offered to women who had "for want of leisure" not yet mastered arithmetic (Taylor, 1954, pp. 128–129). That ladies might find this machine useful for keeping their household accounts does not seem to suggest they were considered less capable than men of learning arithmetic; merely that they may have been too busy to have done so.

However, clearly defined social roles placed women squarely in the home and set limits on the levels of education required by them. Even the 17th century call for the reform of women's education supported the concept that learning for women was primarily a kind of upper class individual accomplishment designed purely for home consumption and decoration.[10]

In the 18th century, a change in attitudes toward women developed; this change would have important implications for setting limits on their intellectual development. When Richard Steele, publisher of the 18th century magazine *The Tatler*, wrote regretfully of "the onset of the age of gallantry which threatened to substitute the *lady* for the *Women*," perhaps he was referring to those changes which, in time, would cause women to be perceived as more artificial, more decorative, and by implication less intelligent. "Lady" sounds more superficial, less serious, than "woman." By the end of the 18th century, we read of the use of strange mechanical devices to stretch and straighten schoolgirls[11] (Armytage, 1964; Gardiner, 1929; Somerville, 1874). Was this too part of the same trend to create artificial, fragile "ladies" instead of healthy "women"?

On the other hand, the spread of mathematics education for men was clearly to enable them to do the work of the world. Taylor cites an early 18th century essay (*Essay on the usefulness of Mathematical Learning in a letter from a Gentleman*) by John Arbuthnot as having had a major influence on the spread of mathematical literacy:

> The writer Arbuthnot had shown the need for mathematics over an ever-widening field of activities, his list including architecture, business affairs, political arithmetic (e.g., vital statistics), theories of change (in gaming, insurance, and investment), astronomy, calendars, mechanics, horology, planetary machines, the making and using of spheres, globes, astrolabes, surveying and gauging, telescopes, microscopes, optics, fortification, gunnery, navigation, engineering, and so on. (Taylor, 1954, p. 146)

It is obvious that this list describes occupations of little interest to 18th and 19th century women.

Thus, at the beginning of the 18th century, women were not generally considered to be intellectually inferior to men. However, women's social roles were defined very differently from men's. The woman's domain was exclusively the home, whereas the man's domain included the larger world as well as the home. The primary goal of education for women was to make women better wives. Sophisticated mathematical training would have had little value to women as homemakers; their only interest in the subject would have been as amateurs. Although the mathematical recreation magazines, of which the *Ladies' Diary* was one, addressed themselves to the non-mathematician, they required a significant degree of mathematical sophistication of their readers, and more so as time went on. At the same time, changes in attitudes toward women throughout the 18th and 19th centuries were evolving an image of women as weaker, less physically capable (than men) of intense brain work, such as that required in the study of mathematics. Thus, the increasing gap in the mathematics education of men and women was not caused by the belief that mathematics is not a womanly subject. Rather, it seems to have been a consequence of the growth of mathematics itself and its evolution, particularly in England, as a useful tool. As a tool it became increasingly an object of the male domain, for it was the men who did the work of the world. Later, as mathematics became more complex, it became less accessible to the amateur, both man and woman. Women, as long as they were exclusively homemakers, could only be amateurs in mathematics. Thus, as time went on, women were to find themselves increasingly excluded from mathematical literacy. In this sense then, the stereotype seems to be an *effect* rather than a *cause.*

At the beginning of the 18th century, mathematics was equally new and equally accessible to both men and women. The popular interest in mathematics was part of the 18th century fascination with the promise of the new science and technology of Newton's universe. However, as mathematics became more complex and the efforts to spread mathematical literacy merged with the call to train young men for jobs in the growing technological society, women, not part of this need, were left behind. As the stereotypes we live with today became fixed, the existence of a popular magazine like the *Ladies Diary* became a wondrous puzzle, but only if seen in relation to these stereotypes (Smith, 1925; Leybourn, 1817). In the light of this analysis, the popularity of the *Ladies' Diary* in its day is not at all surprising.

NOTES

1. The exact title of the *Ladies' Diary* changed over the 140 years of its publication. The following is a list of titles as they appear on the covers of the magazine over the years.

1706–1710	The Ladies Diary: Or, The Womens ALMANACK
1711–1750	The LADIES Diary; Or, The Woman's ALMANACK
1751–1760	The LADIES Diary: or, Woman's ALMANACK
1761–1829	The LADIES' Diary: or Woman's ALMANACK
1830–1839	THE LADIES' DIARY, or COMPLETE ALMANACK
1840	THE LADIES; DIARY

2. This was a typical size for magazines of that time. *The Gentleman's Diary, The Female Spectator,* several almanacs such as *Vox Stellarum* by Francis More, *Merlinue Liberatus* by John Partridge, *Old Poor Robin, Season on Seasons* by Henry Season, one by Tycho Wing, and *The Celestial Atlas* or *Ephemeris* by Robert White were 18th century publications that were the same size as the *Ladies' Diary.*

3. *The Mathematical Repository,* edited by Laybourn, was one of these. Others were *Athenian Oracle* (around 1703), *Gentleman's Diary* (1776–1840 at which time it merged with the *Ladies' Diary*), *Paladium* (edited by Robert Heath who also edited the *Ladies' Diary* from 1745–1753), and the *Mathematical Correspondent.*

4. One may argue with this conjecture, citing cases such as that of Sophie Germain, an outstanding French woman mathematician (1776–1831). Stories tell of great parental opposition to her study of mathematics. However, she did have access to her father's fine library and is an example of an extraordinarily motivated and gifted individual. My understanding also is that the opposition of her parents was a reaction to the *intensity* of her interest in the subject. This seems to fit the hypotheses raised in this paper. In certain circles women are

encouraged to be knowledgeable. It is part of their adornment. Opposition arises, however, only when this interest extends beyond the requirements of the amateur.

5. Some examples of the aliases listed by Leybourn are:

Simpson Patrick O'Cavanaugh, Kubernetes, Anthony Shallow, Hurlothumbo, Timothy Doodle Esp., Marmaduke, Hodgson, (probably others).

Landen Stiff, Walton, Waltoniensis, Bumpkin, Puzzlem

Emerson Merones

Mr. W. Wales, Ann Nichols, Nauticus, Cetii, Gemini, Geometricus

Crakelt Dargenes, Chartreux, Pamphagus

John Holt Mancunlensis

Heath Upnorensis

Mr. Wildbore Amicus

Th. Todd Philalethes Clemsbyensis

6. Subject categories and contributions in each:

No.	Name of category	Total no. of items	No. of items by women	% of total by women
1	Algebra: purely algebraic	135	17	13
	Algebraic theorems	8	0	0
2	Arithmetic	18	1	6
3	Arithmetic of sines	1	0	0
4	Astronomy, geography, etc.	134	12	9
5	Chances, combinations, permutations, etc.	17	2	12
6	Curves	63	0	0
7	Dynamics	152	3	2
8	Fluxions	28	0	0
9	Geodesic problems and theorems	5	0	0
10a	Geometry problems solved by pure geometry	175	7	4
10b	Geometry resolved by algebra	242	10	4
10c	Geometry resolved by arithmetic	96	12	13
10d	Geometry relating to maxima, minima	39	4	2
10e	arithmetic of sines	9	0	0
10f	geometry theorems	35	4	11
11	Gnomics	18	0	0
12	Harmonics	2	0	0
13	Hydrostatics	13	1	8
14	Hydrodynamics	30	0	0
15a	Indeterminate analysis equations of first degree	19	2	11
15b	Indeterminate analysis Diophantine (second and higher)	39	3	8

No.	Name of category	Total no. of items	No. of items by women	% of total by women
16	Interest, annuities, etc.	28	1	4
17	Isoperimetry	6	0	0
18	Navigation	16	0	0
19	Optics	8	0	0
20	Perspective	2	0	0
21	Pneumatics	11	1	9
22	Projections	5	0	0
23	Series	12	0	0
24	Spherics	7	0	0
25	Statics	39	2	5

7. Looking for references to these women in the *Dictionary of National Biography (DNB)* proved fruitless, as expected. However we succeeded in finding men with the same surnames who were involved with mathematics or some subject related to science. This search may have yielded a few strong leads. Although the relationship was verified elsewhere, Henry Beighton is clearly the husband of Caelia Beighton, a contributor who edited the Diary in the interim between her husband's death and the next editor. Another *DNB* entry which looks potentially fruitful is Henry Atkinson. He is listed in the *DNB* as a mathematician, a contributor to the *Ladies' Diary*, and as assisting his father and sister in the managements of schools in various locations around Britain. Perhaps Marie Atkinson, a woman contributor, is his sister. A Peirce Dod(d) is listed in *DNB* as a medical writer who has both a wife and daughter named Elizabeth. Of course, Elizabeth is a common English name, but the dates fit. Perhaps Elizabeth Dod(d) is associated with Peirce Dod(d).

8. Contributions by women to *Ladies' Diary*.

Name	Category	Year	Name	Category	Year
Solution by lady	1	1710	Dod (d), Elix. (Mrs.)	4	1721
Fisher, Lydia (Mrs.)	1	1711		9c	1723
Wright, Mary (Mrs.)	4	1711	Adrastea	1	1722
Sidway, Barbara (Mrs.)	10b,c	1711		7	1722
	5	1712		7	1723
	1	1712		10a	1724
	10a,b	1712	Sylvia	4	1723
	10d	1713		1	1725
	10d	1714	Rosamond Manlove	15b	1748
Wright, Anna (Mrs.)	10c	1712	(Miss)	5	1748
	1	1712	Amanda	15a	1748
	4	1713	Atkinson, Marie	1	1754
Brown, Sarah (Mrs.)	1	1713	(Miss)	1	1755

(continued)

Name	Category	Year	Name	Category	Year
Adway (Mrs.)	4	1714	S—et, T. (Miss)	1	1758
Nelson, Mary (Mrs.)	1	1713	Amaryllis	2	1761
	10a	1713	Nichols, Ann (Mr. W. Wales)	2	1760
	7	1714		1	1761
	10b	1715		4	1761
	10b	1715		4	1763
	10a,b	1716		4	1764
	1	1719		4	1765
Morgan, Ann (Mrs.)	10b	1715		10b	1766
Boydell (Mrs.) Silvia	7	1715		4	1766
	1	1716		10b	1767
	25	1716		4	1768
	10c	1716	Horticultura	10f	1782
	25	1716	Claxton, Betty... (Miss) Benwell nr Newcastle upon Tyne	10b	1791
Heyschia (Mrs.)	1	1717	Mason, Nancy (Miss) Clapham	1	1795
	15a	1720		10a	1796
Heyshot, Lady Mer.	1	1717		10c	1797
Philomathes, Anna	10c	1719	Middleton, Maria (Miss), Eden nr Durham	10b	1797
	10d	1719		10c	1798
	10c	1719		4	1798
	16	1719		21	1799
Cowan, Sarah (Miss)	10c	1797		7	1813
Castieau, M. (Miss) Salop	10c	1800		10f	1814
	10c	1801		10d	1816
Weston, Mary (Miss) Gainsborough	10a	1810		15b	1816
	10b	1811		10f	1816
	15b	1812	Jackson, Susan(nah) (Miss) Mile End	10c	1813
				10c	1814
				1	1814

9. The editors of the *Ladies' Diary* and their dates of editorship are: John Tipper (1704–1713), Henry Beighton (1715–1743), Caelia Beighton (1743–1745), Robert Heath (1745–1753), Thomas Simpson (1754–1760), Edward Rollinson (1761–1763), Charles Hutton (1763–1816+).

10. Perhaps this explains the derision implicit in the English term bluestocking. Compare this with the phrase femmes savants used to describe the comparable group in France (Taylor, 1954, p. 44). The English expression hints at criticism of the women who were not conforming to this design.

11. Mary Somerville writes of her entry into a Miss Primrose's boarding school at the age of 10 "perfectly straight and well made yet was enclosed in stiff stays, with a steel busk, her shoulders drawn back by bands till the shoulder blades met. Then a steel rod with a semi-circle going under the chin was clasped to the steel busk in my stays. In this constrained state I, and most of the younger girls, had to prepare our lessons" (Somerville, 1874). Mrs. Somerville's experience was by no means unique. "Mary Butt (Mrs. Sherwood) about 1780 did her lessons standing in stocks, with an iron collar round her neck, and a back-board strapped over her shoulders" (Gardiner, 1929).

REFERENCES

Armytage, W. H. G. (1964). *Four hundred years of English education*. London: Oxford University Press.

Bayne-Powell, R. (1939). *The English child in the eighteenth century*. New York: Dutton.

Bickley, A. C. (Ed.). (1889). *The gentleman's magazine library: Bibliographical notes*. London: Elliot Stock, 62, Paternoster Row, E.C.

Frank, R. G. Jr. (1977, January 28). Science in the context of Puritan society (book review of Charles Webster's *The Great Insaturation*). *Science, Jan. 28*, 385–386.

Gardiner, D. (1929). *English girlhood at school*. London: Oxford University Press.

Hogan, E. R. (1976). G. Brown and the mathematical correspondent. *Historia Mathematica, 3*, 403–405.

Lee, S. (Ed.). (1903). *Dictionary of national bography*. London: Macmillan.

Leybourn, T. (1817). *The mathematical questions proposed in the Ladies' Diary, 1704–1816*. London: Mawman.

Meyer, G. D. (1955). *The scientific lady in England 1650–1760*. Berkeley: University of California Press.

Ross, R.P. (1975). The social and economic causes of the revolution in the mathematical sciences in mid-seventeenth century England. *The Journal of British Studies*

Smith, D.E. (1925). *History of mathematics*. New York: Dover.

Somerville, M. (1874). *Personal recollections*. Boston: R. Brothers.

Stanton, D.M. (1957). *The English woman in history*. London: Allen & Unwin.

Taylor, E. G. R. (1954). *The mathematical practitioners of Tudor and Stuart England*. England: Cambridge University Press.

Taylor, E. G. R. (1966). *The mathematical practitioners of Hanoverian England*. England: Cambridge University Press.

Thompson, R. (1974). *Women in Stuart England and America*. London and Boston: Routledge and Kagan.

Tipper, J. (1704). *Ladies' Diary*.

White, C. L. (1970). *Women's Magazines 1693–1968*. London: 9M Joseph. (White mentions 70 printing presses reported in London in 1724. This number had increased to between 150 and 200 by 1757.)

ACKNOWLEDGEMENT

I am indebted to Professor Paul Seaver of Stanford University, who shared with me the kinds of questions an historian asks.

CHAPTER 3

CONVERSATIONS OF PARENTS AND CHILDREN WORKING ON MATHEMATICS[1]

Melfried Olson, Judith Olson, Claire Okazaki, and Thuy La
University of Hawai'i

The cognitively demanding language used as parents worked with their children on mathematical tasks was examined in a three-year study funded by the National Science Foundation Research on Gender in Science Technology Engineering and Mathematics Program. The participants in the study consisted of four types of parent–child dyads (mother–daughter, mother–son, father–daughter, father–son). The dyads were videotaped as they worked on three mathematical tasks, one representing each of three content strands, number, algebra, and geometry (National Council of Teachers of Mathematics [NCTM], 2000). Of the 114 parent–child dyads involved in the study, 32 of the dyads consisted of 16 children (eight girls and eight boys) where both mothers and fathers participated with their child in separate sessions. The children were recruited from third and fourth grade classrooms in elementary schools in Hawai'i with low socio-economic status. Hawai'i is diverse with no ethnic majority, and the 32 parents and 16 children in this study reflected this diversity as they reported their ethnicities as

International Perspectives on Gender and Mathematics Education, pages 33–54
Copyright © 2010 by Information Age Publishing

33

Hawaiian, Chinese, Marshallese, Filipino, Indonesian, Nepalese, Mexican, Native American, or Caucasian, and often times indicating a "mixed" ethnicity such as Hawaiian–Chinese, Filipino–Hawaiian, and the like.

The dyads were asked to work together on three tasks in a manner similar to the way they might work together on mathematics at home. Each of the three tasks was timed for 10 minutes. Since the focus of the study was on the language used by the dyads and not the tasks, the parents and children were informed that it was not necessary to complete the tasks in the time given, nor was it important if solutions were correct. After each task was explained to the dyad and the necessary materials provided, researchers remained accessible but did not intervene or interact with the dyads as they worked on the tasks. Following the review of the tasks, the dyads chose the order in which to work on the three tasks.

With each of the 32 dyads completing three different mathematical tasks, there were 96 different videotaped parent–child interactions. The videotapes of each of the 96 sessions were transcribed and then coded using an instrument developed by the researchers based on previous research on the development of coding instruments (Taum & Brandon, 2006; Brandon, Taum, Young, & Pottenger, 2008). The codes were created for levels of cognitively demanding language used by the parents and children. The coded data were analyzed for gender-related differences.

The study focused on the following research question: To what extent are there differences in the use of cognitively demanding language among four types of parent–child dyads (mother–daughter, mother–son, father–daughter, father–son) working together on mathematical tasks in number, algebra, and geometry that initiate high levels of interactions? The theoretical background for the study; the development of the mathematical tasks and the coding instrument; and the analysis of videotaped sessions, results, discussion, conclusions, and implications are presented in this chapter.

THEORETICAL BACKGROUND

Researchers have found that children's perceptions of their abilities in mathematics are directly influenced by parents' beliefs about their children's academic competencies (Bleeker & Jacobs, 2004; Frome & Eccles, 1998; Jacobs & Eccles, 1992) as well as academic performance (Ginsburg & Bronstein, 1993; Yee & Eccles, 1988). In addition, the gender of parents and children has been reported as an important factor contributing to academic performance (Eccles, Frome, Yoon, Freedman-Doan & Jacobs, 2000; Jacobs, 1991, Jacobs & Eccles, 1992). As a result of their longitudinal investigation of the Michigan Study of Adolescent Life Transitions, Frome and

Eccles (1998) reported that parents' perceptions had a stronger influence on children's perceptions than children's own grades. In particular, parents of daughters believed that their daughters had lower mathematical ability than did parents of sons, even when girls received better grades than boys in mathematics. Furthermore, mothers of sons overestimated their sons' mathematical abilities and underestimated their daughters' mathematical abilities. This is particularly relevant given that mothers' achievement-related beliefs were more strongly related to children's achievement than were fathers' beliefs.

Parents also communicate to their children beliefs about their abilities in mathematics by interactions involving homework. Homework often provides an avenue whereby parents can communicate either positive or negative expectations for their child's academic success. While the link between parents' involvement with homework and children's achievement has been found to be both positive and negative (Balli, Demo, & Wedman, 1998; Epstein, 1988; Hoover-Dempsey, Battiato, Walker, Reed, DeJong, & Jones, 2001), one reoccurring finding was that when parents provided uninvited help and monitoring with homework, children felt less competent in their abilities (Grolnick & Slowiaczek, 1994; Ng, Kenney-Benson, & Pomerantz, 2004; Pomerantz & Eaton, 2001). One possible reason children felt this way could be that parents tended to provide unsolicited or intrusive support when they perceived their child as not doing well in school (Ng et al., 2004; Pomerantz & Eaton, 2001). Bhanot and Javanovic (2005) examined homework interactions when parents provided unsolicited help, monitoring or reminding, which the researchers referred to as intrusive support. They found that even though boys were recipients of more intrusive parental support with homework, girls might be more sensitive to these intrusions, specifically when they involved mathematics.

Crowley, Callanan, Tenenbaum, and Allen (2001) found that parents demonstrated gender-related differences in conversations with their children during visits to a science museum. The researchers videotaped the parent–child conversations of 298 families with children from one to nine years of age while they visited interactive science exhibits in a museum. The parents were equally likely to talk to the boys or the girls about how to use the exhibits. There was no significant difference in whether boys or girls initiated engagement. However, boys were three times more likely than girls to hear explanations from their parents, and the type of explanations given was significant. Of the explanations given to boys, 22% were about causal connections, while for girls only 4% were of this type. Parents explained more to boys regardless of their age.

In another study, Tenenbaum and Leaper (2003) investigated parents' teaching language during engagement with science and non-science tasks among 26 sixth-grade and 26 eighth-grade children and their mothers and fathers. The families were predominantly from middle-income European-American backgrounds and were recruited from public schools, summer camps, and after-school activities. The researchers were interested in whether the type of activity was related to how parents used teaching-related speech with sons and daughters. Parents and children were observed interacting in biology, physics, computer, and interpersonal teaching tasks. The biology and physics tasks were selected to examine how the domain of science might influence the interactions between parents and children. The computer task was considered a masculine-oriented task, while the interpersonal task was considered more feminine. While there were no apparent differences between girls and boys in their science-related cognitions or behaviors, there were strong indications of differential treatment by parents. During interactions with their children, fathers tended to act differently within particular tasks depending on the gender of the child. Fathers used more cognitively demanding speech with sons than with daughters when working on a physics task, but not on a biology task. The researchers viewed biology as a more gender-neutral field of study, while physics was viewed as a more masculine task. There were no corresponding gender effects on the level of cognitively demanding language used by either parent during the computer or the interpersonal teaching tasks.

The findings suggest possible biases in what fathers considered appropriate learning activities for sons and daughters. The researchers suggested that perhaps fathers viewed physics as something to encourage in their sons but not their daughters. It is possible that fathers may be encouraging intellectual engagement in science-related activities in sons more than in daughters and therefore may unwittingly be contributing to a gender inequity in science achievement.

DEVELOPMENT OF MATHEMATICAL TASKS

In order to initiate a high level of interaction between parents and children, mathematical tasks were developed that allowed for multiple solutions and/or multiple solution methods. In the first year of the study, parent–child dyads were recruited to assist in the development of the tasks. Six of the parent–child dyads became known as *pilot* dyads. After each pilot videotaping session, parents and children were asked to provide feedback on the clarity of the directions of the tasks, whether the tasks were challenging and enjoyable, and what changes, if any, they would recommend. Researchers examined the video to determine if a desired level of communication was obtained.

The task shown in Figure 3.1 was used as an example (Olson, 1999; Olson, Olson, & Sakshaug, 2000) for designing the geometry task for the research study.

Based on the ideas from the sample task, the first pilot task for geometry was prepared and piloted. It is shown in Figure 3.2.

When reviewing the videotapes it became apparent that parents and the children spent most of the 10-minute session making the shapes in silence or with very little conversation. There was little time left in the session for them to talk about how the shapes were similar or different. It was also noted that the dyads did not grasp the idea of which shapes were appro-

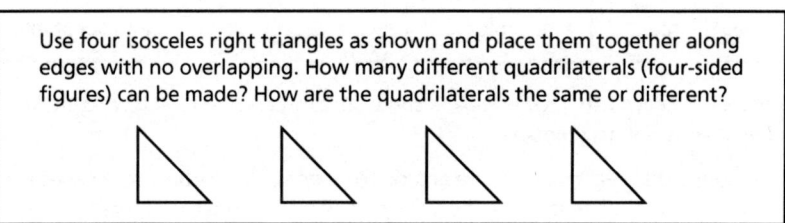

Figure 3.1   Sample task from which pilot geometry task was generated.

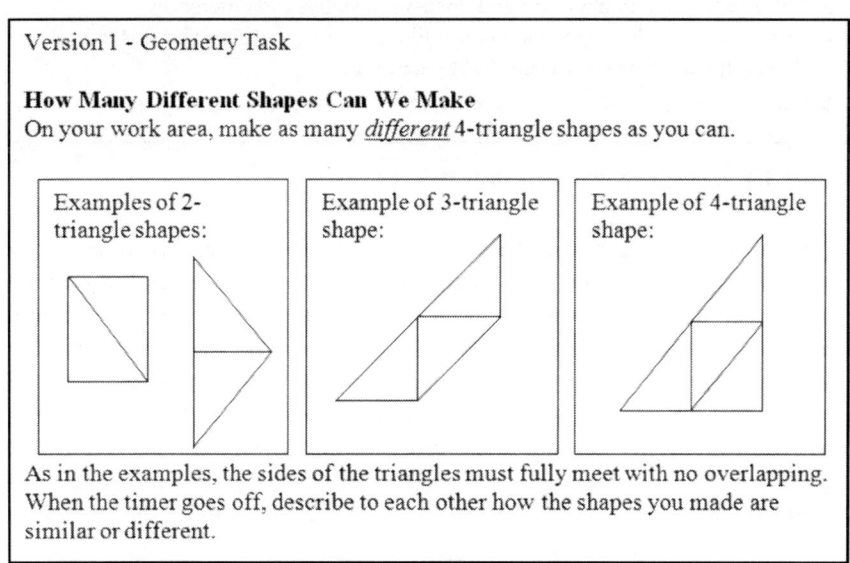

Figure 3.2   First pilot task for geometry.

Final Version – Geometry Task

Making Shapes with 4 Triangles

Your task is to make different shapes where each shape has 4 triangles. Here are acceptable and not acceptable shapes you can make.

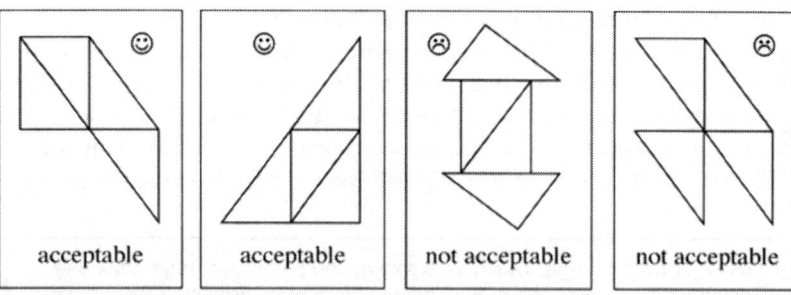

| acceptable | acceptable | not acceptable | not acceptable |

Discuss with each other what makes some shapes acceptable and what makes other shapes not acceptable.

1. On your work area, make 2 more acceptable shapes. Discuss how the shapes are similar to each other as well as how they are different from each other.
2. Leave your shapes on the work area and make another acceptable shape. Discuss how the shape is similar to and different from the ones on your work area.
3. Add another acceptable shape and discuss similarities and differences.
4. Sort the acceptable shapes you made into 2 or more groups. As you sort the shapes, discuss the reasons you made your groups.
5. Make other acceptable shapes. Discuss if they fit into the groups you made.

Figure 3.3 Final version of geometry task.

priate and which were not. Although past research (Leaper, Anderson, & Sanders, 1998) indicated that structured tasks were often associated with less gender differentiation, the interaction patterns between parents and children on pilot videotapes showed the need to revise the tasks to provide more structure to encourage more conversation and interaction between parents and children. Therefore, numbered instructions for items to discuss throughout the session were included. Following several additional iterations, a format for the geometry task generating the desired level of discussion was developed. The seventh and final version of the geometry task is shown in Figure 3.3.

The final version of the geometry task was successful in gaining the desired level of communication. The directions were clear, and the structured questions still allowed opportunities for discussions encouraging dialogue related to mathematical ideas. A parallel task using rhombuses instead of triangles was created for children who worked with the second parent.

As with the development of the geometry task, the number and algebra tasks went through similar revisions to optimize the opportunities for discussion and interaction between parents and children. These were also created with the goal to balance structure with the opportunity for generating discussion. The final versions of the number and algebra tasks are shown in Figures 3.4 and 3.5, respectively. Parallel tasks were also created for children who worked with the second parent. The parallel task for number used 18 in place of 24, while the parallel task for algebra used rhombuses instead of squares.

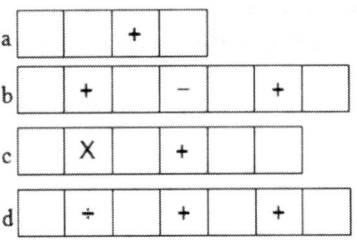

Final Version – Number Task
How Can You Represent 24?

The number 24 can be represented using the digits 0, 1, 2, 3, 4, 5, 6, 7, 8, 9 and operations +, −, X, ÷.

Three examples of representations of 24 are shown:

| 5 | X | 5 | − | 1 | | 6 | X | 2 | + | 1 | 2 |

| 1 | 0 | ÷ | 2 | + | 1 | 0 | + | 9 |

1. Discuss how these are representations of 24.
2. On your work area, use the digit and operation cards to make 3 more examples to represent 24.
3. What digit cards could you use to complete each of the following representation of 24? Make each representation on your work area.

a [ ][ ] + [ ]

b [ ] + [ ] − [ ] + [ ]

c [ ] X [ ] + [ ][ ]

d [ ] ÷ [ ] + [ ] + [ ]

4. Which operation cards could you use to complete the following representation of 24? Make the representation on your work area.

| 8 | | 2 | | 8 |

Figure 3.4 Final version of number task

Final Version – Algebra Task

How Do the Trains Grow? Pattern #1

Here are the first three trains in a pattern:

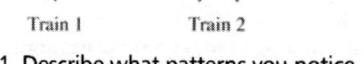

Train 1   Train 2   Train 3

1. Tell each other what patterns you notice.
2. Describe what Train 4 would look like. Build Train 4 on your work area.
3. If this pattern continues and a train has 6 triangles, which train number would that be? Discuss how many squares are in this train.
4. If this pattern continues, describe what Train 10 will look like.
5. Can there be a train in this pattern that has exactly 7 squares? Why or why not?
6. Clear your work area for pattern #2.

How Do the Trains Grow? Pattern #2

Here are the first three trains in a pattern:

Train 1   Train 2   Train 3

1. Describe what patterns you notice.
2. Describe what Train 4 would look like. Build Train 4 on your work area.
3. If this pattern continues and a train has 10 squares, which train number would that be? Discuss how many triangles are in this train.
4. If this pattern continues and a train has 14 triangles, which train number would that be? How many squares are in this train?
5. Discuss how you might find the number of triangles for any train number in this pattern.

Figure 3.5   Final version of algebra task.

DEVELOPMENT OF THE CODING INSTRUMENT

To code the language used by parents and their children for levels of cognitive demand, a team consisting of a researcher with experience coding video and two research assistants was formed to develop a coding instrument. The coding team transcribed the six pilot study videotapes verbatim, studied the videos, and created a list of 10 possible coding headings. The list included how dyads got started, who took control of the tasks; types of questions, reasoning, mathematics vocabulary used; and some aspects of non-verbal communication such as who manipulated the shapes and number cards and how. From this basic coding instrument, the research assistants (coders) individually coded the transcript for each 10-minute session. The two coders then compared and discussed the results of their coding. It became apparent that trying to observe the non-verbal behaviors while attempting to code the language used by the participants was distracting to the coders. Hence, it was decided that only the written transcripts would be coded and that the non-verbal behaviors would be studied separately at a later time. The team developed several drafts of the coding instrument. Each time the coding instrument was refined, the coders re-coded the transcripts using the newly edited coding instruments. As Taum & Brandon (2006) noted, it is the actual coding process that provides the greatest impetus in modifying a coding instrument so that the instrument captures the ideas being studied.

After a year of coding trials and discussions within the project team, a coding guide was finalized. All transcripts of the 32 videotapes, resulting in 96 separate 10-minute tasks, were coded by one of the coders and reviewed by the second coder. The final step, the most time-consuming, required that coders identify and discuss any differences in codes until consensus was reached. Reconciliation between coders was arguably the most demanding aspect of the entire coding process. Any differences in codes were discussed until consensus was reached either among the coders or in consultation with other researchers on the project team.

The final coding instrument consisted of three main sections: Getting Started, Discussion, and Mathematics Vocabulary. The Getting Started section focused on which member of the dyad began the task or began reading from the task card. The Discussion section focused on the types of questions asked, who directed the tasks and how they were directed, types of explanations given, and encouragement provided. The Mathematics Vocabulary section focused on the types of mathematics vocabulary used. The results from the Discussion and Mathematics Vocabulary sections are discussed in this chapter.

The codes for the Discussion section are explained in Table 3.1. Questions (Q), explanations (E), and directing (D) by either parents or children were coded at three levels, while encouragement (CE) was coded with no levels specified. The Mathematics Vocabulary was coded in two levels, as shown in Table 3.2.

TABLE 3.1   Codes and Descriptions for Discussion Section of Coding Instrument

Code	Description
Questioning	
Q1	Questions formulated directly from the task card
Q2	Perceptual questions to elicit concrete or one-word answers
Q3	Conceptual questions to elicit abstract ideas or relationships
Explaining	
E1	Explanation of the task presented on the task card
E2	Explanation of the mathematics
E3	Explanation including reasoning about the mathematics
Directing	
D1	Directing to continue the process of working together
D2	Provide guidance by referring to the task of the mathematics, but not giving explanations
D3	Directing to extend the ideas or mathematics or to prompt for a reasoning response
Encouragement	
CE	Encouragement

TABLE 3.2   Codes and Descriptions for Mathematics Vocabulary Section of Coding Instrument

Code	Description
M2	Procedural or contextualizing familiar words to describe shapes and patterns
M3	Conceptual or more advanced mathematical terms

ANALYSIS

The coded transcripts of the videotapes were entered into ATLAS.ti, a software program used for qualitative data analysis. The software query tool was used to produce quantitative values for the codes for each of the dyads, the gender of the parent, and the gender of the child. The letters, M (mother), S (son), F (father), and D (daughter), were used to identify parent–child dyads. A combination of these letters was used to indicate both the type of dyad and the person who initiated the communication that was coded. These combinations are shown in Table 3.3.

The results are based on the multi-level coding analysis and the quantified codes for each gender or dyad. Identified differences among dyads regard-

TABLE 3.3 Combinations and Labels for Dyads

Dyad	Communication	Label
Mother–Son	Initiated by Mother	MS
Mother–Daughter	Initiated by Mother	MD
Father–Son	Initiated by Father	FS
Father–Daughter	Initiated by Father	FD
Son–Mother	Initiated by Son	SM
Daughter–Mother	Initiated by Daughter	DM
Son–Father	Initiated by Son	SF
Daughter–Father	Initiated by Daughter	DF

ing questioning, explaining, directing, encouragement, and mathematics vocabulary are provided below. In each of the three tasks—algebra, geometry, and number—and the total across tasks (that is, the summation of algebra, geometry, and number tasks), chi-square tests were used to compare the cell frequencies of the codes for each of the following four groupings:

1. M, F
2. MD, MS, FD, and FS dyads
3. D, S
4. DM, SM, DF, and SF dyads.

If there were significant differences found in any of these four comparisons, further chi-square tests were used to determine which pairs of dyads were significantly different. The alpha value level for significance was set at $p < 0.05$.

RESULTS AND CONCLUSIONS

The tables that follow are organized by the categories of gender combinations for the dyads reported in this chapter.

Category 1: Comparisons of M-F and S-D provided comparisons of overall differences between gender of parents or differences between gender of the children.

Category 2: Comparisons of FD-MD, FS-MS, SM-SF, DF-DM showed differences in verbal interactions between mothers and fathers with their daughter or with their son. Since both fathers and mothers were videotaped working with their daughter or with their son in different sessions, this comparison revealed how the language of parents of different genders was used with the same child.

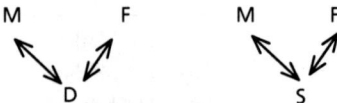

The results reported in this chapter focus on Categories 1 and 2 reporting only on gender-related differences in the language used when both a mother and father worked with the same child. That is, any comparison made was between genders of the parents or when parents were working with the same child. By limiting our analysis to these categories, we examined the best scenario possible to determine the effect of the gender of the parent on language used with a child. Hence, we do not discuss comparisons for other possible gender-related differences that might occur between the following dyad combinations: MD-MS, FD-FS, MD-FS, FD-MS, SM-DM, SM-DF, DF-SM, and DF-SF. Analysis of dyads in these categories will be conducted within the larger sample of 114 dyads and reported separately elsewhere.

Results of Level 3 Codes

Since Level 3 codes represented the highest level of cognitive demanding language used during parent and child interactions, the significant differences found in Level 3 related to questioning, mathematics vocabulary, directing, and explaining, sorted by individual tasks and across tasks are shown in Table 3.4. The differences in codes listed at Level 3 are discussed in the order in which they are listed in Table 3.4.

As shown in Table 3.4, fathers asked significantly more Level 3 questions (Q3) than mothers across tasks. When working with their daughters across

TABLE 3.4 Differences in Level 3 and Encouragement Codes by Mathematical Task Content

Level 3 Codes	Algebra	Geometry	Number	Across Tasks
Q3 (questions)	FD > MD	—	—	F > M FD > MD*
M3 (vocabulary)	—	M > F* MD > FD*	—	—
D3 (directing)	—	—	—	—
E3 (explaining)	—	M > F* MS > FS	—	—

* denotes significance at $p < 0.01$. All differences are significant at $p < 0.05$.

tasks and on the algebra task, fathers asked more Level 3 questions than mothers when working with daughters. Apparently, fathers were asking daughters more cognitively demanding questions that were conceptual in order to elicit abstract ideas or relationships than mothers were asking of daughters.

The only significant differences in mathematics vocabulary were found on the geometry task. Mothers used significantly more conceptual or advanced mathematical terms (M3) when working on the geometry task than fathers used. Mothers also used more Level 3 mathematical vocabulary working with their daughters than fathers used working with their daughters. In light of previous research reporting that females had less well-developed spatial skills than males (Levine, Huttenlocher, Taylor, & Langrock, 1999) and efforts to increase females' spatial skills (Casey, Andrews, Schindler, Kersh, Samper, & Copley, 2008; Medical News Today, 2007), this should be considered very promising. It also is a surprising finding since prior research showed gender differences on spatial skills and geometry (Casey, Nuttall, & Pezaris, 2001; Organisation for Economic Co-operation and Development [OECD], 2004), that girls had fewer out-of-school spatial experiences (Baenninger & Newcombe, 1995), and males in the United States still outperformed females on mathematics/space and shapes scales (OECD, 2004). In addition, Geary and DeSoto (2001) found that adult females were overrepresented at the low end of the Mental Rotations Test while males were over-represented at the high end of the Mental Rotations Test.

Therefore, if mothers were using Level 3 mathematical vocabulary, they were probably having a higher cognitively demanding discussion with their daughters about geometry. Crowley and Callanan (1998) pointed out that as parents used advanced vocabulary, they treated their children in a manner that suggested they believed children were capable of mastering a particular domain. Optimistically, this may indicate that daughters might be receiving a positive message of competence from their mothers related to geometry.

There were no significant differences found in Level 3 directing. Furthermore, almost all directing done by parents did not involve extending the ideas, mathematics, or prompting for a reasoned response (D3), but rather directed to continue the process of working together (D1) or directed the task but did not give any explanations (D2).

As was found with Level 3 mathematics vocabulary (M3), mothers provided significantly more Level 3 (E3) explanations that included reasoning about mathematics than fathers on the geometry task. However, the significant difference on Level 3 explanations indicated mothers used more Level 3 explanations with sons than fathers used with sons on the geometry tasks.

It is interesting to note that on the geometry task mothers used significantly more Level 3 mathematics vocabulary (M3) with daughters than fathers used with daughters, while in the explanation code mothers used significantly more Level 3 explanations (E3) with sons than fathers used with sons.

Results by Coding Areas

While the previous sections focused on Level 3 codes, this section reports significant differences found in Level 2 codes for mathematics vocabulary, questioning, directing, explaining, and encouragement. Level 2 codes provided useful information on the nature of conversations of parents and children as they worked on mathematics. While discussed earlier, Level 3 results are also provided in the following tables as an additional frame of reference.

Mathematical Vocabulary

As can be seen in Table 3.5, when working with children across tasks and on algebra and geometry, mothers used significantly more Level 2 (M2) mathematics vocabulary or mathematical language that was procedural or contextualized familiar words to describe mathematics. Daughters also used significantly more Level 2 mathematics vocabulary than sons used when working on the algebra task.

Further results on Level 2 mathematics vocabulary indicated that on the algebra task and across tasks, mothers used significantly more procedural or contextualizing mathematics vocabulary working with daughters than fathers used working with daughters. Mothers also used significantly more Level 2 mathematics vocabulary with sons than fathers used with sons when working with geometry and across tasks.

When examining these results, it is noted that mothers tended to use more mathematics vocabulary than fathers and with both daughters and sons but at different levels with different tasks. Interestingly, on the geometry task mothers used significantly more Level 3 mathematics vocabulary (M3), which represented conceptual or more advanced mathematical terms, with daughters than fathers used with daughters while mothers used significantly more Level 2 vocabulary with sons than fathers used with sons.

TABLE 3.5   Differences Found in Uses of Mathematical Vocabulary

	Differences found in categories	
	1	**2**
M2—Algebra	D(48) > S*(23) M(43) > F*(16)	MD(37) > FD*(10)
M2—Geometry	M(46) > F(29)	MS(24) > FS*(6)
M2—Across Tasks	M(92) > F*(44)	MS(31) > FS*(12) MD(61) > FD*(32)
M3—Geometry	M(47) > F*(25)	MD(27) > FD*(9)

Note: Numbers in parenthesis indicates usage counts
* denotes significance at $p < 0.01$. All differences are significant at $p < 0.05$

TABLE 3.6 Differences Found in Uses of Questioning

	Differences found per category	
	1	2
Q2—Algebra	M(255) > F*(200)	MS(145) > FS*(94)
Q2—Geometry	—	MD(90) > FD*(58)
Q2—Across Tasks	M(639) > F*(531)	MD(303) > FD*(238)
Q3—Algebra	—	FD(13) > MD(4)
Q3—Across Tasks	F(34) > M(16)	FD(21) > MD*(6)

Note: Numbers in parenthesis indicate usage counts.
* denotes significance at $p < 0.01$. All differences are significant at $p < 0.05$.

Questioning

Questions asked by either parents or children were coded in three levels. Questions formulated directly from the task card were coded as Q1; perceptual questions to elicit concrete or one-word answers were coded as Q2; and conceptual questions to elicit abstract ideas or relationships were coded as Q3. However, only results of Q2 and Q3 codes provided much information about questioning and are reported in Table 3.6.

While fathers asked significantly more Level 3 questions (Q3) and, in particular, more when working with daughters on the algebra task, mothers asked more Level 2 questions (Q2) working with daughters than fathers working with daughters. Mothers asked significantly more Level 2 questions (Q2) when working with sons on the algebra task than fathers working with sons, and also significantly more Q2 level questions working with daughters on the geometry task and across tasks than fathers working with daughters. Therefore, while mothers asked considerably more questions, these were more frequently at Level 2 (Q2).

Explanations

The verbal interactions between parents and children were coded to capture three levels of explanations provided by either parents or children. Whenever parents or children explained the task presented on the card, the language was coded as E1; explanation of the mathematics was coded as E2; while E3 was used when the explanation also included reasoning about the mathematics. The summary of differences found in E2 and E3 is given in Table 3.7.

The significant results for E3 were that on geometry task, mothers explained with reasoning (E3) more than fathers and that mothers explained with reasoning (E3) more with sons than fathers with sons.

TABLE 3.7   Differences Found in Uses of Explaining

	Differences found per category	
	1	2
E2—Algebra	F(80) > M(56)	—
E2—Geometry	S(124) > D(92)	FS(62) > MS*(32)
	F(108) > M(76)	SF(83) > SM*(41)
E2—Across Tasks	F (225)> M(180)	FS(122) > MS*(81)
		SF(225) > SM*(168)
		DM(195) > DF(156)
E3—Geometry	M(33) > F*(13)	MS(19) > FS(7)

Note: Numbers in parenthesis indicate usage counts.
* denotes significance at $p < 0.01$. All differences are significant at $p < 0.05$.

The significant differences for E2 are:

- On algebra, geometry, and across tasks, fathers explained mathematics (E2) more than mothers.
- On geometry, sons explained mathematics (E2) more than daughters.
- On geometry and across tasks, fathers explained mathematics (E2) more with sons than mothers with sons.
- Across tasks, and on geometry tasks, sons explained mathematics (E2) more with fathers than with mothers.
- Across tasks, daughters explained mathematics (E2) more with mothers than with fathers.

It would be expected that when parents asked more questions, children would answer these questions resulting in more explanations on the part of the child. This did indeed occur, however only at the E2 level. As reported earlier, mothers asked significantly more Level 2 questions (Q2) with daughters than fathers asked with daughters across tasks and, in return, daughters explained (E2) significantly more with mothers than with fathers. No similar results were found for Level 3 questioning and explaining.

Directing

Interactions between parents and children were coded to capture three levels of directions provided by either parents or children. Whenever parents or children directed to continue the process of working together, the interaction was coded as D1. Directing by referring to the task or the mathematics, but not explaining, to guide the thinking was coded as D2. D3 was used when the interaction extended the ideas or prompted a reasoned response. As mentioned earlier, there were no significant differences found for D3. In fact, there were few instances of discussions that warranted the

TABLE 3.8   Differences Found in Uses of Directing

	Differences found per category	
	1	**2**
D2—Algebra	D(17) > S*(4)	MS(64) > FS(36)
	M(131) > F*(83)	
D2—Geometry	M(113) > F*(61)	MD(64) > FD*(29)
D2—Number	M(304) > F*(224)	MD(176) > FD*(120)
D2—Across Tasks	D(53) > S(32)	MS(241) > FS*(172)
	M(548) > F*(368)	MD(307) > FD*(196)

Note: Numbers in parenthesis indicate usage counts.
* denotes significance at $p < 0.01$. All differences are significant at $p < 0.05$.

D3 code; therefore, only the summary of differences found for D2 is given in Table 3.8.

Results of the analysis of directing showed mothers directed (D2) more than fathers on each task as well as across tasks, and daughters directed more than sons on algebra and across tasks. This contrasts with explaining, where fathers explained (E2) more than mothers on algebra, geometry, and across tasks. Similarly, across tasks mothers directed (D2) more with sons than fathers with sons, while across tasks fathers explained (E2) more with sons than mothers with sons.

In geometry, number, and across tasks, mothers directed more with daughters more than fathers with daughters. In algebra, mothers directed more with sons more than fathers with sons.

There are several parallels in Level 2 questioning and directing as seen in Tables 3.6 and 3.8. This is not surprising, as there could possibly be some directing referring to the task or the mathematics along with questioning, especially at the perceptual level (Q2) when asking questions that elicit concrete or one-word answers. Mothers tended to do significantly more of both directing and questioning at Level 2. Mothers did significantly more directing and explaining with sons than fathers with sons on the algebra task and significantly more with daughters than fathers did with daughters on the geometry task.

Encouragement

The language used by parents and children as they worked on the task was also coded for encouragement; however, no levels were defined. The summary of differences found in encouragement is given in Table 3.9. While there were no significant differences in encouragement between mothers and fathers or sons and daughters, there were significant differences found among the dyads. Across tasks and on number, fathers provided significantly more encouragement with daughters than mothers provided with daughters, while on algebra and across tasks, mothers provided significantly more

TABLE 3.9   Differences Found in Uses of Encouragement

	Differences found per category	
	1	2
Encouragement—Algebra	—	MS(36) > FS(17)
Encouragement—Number	—	FD(23) > MD(8)
Encouragement—Across Tasks	—	MS(86) > FS(56)
		FD(44) > MD(20)

Note: Numbers in parenthesis indicate usage counts
All differences are significant at $p < 0.05$

encouragement with sons than fathers provided with sons. This suggests that parents tended to provide more encouragement with their children of the opposite gender.

CONCLUSIONS AND IMPLICATIONS

While mothers overall appeared to talk more with their children than fathers, there were different ways that parents used language with their children when working with mathematics. These differences were spread out among the different levels of codes for mathematics vocabulary, questioning, explaining, directing, and encouragement as well as with the three different tasks. There were three significant gender differences at Level 3 that stood out: (1) Mothers used significantly higher levels of mathematics vocabulary (M3) with daughters than fathers used with daughters when working on the geometry task, (2) mothers provided more high cognitive level explanations (E3) with their sons than fathers did with their sons, and (3) fathers asked significantly higher levels of cognitively demanding questions (Q3) with daughters than mothers asked with daughters on the algebra task.

By listening to parents' use of cognitively demanding language while working on mathematics tasks, children might emulate this type of talk. The results given above provide an important start for understanding how parents interact with children, including the type of language used when working together on mathematical tasks. The language that parents use can help children develop an interest in mathematics as well as to think about mathematical ideas at a deeper level.

Some of the identified gender-related differences could be attributed to the way the mathematical tasks were structured. The tasks were designed to be accessible to all participants so they could easily engage in the mathematics involved in the task. However, the directions and questions in the tasks provided opportunities for parents and children to use language at higher cognitive levels. The nature of the content of each task provided for distinct types of conversations. Therefore, the mathematical tasks determined the activity, and then the parents and children within this setting constructed their interpretation of the activity. Closer examination of conversations that occurred at various points in the tasks provided further insights into the gender-related differences that were found. At this time the data have not yet been analyzed at this depth. The fact that gender differences were found with structured tasks is noteworthy since in previous research (Leaper et al., 1998) structured tasks were found to show less gender differentiation.

One aspect of the findings in this study important to the field of gender studies concerns the encouragement fathers and mothers gave to their children. While encouragement was not coded at different levels, it was found that mothers encouraged sons more than fathers encouraged sons on the algebra task. Paralleling these findings were the results that mothers also asked more Level 2 (Q2) questions with sons than fathers asked with sons on the algebra task. This could indicate that asking perceptual questions requiring short answers was linked to the encouragement mothers gave when working with sons. It was also found that fathers encouraged their daughters more than mothers encouraged their daughters on the number task.

Children form their perceptions of their self-efficacy in mathematics through interactions with more advanced members of their community, such as parents. Parents who convey to children that they can understand mathematics likely produce self-efficacious children and influence subsequent academic and career choices of children (Bandura, 1997; Harter, 1992). Crowley and Callanan (1998) remind us that when parents take the time to explain to their children and ask conceptual questions they are providing their children with skills to become more competent in mathematics. Therefore, when engaging in social interactions such as homework, the encouragement conveyed and language used by parents play important roles. Thirteen mothers reported being the primary person in the family who helped children with homework, while 12 parents reported that both mother and father helped with homework. Only five fathers reported that they were the primary person who helped with homework. With mothers playing such a prominent role in working with children on their homework, they are likely to influence their children in the way their children view mathematics. Bhanot and Javanovic (2005) suggested that girls' confidence might be eroded through the interactions they have with parents regarding mathematics homework since they were susceptible to even the most subtle and seemingly innocuous parental behaviors including the lack of encouragement.

While researchers in this study made efforts to be unobtrusive as parents and children worked together, it is unclear to what extent the language used by parents and children while they were being videotaped was natural and represented the way they worked together on mathematics at home. The snapshot provided here might not occur on a regular basis when parents and children work together on mathematics.

The choices researchers made about what and how to code language was based on previous research and experience with coding; however, other choices may have resulted in different gender-related findings. It is possible that language used by parents and children differed in other ways than those that were identified.

The importance for parent involvement in the mathematics learning of their children has been described in the literature (Burrill, 1996; Civil, 2001; Perissini, 1998). However, defining the nature of the parents' role remains a question for leaders in mathematics education (Perissini, 1998). Understanding the ways parents and children work together on different mathematics content areas will provide needed information for all stakeholders in children's mathematics learning and in gender-related studies. By examining different parent–child contexts, researchers can come to understand how parents help children make important decisions related to their mathematics achievement and their future study of mathematics. While the sample size was small, the findings from this research could be useful for the development of parent materials and parent involvement programs to address the gender differences in language and actions of parents and their children when they work on mathematics.

NOTE

1. The research reported in this chapter was generated by the grant, *The Role of Gender in Language Used by Children and Parents Working on Mathematical Tasks (#0522946)* funded by National Science Foundation Research in Gender in Science and Engineering (GSE) program. The views expressed in this chapter are the views of the authors and do not necessarily represent the views of the National Science Foundation.

REFERENCES

Baenninger, M., & Newcombe, N. (1995). Environmental input to the development of sex-related differences in spatial and mathematical ability. *Learning & Individual Differences, 7,* 363–379.

Balli, S. J., Demo, D. H., & Wedman, J. F. (1998). Family involvement with children's homework: An intervention in the middle grades. *Family Relations, 47*(2), 149–157.

Bandura, A. (1997). *Self-efficacy: The exercise of control.* New York: W. H. Freeman

Bhanot, R. & Javanovic, J. (2005). Do parents' academic gender stereotypes influence whether they intrude on their children's homework? *Sex Roles, 52*(9/10), 597–607.

Bleeker, M. M., & Jacobs, J. E. (2004). Achievement in math and science: Do mothers' beliefs matter 12 years later? *Journal of Educational Psychology, 96*(1), 97–109.

Brandon, P. R., Taum, A. K. H., Young, D. B., & Pottenger, F. M. (2008). The development and validation of the Inquiry Science Observation Coding Sheet. *Evaluation and Program Planning, 31,* 247–258.

Burrill, G. (1996, May/June). President's message: We are making progress. *National Council of Teachers of Mathematics News Bulletin, 10,* 3.

Casey, B. M., Andrews, N., Schindler, H., Kersh, J. E., Samper, A., & Copley, J. (2008). The development of spatial skills through interventions involving block building activities. *Cognition and Instruction, 26*(3), 269–309.

Casey, M. B., Nuttall, R. L., & Pezaris, E. (2001). Spatial-mechanical reasoning skills versus mathematics self-confidence as mediators of gender differences on mathematics subtests using cross-national gender-based items. *Journal for Research in Mathematics Education, 32*(1), 28–57.

Civil, M. (2001, April). *Redefining parental involvement: Parents as learners of mathematics.* Paper presented at the National Council of Teachers of Mathematics research pre-session, Orlando, FL.

Crowley, K. & Callanan, M.A. (1998). Identifying and supporting shared scientific reasoning in parent–child interactions. *Journal of Museum Education, 23*(1), 12–17.

Crowley, K., Callanan, M. A., Tenenbaum, H. R., & Allen, E. (2001). Parents explain more often to boys than to girls during shared scientific thinking. *Psychology of Mathematics Education PME–XI, 12*(3), 258–261.

Eccles, J. S., Frome, P., Yoon, K. S., Freedman-Doan, C., & Jacobs, J. (2000). Gender-role socialization in the family: A longitudinal approach. In T. Eckes & H. M. Trautner (Eds.), *The developmental social psychology of gender* (pp. 333–360). Mahwah, NJ: Lawrence Erlbaum Associates.

Epstein, J. (1988). *Homework practices, achievements, and behaviors of elementary school students* (ERIC Document Reproduction Service No. PS 017 621). Baltimore, MD: Center for Research on Elementary and Middle Schools, Johns Hopkins University

Frome, P. M., & Eccles, J. S. (1998). Parents' influence on children's achievement-related perceptions. *Journal of Personality and Social Psychology, 74*(2), 435–452.

Geary, D. C. & DeSoto, M. C. (2001). Sex differences in spatial abilities among adults from the United States and China. *Evolution and Cognition, 7*(2), 172–177.

Ginsburg, G. S., & Bronstein, P. (1993). Family factors related to children's intrinsic/extrinsic motivational orientation and academic performance. *Child Development, 64*(5), 1461–1474.

Grolnick, W. S., & Slowiaczek, M. L. (1994). Parent's involvement in children's schooling: A multidimensional conceptualization and motivational model. *Child Development, 65*(1), 237–252.

Harter, S. (1992). The relationship between perceived competence, affect, and motivational orientation within the classroom: Processes and patterns of change. In A. K. Boggiano & T. S. Pittman (Eds.), *Achievement and motivation. A social developmental perspective* (pp. 77–114). New York: Cambridge University Press.

Hoover-Dempsey, K. V., Battiato, A. C., Walker, J. M. T., Reed, R. P., DeJong, J. M., & Jones, K. P. (2001). Parental involvement in homework. *Educational Psychologist, 36*(3), 195–209.

Jacobs, J. E. (1991). The influence of gender stereotypes on parent and child mathematics attitudes. *Journal of Educational Psychology, 83*(4), 518–527.

Jacobs, J. E., & Eccles, J. S. (1992). The impact of mothers' gender-role stereotypic beliefs on mothers' and children's ability perceptions. *Journal of Personality and Social Psychology, 63*(6), 932–944.

Leaper, C., Anderson, K. J., & Sanders, P. (1998). Moderators of gender effects on parents' talk to their children: A meta-analysis. *Developmental Psychology, 34*(1), 3–27.

Levine, S.C., Huttenlocher, J., Taylor, A., & Langrock, A. (1999). Early sex differences in spatial ability. *Developmental Psychology, 35*(4), 940–949.

Medical New Today. (2007). *Playing Video Games Reduces Sex Differences In Spatial Skills.* Retrieved October 31, 2008, from http://www.medicalnewstoday.com/articles/84625.php

National Council of Teachers of Mathematics (NCTM). (2000). *Principles and standards for school mathematics.* Reston, VA: NCTM.

Ng, F, F., Kenney-Benson, G. A., & Pomerantz, E. M. (2004). Children's achievement moderates the effects of mothers' use of control and autonomy support. *Child Development, 75*(3), 764–780.

Organisation for Economic Co-operation and Development (OECD). (2004). *Learning for tomorrow's world: First results from PISA 2003.* Paris, France: OECD.

Olson, J. (1999). What shapes can you make? *Teaching Children Mathematics. 5*(6), 330–331.

Olson, J., Olson, M., & Sakshaug, L. (2000). Responses to the "What shapes can you make?" problem. *Teaching Children Mathematics, 6*(5), 308–309, 321–323.

Peressini, D. D. (1998). The portrayal of parents in the school mathematics reform literature: Locating the context for parental involvement. *Journal for Research in Mathematics Education, 29*(5), 555–582.

Pomerantz, E. M., & Eaton, M. M. (2001). Maternal intrusive support in the academic context: Transactional socialization processes. *Developmental Psychology, 37*(2), 174–186.

Taum, A. K. H., & Brandon, P. R. (2006, April). *The iterative process of developing an inquiry science classroom observation protocol.* Paper presented at the annual meeting of the American Educational Research Association, San Francisco, CA.

Tenenbaum, H. R., & Leaper, C. (2003). Parent–child conversations about science: The socialization of gender inequities? *Developmental Psychology, 39*(1), 34–47.

Yee, D. K., & Eccles, J. S. (1988). Parent perceptions and attributions for children's math achievement. *Sex Roles, 19*(5/6), 31–333.

CHAPTER 4

OUT-OF-SCHOOL-TIME (OST) PROGRAMS AS MATHEMATICS SUPPORT FOR FEMALES

Lynda R. Wiest
University of Nevada, Reno

Females have gained ground in mathematics in relation to males, yet many areas of concern remain that might be addressed, in part, by learning opportunities that take place outside of school hours. Some recent research findings in the U.S. indicate that girls now perform as well as boys on standardized mathematics tests (e.g., Hyde, Lindberg, Linn, Ellis, & Williams, 2008). Others show that females score below males on the mathematics sections of important standardized national and international tests. For example, males significantly outperformed females in 35 of 57 countries on the 2006 Programme for International Student Assessment (PISA) (Organisation for Economic Co-operation and Development, 2007). Gender gaps favoring males are particularly pronounced at the highest achievement levels (e.g., Andreescu, Gallian, Kane, & Mertz, 2008). Females' participation in mathematics (e.g., mathematics clubs/contests, college majors, and

International Perspectives on Gender and Mathematics Education, pages 55–86
Copyright © 2010 by Information Age Publishing
All rights of reproduction in any form reserved.

careers) also remains low. In the United States, for instance, females earned 57.5% of all 2005–06 bachelor's degrees but only 45.1% of those conferred in mathematics and statistics (Institute of Education Sciences, 2008). In the world of work, women comprise only 25% of U.S. workers classified into Computer and Mathematical Occupations (Bureau of Labor Statistics, 2008). Moreover, females display a host of negative affects toward mathematics in comparison with males, such as weak self-concept and self-confidence, anxiety, and lower interest (e.g., Frenzel, Pekrun, & Goetz, 2007; Halpern, Aronson et al., 2007; Ma & Cartwright, 2003; McGraw, Lubienski, & Strutchens, 2006).

Making better use of potential learning time outside of school offers one strategy for supporting and encouraging females in mathematics. Thus, the purpose of this chapter is to provide an overview of the concept of out-of-school-time (OST) academic opportunities with a particular focus on those designed and conducted for females in mathematics. OST programs, structured learning opportunities that occur outside of regular school hours, are increasing in number as well as in the professional and public attention they receive (Bouffard, Little, & Weiss, 2006; Cavanagh, 2007; Phi Delta Kappa International, 2006). In reporting the rise of U.S. academic camps, Cavanagh (2007) noted that the number of such programs for girls increased 140% in less than a decade. Reasons for increased focus on OST programs in the U.S. are legislative attention from the reauthorized No Child Left Behind (NCLB) Act of 2001[1] and the reformed Title IX,[2] as well as mounting research on the academic and other benefits that can result from participation in these programs (Bouffard et al., 2006; Froschi, Sprung, Archer, & Fancsali, 2003; Hughes, 2006–07; Lauer et al., 2006).

Research indicates that students make greater cognitive gains in mathematics and reading (the subjects studied most often) during the school year than during the summer (Alexander, Entwisle, & Olson, 2004) and may in fact show declined achievement test scores after the summer break (Cooper, Nye, Charlton, Lindsay, & Greathouse, 1996).[3] These tendencies are stronger for mathematics than language skills, thus making mathematics learning more dependent on structured opportunities (Alexander et al., 2004; Cooper et al., 1996). For such reasons, many researchers contend that OST programs in mathematics, as well as technology, are particularly important (Bouffard et al., 2006; Cooper et al., 1996; Wimer, Hull, & Bouffard, 2006).

Summer learning loss strikes students with greater learning needs, especially those from lower socioeconomic status (SES) backgrounds, harder than more advantaged students (Alexander et al., 2004; Burkam, Ready, Lee, & LoGerfo, 2004; Cooper, 2003; Phi Delta Kappa International, 2008). Alexander et al. (2004) state, "Upper SES children forge ahead, whereas lower SES children stagnate. This is 'summer slide' made vivid" (p. 34). Wimer and Gunther (2006) add, "[OST] programs can make the summer months an opportunity for [academic] growth, rather than a time when at-risk youth fall further behind" (p. 12). Thus, many OST programs focus on struggling or underrepresented student groups, such as racial/ethnic minorities, females, and students of low SES; gifted students may also be targeted due to perceived need for greater academic challenge than they may receive in regular school settings (e.g., Alexander et al., 2004; Andreescu et al., 2008; Reid & Roberts, 2006; Wimer & Gunther, 2006).

Given their potential educational benefits, OST programs need greater investment and research (Bouffard et al., 2006; Lauer et al., 2006). Research on females and the STEM (science, technology, engineering, and mathematics) disciplines is especially lacking (Froschi et al., 2003). Accordingly, this chapter provides a synthesis of professional literature on academic out-of-school-time programs, specifically, their structure, purpose, features, outcomes, barriers, and challenges, with special attention to mathematics programs for females. Data are also presented on a sample program, the Northern Nevada Girls Math & Technology Program.

Although data presented in this chapter are based on U.S. programs,[4] the important OST characteristics discussed crosscut different types of programs and thus have merit for readers anywhere who have an interest in this topic. Although the concept of summer typically means a 12-week break in the U.S., that is not always the case. As of 2007, 3000 schools were year-round (National Association for Year-Round Education, 2007), which means they have much shorter breaks. Moreover, the literature summarized here includes after-school programs and programs as short as one day in duration. Therefore, the ideas addressed in this chapter offer general concepts likely to be useful to most countries interested in maximizing out-of-school-time academic learning, regardless of their school-calendar structure.

ACADEMIC OUT-OF-SCHOOL-TIME (OST) PROGRAMS

In this section, common characteristics of academic OST programs are described, as well as program outcomes, barriers, and challenges. Literature for OST programs in general is integrated with specific data on the narrower topic of intervention programs for females in the STEM disciplines, particularly mathematics.

Structure of OST Programs

OST programs occur most often after school and in the summer (Lauer et al., 2006), although Saturday programs also exist. They tend to take place in school or university settings. Summer programs may be residential or daytime-only, and program length might range from one day to several months. OST programs range from remedial services to academic acceleration or enrichment. Intervention programs, a specific type of OST program, are early attempts to support special populations before potential needs become manifest, rather than serving as "after-the-fact" remedial efforts (National Council of Teachers of Mathematics, 2008). OST programs in the form of interventions typically target a specific population, such as low-income children or females in the STEM disciplines.

OST programs span grade levels from elementary through higher education, but STEM programs seem particularly prevalent in middle and high school. This makes sense in that Lauer et al.'s (2006) meta-analysis of mathematics achievement in 22 OST programs revealed that the main benefits occur at the secondary level. Moreover, many contend that the middle-grades level is the lynchpin for addressing mathematics achievement, attitudes, and participation, especially for females (Andreescu et al., 2008; Blue & Gann, 2008; Halpern, Aronson et al., 2007; Lawhead, Loyd, Schep, Laws, & Price, 2005; Ma & Cartwright, 2003; Stevens, Wang, Olivárez, & Hamman, 2007). This is the time when adolescents become particularly sensitive to societal and peer perspectives, as well as when many make decisions about future course enrollments and career paths (Andreescu et al., 2008; Stevens et al., 2007).

Purpose of OST Programs

Academic OST programs promote out-of-school-time learning for all students or particular populations, as noted. The foremost goal is to develop and extend knowledge and skills in a particular domain, often with attention to building conceptual understanding and applying learning meaningfully (e.g., Fiebrink & Alcott, 2003; Morrow, 2006; Reid & Roberts, 2006; Wiest, 2004). An important goal of many STEM OST programs is participation and persistence—that is, to encourage a particular population (e.g., females or African Americans) or students in general to continue taking STEM coursework and consider STEM careers (Edwards, Kahn, & Brenton, 2001; Morrow, 2006; Pollock, McCoy, Carberry, Hundigopal, & You, 2004; Reid & Roberts, 2006; Roberts-Ohr, 2008; Wiest, 2004; Wolbach, 2007). Motivating students to attend college in general may be an objective (Wolbach, 2007). Another key intent of OST programs, especially interventions, is to improve attitudes and beliefs related to the subject matter of focus (Lawhead et al., 2005; Pollock et al., 2004; Reid & Roberts, 2006; Wiest, 2004). Finally, a goal articulated for OST interventions for females in STEM is providing female role models and mentoring (Edwards et al., 2001; Pollock et al., 2004; Roberts-Ohr, 2008; Wiest, 2004).

Features of Effective OST Programs

Out-of-school-time programs must be of high quality to be effective (Bouffard et al., 2006; Miller, 2003). Effective programs are those that yield positive results in relation to the types of worthwhile educational goals delineated earlier, in particular, intent to improve participant content knowledge and skills, dispositions, and participation and persistence in one or more disciplines.

Key features of successful OST programs for females in STEM (especially mathematics), as gleaned from recent professional literature, are discussed below. These features include:

- sufficient duration and intensity
- good planning, structure, administration, and organizational support
- adequate facilities, equipment, materials, and funding
- clearly articulated program mission, goals, and objectives, with high expectations for all participants
- strong, focused academic curriculum with enrichment opportunities
- social/recreational aspect
- concerted attention to attitudes and beliefs
- single-sex setting
- positive, supportive climate, with a cohesive participant/staff community
- high-quality, well-trained, diverse staff
- student-centered instruction (especially hands-on learning and small-group work)
- female role models and mentors
- career and academic preparation information
- meaningful connections with families, academic and non-academic entities, and the community
- ongoing evaluation and revision
- program promotion and publicity
- selected other features (e.g., residential versus day settings and inclusion of a culminating experience)

Sufficient Duration and Intensity

The timeframe when OST programs are offered—summer, after school, or weekend—does not seem to significantly influence impact, but maximally effective programs appear to be of longer duration (Froschi et al., 2003; Lauer et al., 2006; Miller, 2003). As a result of their meta-analysis of OST programs, Lauer et al. (2006) set the minimum amount of time at 45 hours and, based on a review of the research, Miller (2003) adds that program intensity also matters. Wiest's (2008a) recommendation for continued participant contact/support through, for example, follow-up sessions or a program web site, resonates with the idea of sufficient program duration. In addition, Wong (2006) suggests that Expanding Your Horizons' one-day conferences may have a stronger impact on participants if they were extended in some manner.

Good Planning, Structure, Administration, and Organizational Support

Effective OST programs require good planning, structure, and administration within a wider system of organizational support (Miller, 2003; NCTM, 2008; Wiest, 2008a). Program planning should include researching and incorporating key elements of successful, related OST programs.

Adequate Facilities, Equipment, Materials, and Funding

Sustainable programs need adequate facilities, equipment, and materials, as well as stable funding (Miller, 2003; Wiest, 2008a). The instruction room for one STEM program for females, for example, was described as "large with various seating configurations in different areas, lots of board space, a computer projection system, and lots of mathematical equipment and 'toys'" (Morrow, 2006, p. 24).

Clearly Articulated Program Mission, Goals, and Objectives with High Expectations for all Participants

Effective OST programs clearly delineate and make public their program mission, goals, and objectives (Miller, 2003; Phi Delta Kappa International, 2006; Roberts-Ohr, 2008; Wiest, 2008a). For example, the mission of the Northern Nevada Girls Math & Technology Program (see http://www.unr.edu/educ/girlsmathcamp) is:

> to support and encourage girls from varied backgrounds to increase their knowledge, skills, and confidence in mathematics, as well as technology use for mathematics learning. The program provides learning opportunities, resources, and participation in a community of support to better prepare girls to enhance the quality of their academic, vocational, and everyday lives and to contribute to advancement of the wider society.

Specific goals and objectives are established to serve the broader program mission. Examples for a mathematics program for girls might be:

- Program goal: To increase the number of females who seriously consider and who enter mathematics-related careers
- Participant objective: To gain greater knowledge of mathematics-related careers and the academic preparation needed to pursue them

In striving to meet articulated ends, high-quality OST programs set, and aim to maintain, high expectations for all participants (Miller, 2003; Phi Delta Kappa International, 2006).

Close Supervision with Clear Expectations and Consequences

Participants should be closely supervised, and expectations for behavior, as well as consequences for noncompliance, should be clear and enforced (Miller, 2003; Wiest, 2008a). Girls who attend the Northern Nevada Girls Math & Technology Program, as well as their parents, sign a participation agreement that details expected participant behaviors and indicates possible dismissal from the program for failure to uphold these expectations. One safety measure in this agreement is that participants are never to go anywhere without the group, which always has staff supervision. Staff are briefed initially, and reminded regularly, of the importance of consistently monitoring and enforcing the girls' behavior according to the participation agreement.

Strong, Focused Academic Curriculum with Enrichment Opportunities

Successful OST programs have a strong, focused academic curriculum designed to increase content knowledge and build skills (Alexander et al., 2004; Bouffard et al., 2006; Brown, 2008; Chacon & Soto-Johnson, 2003; Fiebrink & Alcott, 2003; Froschi et al., 2003; Lauer et al., 2006; Miller, 2003; Morrow, 2006; Wiest, 2004; Wimer & Gunther, 2006). Some researchers and OST program providers contend that enrichment learning and exposure to new ideas should be incorporated into all OST programs (Alexander et al., 2004; Chin & Phillips, 2005; Cooper, 2003; Froschi et al., 2003; Miller, 2003). Some also promote strong infusion of technology use due to its increasing importance in all aspects of life (e.g., Reid & Roberts, 2006; Wiest, 2004; Wimer et al., 2006). This might include using computers and graphing calculators as mathematics learning tools (e.g., Morrow, 2006; Wiest, 2004). SummerMath (Morrow, 2006), an intensive four-week mathematics, science, and computing program for high school girls, extends its learning opportunities beyond its express focus of STEM content by including workshops on important health-related topics. Expanding Your Horizons, a one-day conference, includes STEM professionals addressing how to balance career and family (Wong, 2006).

Social/Recreational Aspect

Some program directors emphasize the importance of social/recreational activities, and programs that include this aspect have been found to have a more positive influence on participants than those that involve academics alone (Chacon & Soto-Johnson, 2003; Lauer et al., 2006; Morrow, 2006; Pollock et al., 2004; Wiest, 2004). Examples include going roller skating, doing beadwork, and having a movie night. In reference to their summer mathematics camps, Chacon and Soto-Johnson (2003) say: "Social activities provided a much-needed break after the grueling hours of work" (p. 281). Besides creating a more balanced and thus desirable setting for young women, social and recreation time encourages the building of more complete relationships with peers and staff. In an OST environment, social time can also encourage academically oriented conversations among peers or between peers and staff that may not take place otherwise.

Concerted Attention to Attitudes and Beliefs

Ways to positively influence girls' confidence, self-efficacy, and interest in STEM, addressing stereotypes and misconceptions about these fields, are often implemented in STEM intervention programs for females (e.g., Fiebrink & Alcott, 2003; Froschi et al., 2003). This effort may also extend to academic behaviors. For example, in their Saturday GO-GIRL (Gaining Options: Girls Investigate Real Life) mathematics program, Reid and Roberts (2006) seek to help seventh-grade girls gain confidence to "speak up" in classes through improved interpersonal and communication skills.

Single-Sex Setting

Some researchers and practitioners contend that some intervention programs, such as STEM programs for females, should be single-sex to achieve maximum benefits (e.g., Lawhead et al., 2005; Morrow, 2006; Wiest, 2004). In coeducational settings, girls may be relegated to sharing the results of hands-on work that boys have conducted (Cavanagh, 2007). In all-female learning environments, girls have more opportunity to speak and are more comfortable reporting ideas and doing problems in front of others (Morrow, 2006). In general, they show better attitudes, focus, leadership, and academic gains (e.g., Spielhagen, 2006; Stutler, 2005). Across subject areas, single-sex education may be especially important in the middle grades (Spielhagen, 2006).

Positive, Supportive Climate with a Cohesive Participant/ Staff Community

An effective OST environment is physically and emotionally safe with positive peer relations developed through team-building and other activities; participants have a sense of belonging, relating to others, and mattering (Alexander et al., 2004; Bouffard et al., 2006; Lawhead et al., 2005; Miller, 2003; Phi Delta Kappa International, 2006; Reid & Roberts, 2006; Roberts-Ohr, 2008; Wiest, 2008a; Wimer & Gunther, 2006). Together, staff and participants form a sense of community (Morrow, 2006; Reid & Roberts, 2006).

High-Quality, Well-Trained, Diverse Staff

A vital component of good OST programs is a high-quality, well-trained staff that is both duly compensated and maintains some measure of continuity over time (Miller, 2003; Reid & Roberts, 2006; Wiest, 2004; Wimer & Gunther, 2006). Staff should know how to support the developmental growth of youth from a wide variety of backgrounds (Miller, 2003). Staff diversity itself (e.g., racial/ethnic) can enhance programs by, for example, providing a range of types of female role models. Positive staff–participant relationships and low adult-to-youth ratios are aspects of a strong program staff (Bouffard et al., 2006; Miller, 2003; Wiest, 2004). The staff at Summer-Math meet a full week before participants arrive in order to prepare for the program (Morrow, 2006).

Student-Centered Instruction (especially, hands-on learning and small-group work)

Topping the list of effective instructional techniques is the use of small-group work and hands-on learning. Small-group work, which emphasizes cooperation and communication, is a common arrangement in many OST programs (Chacon & Soto-Johnson, 2003; Lawhead et al., 2005; Miller, 2003; Morrow, 2006; Pollock et al., 2004; Reid & Roberts, 2006; Wiest, 2004). In their summer mathematics camps for academically talented high school girls, Chacon and Soto-Johnson (2003) used teams of four that were different in the morning and afternoon sessions, and Wiest (2004) organized the middle school girls in her summer camp into randomized groups of four that changed daily. The hands-on and physical tasks used in many programs, which may or may not be project-based, emphasize experiential, investigative explorations that require active engagement and reflection (Alexander et al., 2004; Chacon & Soto-Johnson, 2003; Froschi et al., 2003; Miller, 2003; Morrow, 2006; Pollock et al., 2004; Roberts-Ohr, 2008; Wiest, 2004; Wimer & Gunther, 2006). Discussing and sharing ideas in small- and whole-group settings are common practices. These active methods may be particularly important in out-of-school-time settings for engaging youth in extended academic work. They also encourage and accelerate interpersonal connections and community building in unfamiliar settings with new peers.

Female Role Models and Mentors

Many OST programs articulate the importance of staff members who engage frequently in positive, supportive interactions with participants (Miller, 2003; Wimer & Gunther, 2006). STEM intervention programs for girls typically provide diverse historical and/or contemporary female role models, female mentors, and opportunities to network with peers (Brown, 2008; Fiebrink et al., 2003; Froschi et al., 2003; Halpern, Aronson et al.,

2007; Lawhead et al., 2005; Pollock et al., 2004; Reid & Roberts, 2006; Roberts-Ohr, 2008; Wiest, 2004).

Career and Academic Preparation Information

Career-oriented information is a key element of effective STEM intervention programs for girls (Brown, 2008; Froschi et al., 2003; Lauer et al., 2006; Lawhead et al., 2005; Morrow, 2006; Pollock et al., 2004; Reid & Roberts, 2006; Roberts-Ohr, 2008; Wiest, 2004). Brown (2008) notes that students sometimes do not know the formal education required for particular careers and that this "may be particularly problematic for students from underserved populations or for those whose family members have not attended college" (p. 13).

Meaningful Connections with Families, Academic and Non-Academic Entities, and the Community

OST programs should build meaningful connections with participants' families and with schools and universities, non-school organizations and programs (e.g., similar OST programs), and the community at large (Bouffard et al., 2006; Froschi et al., 2003; Miller, 2003; Phi Delta Kappa International, 2006; Wiest, 2008a; Wimer & Gunther, 2006). School connections might involve sharing resources or aligning program curricula with those of state or local content standards (Bouffard et al., 2006; Wiest, 2004). Parents might be offered information, resources, or workshops on understanding the program's academic content, supporting girls at home, and obtaining career-related information (Wiest, 2008a; Wimer & Gunther, 2006). Parent sessions can help address the stereotypes that females may face at home and beyond, as well as the lower expectations family and society have for girls, especially in the STEM disciplines (Brown, 2008; Froschi et al., 2003). For these reasons, in addition to girls' tendency to have greater home responsibilities than boys, as well as the voluntary, self-selected nature of OST programs, outreach to girls and families is crucial (Froschi et al., 2003). Other ways to assist families include helping to facilitate carpooling or providing transportation, accommodating parent schedules, providing or assisting with childcare arrangements (e.g., facilitating daycare exchanges), and offering scholarships to participants in financial need (Chin & Phillips, 2005; Wiest, 2004; Wimer & Gunther, 2006). OST programs should also consider offering program materials, especially those provided to parents, in several dominant local languages. Likewise, linguistic diversity among staff can improve communication with families and the public at large.

Ongoing Evaluation and Revision

Effective OST programs are research-based and conduct ongoing evaluation and revision (Bouffard et al., 2006; Miller, 2003; Morrow, 2006; NCTM,

2008; Phi Delta Kappa International, 2006; Wiest, 2008a). Program assessment may include external observation but must also include both short- and long-term input from participants, parents, staff, and other important stakeholders. For longer programs, assessment of the program and/or its participants, with associated staff discussion and possible remedial action, may take place while the program is in progress. Young people appreciate an opportunity to voice their opinions and help shape programs. For example, on a recent exit survey for the Northern Nevada Girls Math & Technology Program's summer camp, a middle school girl wrote, "It's good that you are asking for what we think about this camp (other camps usually don't)."

Program Promotion and Publicity

On a practical note, promotion and publicity are important to reach potential participants, bring programs greater public prominence, and help attract funding (e.g., Chin & Phillips, 2005; Wiest, 2008a). Special efforts should be made to reach populations who may need these supplemental programs the most, such as low-income youth and particular racial/ethnic minorities. A program might be publicized, for example, to selected local youth service providers (e.g., in the U.S., the Boys and Girls Clubs of America).

Selected Other Features

Several other recommendations have been offered for successful OST program design and operation. Wiest (2008a) recommends residential rather than day programs to accommodate non-local participants, allow greater social/recreational opportunities, encourage community-building among participants and staff, and provide a more expansive college experience (if held in a university setting) that promotes familiarity with, and interest in, college in general. Similarly, Morrow (2006) states, "The residential program is critical to creating a strong, close-knit community at SummerMath" (p. 23). Engaging educational field trips might be part of an OST program (Wimer & Gunther, 2006). These might help serve program goals such as providing career awareness. Participants might also be given an item related to the program's academic focus to serve as a memento and to encourage continued exploration of the program's learning objectives. For example, the younger camper group in the Northern Nevada Girls Math & Technology Program receives a set of tangrams with a packet of tangram puzzle sheets, and the older group receives a puzzle-cube spatial task. The seventh-grade girls in Reid and Roberts' (2006) summer mathematics program are given graphing calculators. In both cases, participants use these products during the program and then keep them.

Besides their strong positive response to staying overnight in university dormitories, the middle school girls in the Northern Nevada Girls Math & Technology Program commented frequently on the amount and quality of food provided for meals and snacks during the summer camp (Wiest, 2008a). These types of experiences and provisions cannot be overlooked in maximizing an OST program's appeal. Finally, some programs end with an event where participants share their learning in such forms as demonstrations for parents or a "conference" where participants share and discuss the results of their work (Morrow, 2006; Wiest, 2008a).

Key OST Program Features for Females in STEM

Effective intervention programs for females in mathematics incorporate the features described above for both OST programs in general and, more specifically, for females in the STEM disciplines. As noted, OST programs should be of sufficient duration (at least 45 hours according to Lauer et al., 2006), and they appear to be more important for mathematics than for linguistic (e.g., reading) achievement. STEM programs have been found to be most effective at the secondary level, with middle school a particularly important crossroads. Single-sex settings are often advised. Some of the most salient mathematics OST program features for females include a focus on increasing participation and persistence, as well as on improving attitudes and beliefs. Girls need female role models and mentors, and they thrive best in supportive, communal instructional climates that emphasize hands-on tasks and cooperative group work. Residential programs that include social and recreational opportunities may especially serve some of these goals. Finally, mathematics OST programs for girls should provide career and career-preparation information and incorporate strategies for gaining active parent support to help validate, value, and encourage females as mathematicians.

Outcomes of OST Programs

Increased knowledge and skills gained through OST programs have been shown to translate to higher achievement (Miller, 2003; Phi Delta Kappa International, 2006; Reid & Roberts, 2006; Wiest, 2004; Wimer et al., 2006). These programs have also been associated with improved participant attitudes—in particular, confidence, as well as interest and enthusiasm, motivation and effort, and a sense of academic competence and of self as learner (Chacon & Soto-Johnson, 2003; Froschi et al., 2003; Miller, 2003; Morrow & Schowengerdt, 2008; Pollock et al., 2004; Reid & Roberts, 2006; Wiest, 2004; Wimer et al., 2006; Wong, 2006). High school girls who attended a single-sex mathematics OST program showed increased confidence and skill in explaining their mathematical thinking, as well as greater attention to and valuing of mathematical processes (Morrow & Schowengerdt, 2008).

Personal and social benefits also accrue from OST program participation, including self-confidence, independence, personal and social responsibility, social/collaboration skills, and exposure to new and/or nonacademic experiences (Froschi et al., 2003; Wiest, 2004; Wimer et al., 2006). Among the prosocial behaviors gained, girls have been shown to develop a greater propensity to assist others (e.g., friends, families, teachers) with their new learning (Wiest, 2004). An especially powerful outcome of participation in an OST program is a broader and stronger sense of community. This network involves "membership" in a group of peers who like STEM, relationships with female role models and mentors, and increased family involvement in participants' lives (Chacon & Soto-Johnson, 2003; Miller, 2003; Wiest, 2004; Wolbach, 2007).

Attending STEM intervention programs appears to increase girls' participation in STEM activities both in and out of school. Examples include joining a mathematics competition team, playing mathematics-oriented games, taking more STEM coursework, and pursuing STEM careers (Chacon & Soto-Johnson, 2003; Pollock et al., 2004; Wiest, 2004; Wolbach, 2007). Chacon and Soto-Johnson (2003), for example, observed that high school girls attending their summer mathematics camps spent time doing mathematics even during social/recreational time. OST program participants have also shown greater awareness of, and interest in, college and career options (Froschi et al., 2003; Morrow & Schowengerdt, 2008; Reid & Roberts, 2006; Wiest, 2004; Wimer et al., 2006; Wolbach, 2007; Wong, 2006).

Participants in OST programs tend to engage more fully and perform better in school compared with their previous behaviors and/or to that of peers. They take more STEM coursework, are better prepared for school, earn higher test scores and grades, display better academic skills and behaviors, and have better attendance, grade-promotion, and graduation rates (Froschi et al., 2003; Miller, 2003; Wiest, 2004). OST programs may also help participants bridge home and school cultures (Miller, 2003).

In sum, OST programs have been associated with positive outcomes in the form of academic achievement, affective attributes, personal and social behaviors and relationships, participation in STEM activities and career paths, college and career awareness and interest, and school engagement and performance. Miller (2003) notes that families and society, as well as individual participants, benefit from these gains.

Program Barriers and Challenges

Parents of all social classes value the opportunity to enroll their children in quality OST programs (Chin & Phillips, 2005). Nevertheless, a number of barriers make OST programs less accessible to students who are likely to need them most. Students from higher-income and better-educated families are more likely to enroll in OST programs than other youth (Chin & Phillips, 2005; Miller, 2003). Girls and students whose first language (or that of their parents) differs from the language of instruction are also less likely to participate in OST programs. Program participation barriers for these populations include finances, time, schedule conflicts, transportation, safety concerns, cultural/language mismatch, lack of cultural[5] or academic knowledge, greater out-of-school responsibilities for girls compared to boys, and stereotypes that yield lower expectations for girls in some pursuits, including the STEM fields (Chin & Phillips, 2005; Froschi et al., 2003; Miller, 2003). Compared with their counterparts, low-income and minority[6] parents report much less satisfaction with the quality, availability, and affordability of their OST options, as well as substantially stronger interest in academic compared to nonacademic programs (Duffett & Johnson, 2004).

Several challenges have been documented in relation to conducting OST programs beyond the amount of work required to plan and operate these programs and reliance on needed facilities, equipment, and materials. Securing ongoing funding is one such concern (Miller, 2003; Wiest, 2008a; Wimer & Gunther, 2006). The broad ability range of participants drawn from a wide geographic area, participant difficulty adjusting to program settings (especially residential programs), and interpersonal conflict among participants are other challenges (Wiest, 2008a). Finally, satisfying increasing accountability demands for demonstrating results can tax program staff (Wimer & Gunther, 2006).

A SAMPLE OST PROGRAM: THE NORTHERN NEVADA GIRLS MATH & TECHNOLOGY PROGRAM

In this section, an example of an OST program—specifically, a mathematics-technology intervention program for middle school girls—is described. Selected results from data collected on this program are shared. This program was designed to embody all of the important OST features detailed above, as the brief description below illustrates in limited measure. The "program promotion and publicity" feature is not addressed below. In short, local print, electronic, and visual media are routinely notified about the program's activities, presentations about the program are made to community groups and at professional meetings, papers are published about the program, a web site is under development, and production of a short film about the program is under consideration. The program illustrates the important OST features discussed previously, although there are several areas considered to be weaker aspects of the program. One is the feature termed "sufficient duration and intensity." This five-day residential program is intense, but it might be more effective if it were to involve extended contact with, and support for, participants after they have attended the program. The extensive, interactive web site under development is intended to serve this purpose by providing year-round services and resources to participants and their parents. The program has had adequate funding to date, but seeking financial support is time-consuming and attaining it is perpetually uncertain. Finally, the "career and academic preparation" feature of this program is limited due to the program's short duration. This aspect consists mainly of one guest speaker discussing her job, a session on famous female mathematicians/computer scientists, and some handouts. The forthcoming web site will help provide more information in this area. Despite these somewhat weaker program features, the program, described in more detail below, has been highly successful due to its overall strong observance of the features deemed important for effective OST programs.

The Northern Nevada Girls Math & Technology Program

Each year about 60 Northern Nevada middle school girls are selected randomly from a wide variety of applicants to participate in the Northern Nevada Girls Math & Technology Program, which began in 1998. The program's purpose is to increase girls' knowledge, skills, and confidence in mathematics and technology in order to enhance mathematical and technological competence in girls' personal, academic, and occupational lives. The main program component is a five-day residential summer camp held at the University of Nevada, Reno. Two full-day Saturday follow-up sessions that used to be held, one each in autumn and spring, were discontinued in 2007 in favor of efforts to complete a program web site to provide year-round services. The typical class size is 30 girls who work within their own grade level—those who will enter grade 7 or grade 8 in the next school year (here termed "rising" seventh or eighth graders). Scholarships are available to participants with demonstrated financial need.

Instructional topics include problem solving, geometry, spatial skills, data analysis and probability (younger group only), algebra (older group only), and more powerful uses of technology. Technology includes use of graphing calculators as well as the computer programs *Terrapin Logo* (Terrapin Software, http://www.terrapinlogo.com/terrapin-logo.php) and *The Geometer's Sketchpad* (Key Curriculum Press, http://www.keypress.com/x5521. xml). These components are deemed to be particularly important to girls' academic and vocational futures and may be areas that receive insufficient attention in the elementary and middle grades. There have, for example, been increasing calls for improving all U.S. students' algebra and geometry skills (e.g., Blume, Galindo, & Walcott, 2007; Kortering, deBettencourt, &

Braziel, 2005). Females underperform in relation to males in some of these topics. Particular concern has been expressed for girls' geometry and spatial skills, which lag behind those of boys (e.g., Halpern, Benbow et al., 2007; Liu, Wilson, & Paek, 2008). Technology use is intended to address females' low representation in computing degree programs and careers (Bureau of Labor Statistics, 2008; Institute of Education Sciences, 2008) and to serve as a powerful tool for learning mathematics (cf. Wiest, 2008b). Program participants also gain exposure to female role models in mathematics and technology through hearing a female guest speaker discuss the use of mathematics and/or technology in her job, learning biographical information about contemporary and/or historical female mathematicians and computer scientists, and working with an all-female staff.

The program staff consists of main instructors—highly recommended, experienced, local teachers who are active in mathematics education—and instructional assistants who are either teacher education majors near the end of their degree programs or beginning teachers.[7] Two to three instructors are present in each classroom at all times. An instructor who has special expertise in the mathematics software used in the program conducts the computer instruction segments. The instructors base their lessons on the Nevada Mathematics Standards (Nevada Department of Education, 2006) established for the grade level the girls will enter in the next school year. For most class sessions, the girls work collaboratively at tables in randomized groups of four that change daily. This practice encourages girls to meet new peers across the week, as well as to hear different perspectives and learn how to work with a variety of girls in a collaborative, communicative environment.

Instructors are asked to use instructional methods considered especially beneficial to girls. These approaches include a constructivist orientation to learning, which is based on a belief that each individual actively constructs her own knowledge through interacting with and reflecting on materials and ideas; this process is enhanced by interacting with a variety of other thinkers, including more knowledgeable others. The key pedagogical strategies employed are supported by the professional literature detailed earlier and include hands-on activities, mixed-ability cooperative group work, real-world applications, and problem solving and investigations in a supportive learning environment.

Instructional sessions take place in the College of Education. Participants and overnight staff engage in recreational activities each evening (e.g., roller skating and a movie night), stay in the university dormitories, and eat in the campus cafeteria. Parents are invited to participate in introductory and concluding activities during the summer camp.

Figure 4.1 shows a sample schedule (Tuesday) for the youngest participant group, rising seventh graders. Information about the Northern Ne-

8:45 a.m.	Problem solving/reasoning
9:15 a.m.	Guest speaker on her use of mathematics/technology on the job
10:00 a.m.	Geometry
10:30 a.m.	Snack break
10:45 a.m.	Geometry (continued)
11:55 a.m.	Spatial-reasoning task
12:15 p.m.	Lunch
1:15 p.m.	Data analysis and probability
3:00 p.m.	Snack break
3:15 p.m.	Using geometry software (*Terrapin Logo*)
4:15 p.m.	Spare time
5:30 p.m.	Dinner
6:30 p.m.	Spare time
7:45 p.m.	Movie night

Note: Printed schedules for program participants include instructor/speaker names for each segment, room locations, and options available for spare-time segments.

Figure 4.1 Sample Tuesday schedule for rising 7th graders.

vada Girls Math & Technology Program and general resources for the public can be found at http://www.unr.edu/educ/girlsmathcamp.

Securing ongoing funding for this program is a challenge in that residential programs require much greater cost than daytime-only programs. To date, most funding has consisted of university support for a program assistant, participant registration fees, and grant monies from local foundations. It may become more difficult to continue the program in its current form if local funding sources begin to "dry up."

Participants

Various short- and long-term assessment measures have been used to gather data on participants, including participant surveys given at camp beginning, camp end, and several months later; parent surveys conducted several months after the summer camp; and participant interviews. The sample for the data shared here includes 297 Northern Nevada middle school girls who attended the program in the years 2001 to 2007. Of these, 213 first participated in the program as rising seventh graders, and 84 first did so as rising eighth graders. The sample was 65% White, 15% Hispanic, 6% American Indian, 3% Black, 2% Asian, and 6% biracial/multiracial; 3% did not disclose this information. Almost one-third of the rising seventh graders and two-fifths of rising eighth graders were classified as low SES, as determined by participation in their schools' free/reduced-price lunch programs.[8] Additionally, 65 parents of girls who attended this program during the years 2005 to 2007 participated in the research on the program.

Program Evaluation

Selected quantitative and qualitative data are shared here to provide insight into the program's impact. For the years 2001 to 2007, the quantitative data come from the Modified Fennema-Sherman Mathematics Attitude Scale (Doepken, Lawsky, & Padwa, 1993). (See Wiest, 2004, for data from earlier years.) Participants completed this instrument at the beginning and at the end of the summer camp, and again about three months later at either the autumn follow-up session or by mail. The instrument contains four subscales: personal confidence about mathematics, usefulness of mathematics, perception of mathematics as a male domain, and perception of teachers' attitudes toward the respondent in mathematics. On this instrument, students respond to 47 intermingled positively and negatively phrased statements using a five-point Likert-type response format that ranges from "strongly agree" to "strongly disagree." Sample statements include:

Confidence:
> I am sure that I can learn math.
> Math is hard for me.

Usefulness:
> Knowing mathematics will help me earn a living.
> I don't expect to use much math when I get out of school.

Male domain:
> Males are not naturally better than females in math.
> Studying math is just as good for women as for men.

Teachers' attitudes:
> My teachers think I'm the kind of person who could do well in math.
> Getting a teacher to take me seriously in math is a problem.

Scores on the Confidence, Usefulness, and Teacher's Attitudes subscales range from 12–60 each for the 12 items, and scores for the Male Domain subscale range from 11–55 for its 11 items.[9]

Participants and their parents completed a number of open-ended questions on written surveys. Of interest here are those that participants completed at the end of the summer camp and the autumn surveys that participants and their parents completed (separately) several months later. The qualitative data addressed here included items for which respondents were asked to list up to three program features that they considered to be most important, to compare camp and school settings/instruction, to report whether the program had encouraged greater mathematics/technology participation, and to indicate whether they would return to the camp the next year if eligible.

Selected Quantitative Data Results

Compared to their scores at the beginning of the summer camp, the participants' Modified Fennema-Sherman Mathematics Attitude Scale scores increased for all four survey subscales at both summer camp exit and again several months later. (Increases indicate improved attitudes. For example, in the Male Domain category, this means that respondents were less likely to stereotype mathematics as a male field.) Table 4.1 shows these data only for the first year girls had participated in the program in order to remove previous program participation as a confounding variable for those who had attended both years. Because attitudes and beliefs tend to be stable over time, camp exit data may be unreliable. Therefore, a one-tailed Wilcoxon signed-ranks test was performed only for the "All Participants" category of the final (autumn follow-up) survey results in comparison with camp entry. These four analyses each included 127–134 valid pairs of scores. (Note that the sample size for the autumn follow-ups is about half that of the summer camp sample due to the smaller number of participants who attended these sessions or returned mail surveys.) Other missing data result, to a lesser degree, from participants inadvertently skipping items. Also, statistical tests were limited to more global analyses for this paper in order to necessarily limit its scope.) Participant scores increased significantly on both the Confidence ($p < .001$) and Male Domain ($p < .05$) subscales. The effect sizes were .11 (medium) and .03 (small), respectively. Table 4.1 also shows paired scores for selected participant subgroups by race/ethnicity,[10] SES, and grade level.

In Table 4.2 the scores for rising eighth graders are compared by first-time attendees versus return participants. Both sets of campers had been randomly drawn from the program's applicant pool in either their first or second eligible year, with return campers having already participated in the program the previous year. The "new" campers were those who had either not applied to the program or were not selected in the random draw as ris-

TABLE 4.1 Modified Fennema-Sherman Mathematics Attitude Scale Scores for Six Camper Groups' Initial Year of Program Attendance, 2001–2007 (program hiatus 2003)

	N	Confidence		Usefulness		Male Domain		Teacher's Attitudes	
		M	SD	M	SD	M	SD	M	SD
Completion Point									
Summer Camp Entry									
All Participants	276	48.49	8.83	53.24	6.17	46.75	5.21	48.74	6.64
White	182	49.71	8.73	53.56	6.22	47.43	4.94	49.29	6.75
Non-White	87	46.14	8.71	52.55	6.15	45.20	5.56	47.48	6.41
Medium/High SES[a]	135	49.33	8.08	53.75	6.03	47.49	4.98	49.24	6.36
Low SES	60	50.17	8.32	54.02	5.40	46.36	6.49	49.73	6.34
Rising 7th Graders	193	49.65	8.17	53.94	5.72	47.20	5.31	49.38	6.26
Rising 8th Graders	83	45.74	9.73	51.63	6.86	45.71	4.85	47.24	7.27
Summer Camp Exit									
All Participants	263	50.36	8.34	53.77	6.30	47.60	5.14	49.27	7.65
White	174	51.32	7.99	54.23	6.35	48.23	4.82	50.04	7.19
Non-White	82	48.35	8.55	53.17	5.97	46.30	5.62	47.51	8.36
Medium/High SES[a]	127	50.75	7.70	54.64	5.86	48.52	4.71	50.07	6.76
Low SES	57	51.98	7.33	54.37	5.01	46.06	6.42	49.18	8.78
Rising 7th Graders	181	51.21	7.58	54.70	5.55	47.87	5.25	49.84	7.43
Rising 8th Graders	83	48.42	9.62	51.75	7.31	47.01	4.87	47.97	8.04
Autumn Follow-Up									
All Participants	136	51.95***	7.84	53.56	6.61	48.13*	4.58	50.10	7.13
White	93	51.94	7.90	53.77	7.05	48.38	4.67	49.67	7.85
Non-White	40	51.92	7.90	52.89	5.69	47.33	4.43	51.00	5.19
Medium/High SES[a]	66	53.51	6.83	54.33	6.59	48.86	4.79	50.11	7.51
Low SES	30	52.17	7.03	54.48	4.85	47.55	4.76	50.87	5.53
Rising 7th Graders	96	53.26	6.60	54.73	5.33	48.61	4.06	50.68	6.34
Rising 8th Graders	40	48.40	9.74	50.66	8.45	47.00	5.51	48.72	8.66

Note: Data represent participants' first year in the program: summer after either sixth or seventh grade. Scores for each of the Confidence, Usefulness, and Teacher's Attitudes subscales range from 12-60 for the 12 items with a "neutral" score being 36, and scores for the 11 Male Domain items range from 11–55 with a "neutral" score being 33 (the higher the score, the more favorable the response).
[a] Participants not identified as low SES based on free/reduced-price lunch status, the only SES information collected.
*** $p < .001$. * $p < .05$.

ing seventh graders. The data show that among the rising eighth graders, the return campers had higher scores than the new campers on each of the four subscales at each of the three survey administrations.

TABLE 4.2 Modified Fennema-Sherman Mathematics Attitude Scale Scores: Rising Eighth Graders, 2001–2007

	N	Confidence		Usefulness		Male Domain		Teacher's Attitudes	
	N	M	SD	M	SD	M	SD	M	SD
Summer Camp Entry									
New Camper	83	45.74	9.73	51.63	6.86	45.71	4.85	47.24	7.27
Return Camper	60	50.37	8.67	54.12	5.88	48.23	4.16	49.53	7.85
Summer Camp Exit									
New Camper	83	48.42	9.62	51.75	7.31	47.01	4.87	47.97	8.04
Return Camper	57	51.09	7.57	54.13	6.36	48.20	4.93	49.51	7.58
Autumn Follow-Up									
New Camper	40	48.40	9.74	50.66	8.45	47.00	5.51	48.72	8.66
Return Camper	29	52.90	5.96	54.24	6.13	49.31	4.45	51.28	6.05

Note: See Table 4.1 note for interpretation of scores.

Score changes from camp entry to camp exit and camp entry to the autumn follow-up are shown in Table 4.3 for various student subgroups. Non-Whites made greater gains than Whites on later surveys compared to their entry survey scores, especially on the Confidence subscale and in the long-term. Girls of higher SES tended to improve their scores more than girls of lower SES for both follow-up survey administrations. The results by grade level are mixed but tend to show increasing scores over time for the younger group (rising seventh graders).

Selected Qualitative Data Results

On the surveys at the end of the summer camp week and several months later, participants were asked to name up to three program features they found most important. About 75% of the 177 camp-end and 89 autumn follow-up survey respondents named one or more academic features related to specific mathematics or technology topics (e.g., geometry, data analysis, or use of *Terrapin Logo* software) or to mathematics/technology learning in general. Secondary comments centered on the social and recreational camp aspects, as well as campus life (e.g., staying in a university dormitory). Parents were also asked this question on the autumn follow-up survey.[11] About half of the 67 parents named academic topics as one or more of their top three program features. Parents were equally inclined to name personal, social, and academic/career skills and supports, such as helping girls gain confidence, interest, and motivation in relation to mathematics and technology; provid-

TABLE 4.3 Change in Modified Fennema-Sherman Mathematics Attitude Scale Scores: First Camp Year from Entry to Exit and Entry to Autumn Follow-Up, 2001–2007

	Confidence	Usefulness	Male Domain	Teacher's Attitudes
Completion Point				
Summer Camp Exit				
All Participants	+1.87	+0.53	+0.85	+0.53
White	+1.61	+0.67	+0.80	+0.75
Non-White	+2.21	+0.62	+1.10	+0.03
Medium/High SES[a]	+1.42	+0.89	+1.03	+0.83
Low SES	+1.81	+0.35	−0.30	−0.55
Rising 7th Graders	+1.56	+0.76	+0.67	+0.46
Rising 8th Graders	+2.68	+0.12	+1.30	+0.73
Autumn Follow-Up				
All Participants	+3.46	+0.32	+1.38	+1.36
White	+2.23	+0.21	+0.95	+0.38
Non-White	+5.78	+0.34	+2.13	+3.52
Medium/High SES[a]	+4.18	+0.58	+1.37	+0.87
Low SES	+2.00	+0.46	+1.19	+1.14
Rising 7th Graders	+3.61	+0.79	+1.41	+1.30
Rising 8th Graders	+2.66	−0.97	+1.29	+1.48

Note: All scores changes are based on comparison with summer-camp entry scores. Sensitivity to change is slightly lower for the Male Domain category, which has 11 items and a possible total score of 55, whereas the other three categories each have 12 items and a possible total score of 60.

[a] Participants not identified as low SES based on free/reduced-price lunch status, the only SES information collected.

ing female role models and career information or inspiration; interacting with like-minded peers; and spending time in a college setting.

The girls themselves also showed awareness of their affective gains. For example, one seventh grader wrote on her autumn survey, "This program made me relize [*sic*] that I am better at math than I thought." However, parents were even more likely to comment on their daughters' improved confidence:

> I feel my daughter is more comfortable asking for help and more confident when facing challenging problems. (parent of 7th grader)

> Your program increased her self-confidence in the mathematic[s] sphere and that's priceless as she goes forward into high school. (parent of 8th grader)

> It is a gained confidence level that we do not receive in our public school setting. (parent of 8th grader)

Parents valued their daughters "interacting with other girls with the same interests" and having an "introduction to a scholarly environment." In keeping with their greater attention to and appreciation of contextual factors, parents were more likely than their daughters to include the program's single-sex setting among their top three "most important" features on the autumn follow-up surveys (9% of girls, 21% of parents). Interestingly, only 1% of girls had listed this feature on the camp-end survey—that is, it rose in prominence retrospectively. It should be noted that these figures only reflect the top-rated features, and not whether individual features were important in their own right.

Girls and their parents were asked to compare the camp content and instructional approach with school classes in the same subject areas. Most said that camp and school tended to differ, particularly in two main areas. One was instructional orientation (for example, a greater use of active learning and group work in the camp setting). One rising seventh grader wrote, "We get to work in groups to explain our thinking (at math camp) but at school usually we don't." The other divergence was in the degree and role of technology used at the camp. Many campers and parents stated that computers were not used at all or were used in a limited sense at school. When computers were used, they tended to be used in less powerful ways, such as typing or seeking information. One eighth grader reported, "In my computer class at school we just learn how to do regular stuff like slide shows but here we learned more sofisticated [*sic*] things." Another wrote, "A lot of the technology that was used in helping the girls at Math Camp was very advanced. I never saw a lot of the stuff that was used at camp."

Survey responses indicated that some girls increased their participation in mathematics and/or technology as a result of having attended the program, such as seeking or gaining entry into more advanced classes or joining relevant clubs or competitions (e.g., MATHCOUNTS, a U.S. national mathematics competition for middle school students). Sample comments on the autumn follow-up survey in relation to perceived camp impact include:

[When] I first started school, I asked my math teacher . . . to give me advanced work. (7th grader)

She wasn't doing well in Pre-algebra and her teacher was going to pull her out and she went and talked to him and got help to stay in Pre-algebra. (parent of 7th grader)

The computer stuff [at camp] got me very interested. I've been spending a lot more time working with computers. (8th grader)

Last year in 7th grade, my teacher said some of the people in my class could go to Algebra I. I worked really hard to be one of those people. And I made it. All because I attended the Math Camp. (8th grader)

It has influenced me [to] take part [in] tutoring other people in my class after school. (8th grader)

Program participants were asked whether they would return to the camp the following year if they were eligible. The majority (92% camp-end, 93% autumn follow-up) said a resounding yes, often accompanied by strong positive commentary. Some of those who said no or they were unsure indicated reasons unrelated to the program, such as plans to move or a desire to spend more time with parents.

Discussion of Program Data

The limited quantitative and qualitative data shared here to shed some insight into an example of an OST (here, an intervention) program support earlier literature and contentions that quality out-of-school-time programs can positively impact participants. Three examples are evident in these data. First, participants' attitude scores improved from program beginning to two later survey administrations, achieving statistical significance in two of four measures tested on the autumn follow-up survey (Confidence, Male Domain). Scores on the Usefulness and Teacher's Attitudes subscales improved but did not achieve statistical significance, perhaps because they already had higher mean ratings than the other two categories at the outset. In some research, affective measures have been found to correlate positively with performance and participation (e.g., Antunes & Fontaine, 2007; Halpern, Aronson et al., 2007; Ma & Johnson, 2008; Meelissen & Luyten, 2008). Ma and Johnson (2008) say, "Fostering a positive attitude toward mathematics could hold the key to retaining both females and males in advanced mathematics coursework and eventually attract them to the STEM fields" (p. 77). The greatest gains in the Northern Nevada Girls Math & Technology Program appeared in the area of confidence, which might be particularly important for future performance, participation, and quality of experience for females in the STEM disciplines.

Other student demographic factors have also been found to interact with gender in research results obtained in areas such as mathematics achievement, attitudes, and participation (e.g., Andreescu et al., 2008; McGraw et al., 2006). McGraw et al. (2006) argue, "Any report on gender differences in achievement would be incomplete without discussion of race/ethnicity and SES" (p. 140). The same is true for other important variables, including attitudes. Accordingly, the disaggregated data in this study show that some student subgroups responded differently than others to the Northern Nevada Math & Technology Program. The fact that attitude scores for non-White students in the sample tended to increase over time more than those of White students may simply reflect the fact that these students had lower initial scores and thus more room for growth, or it may mean that the inadequately served populations that comprise this category of student are in particular need and respond well to additional mathematics support. Because girls of higher SES had greater score increases over time than lower-income girls, the latter may need stronger and more sustained assistance in seeking to reduce academic-oriented gaps among students from different socioeconomic backgrounds. Similarly, more concerted and ongoing support may be especially important for older girls. As reported in the literature discussed earlier in this chapter, issues concerning girls in relation to the STEM disciplines begin to appear by at least the middle grades. Indeed, the attitudinal data reported here appear to support those findings. The scores of the younger camper groups tended to continue to improve from camp exit to the autumn follow-up, whereas the older camper group had relatively stable or decreased scores on three of the four subscales (see Table 4.3).

A second set of data reported here also supports the positive impact of this program. There was clear evidence that girls who had participated in the program entered the second eligible camp year as rising eighth graders with higher attitude scores than new participants, those of the same age coming for the first time. This advantage was maintained at the two later administrations of the attitude survey on each of the four subscales (see Table 4.2).

Third, in their written commentaries, program participants and their parents reported strong positive experiences and outcomes associated with the camp, both during and after the program. According to the respondents, program benefits included improved affect, knowledge, and skills, as well as greater participation in mathematics and technology. Participants and, especially, their parents noticed the girls' improved confidence levels. Parents may be uniquely positioned to observe this type of change in their children, compared with young people's own abilities to identify such transformations. Thus, parent perspectives strengthen the claims of positive program outcomes.

In addition to data supporting the positive impact of this program on mathematics attitudes and participation (and likely other areas), it is encouraging to note that on the retrospective surveys the girls who participated in the program generally found academic program features—namely, specific mathematics topics such as geometry and data analysis or the mathematics-oriented computer segments—most important, despite their keen and often-vocalized interest in the nonacademic aspects of the program, such as staying in university dormitories, eating in the campus cafeteria, and engaging in evening recreational activities. This demonstrates that girls have the ability to identify and appreciate worthwhile learning opportunities. A clear focus on discipline-based content (here, mathematics) and use of academic physical settings (e.g., here, a university) can help foster a scholarly climate for OST programs. These two features were prominent in this program and may have contributed to the girls' valuing of the academic aspects of the summer camp experience.

Another participant response that merits attention is the limited and low-level use of computers that was evident in these middle school girls' scholastic lives. This is a serious shortcoming in terms of developing familiarity and skill with technology, an increasingly integral part of personal, academic, and occupational life. This neglect also minimizes or omits technology's potential as a tool for learning mathematics (cf. Wiest, 2008b). This finding supports the contention that technology should be strongly infused into OST programs, especially for girls (e.g., Reid & Roberts, 2006; Wiest, 2004; Wimer et al., 2006).

FINAL COMMENTS

Out-of-school-time (OST) programs are increasing in number and importance in the United States and may have important learning implications in the U.S. and beyond. There is an increased call for OST programs, sometimes in the form of interventions, to address the needs of all students or particular groups in specific subject areas. Here, the focus was on females in mathematics. Much more research is needed on OST programs to inform policy and practice. Research should address such areas as effective characteristics and longitudinal impacts (e.g., on persistence in particular fields) of OST programs. Such studies should also disaggregate data by student subgroups, especially gender, race/ethnicity, and socioeconomic status (e.g., Froschi et al., 2003).

Results from OST programs to date, such as those shared in this chapter, reveal some consistent results and recommendations for practice. The information reported here combines general research on OST programs with that of intervention programs for females in the STEM disciplines. Females continue to be underrepresented in mathematics degree programs and careers, and they display less positive attitudes toward mathematics than males. OST programs can be one effective response to help support and encourage females in mathematics. Some contend that middle school is a key crossroads where OST programs are especially important. Accordingly, Miller (2003) states:

> We need to increase the number of [afterschool] programs available to middle school students, and take the steps necessary to ensure that these programs have the resources, including knowledgeable and skilled staff, to effectively build the developmental assets of young people. (p. 83)

However, these initiatives require expertise, time, and funding (e.g., Andreescu et al., 2008). Thus, national policy may be especially important in driving the OST-program movement to a place of prominence and maximum impact (cf. Marks, 2008). As indicated earlier, recent changes in federal policies have led to a surge of interest in, support for, and thus proliferation of OST programs in the United States. Nevertheless, with or without the support of national policy, carefully planned and evaluated OST programs offered by educational institutions or other qualified entities are recommended as one measure in ongoing efforts to raise the status of females in mathematics.

NOTES

1. This Act is a federal law that affects elementary and secondary education (grades K-12). Its purpose is to increase accountability for results, use of research-based methods, local control and flexibility, and parent choice. For more information, see http://www.ed.gov/nclb/.
2. This law prohibits sex discrimination in educational programs and activities that receive federal funds. See, for example, http://www.usdoj.gov/crt/cor/coord/titleix.php.
3. Potential summer gains relate to students' family and community experiences, such as academically oriented efforts and activities in the home, participation in optional academic summer programs, and both structured and informal individual learning experiences.
4. An extensive search was made to find OST programs in other countries. This effort was unsuccessful other than determining that the U.S.-based Expanding Your Horizons program (http://www.expandingyourhorizons.org/) has recently extended its one-day STEM programs for girls to five other countries: China, Malaysia, Singapore, Thailand, and Switzerland. The author is interested in being informed about STEM OST programs for girls being conducted both within and outside of the U.S.
5. Cultural knowledge refers to educational content gained through exposure to enriching cultural, historical, and scientific experiences through travel and home opportunities (e.g., sites visited and resources available, such as books, educational games, and computers). It also includes an understanding of the behavioral, social, communicative, and work styles preferred by the dominant school culture.
6. As used in this document, "minority" means non-White racial/ethnic groups, all of which have lower societal status than Whites.
7. Staff names are obtained by cross-referencing recommendations made by the Mathematics Coordinator(s) for the local school district, mathematics professional development specialists for the region in which the program is conducted, the President of the Northern Nevada Mathematics Council, and mathematics education faculty members at the local university. Instructors chosen to work with the program have a reputation for teaching in the

spirit of the *Principles and Standards for School Mathematics* (National Council of Teachers of Mathematics, 2000). For example, they emphasize a conceptual approach. Typically, two or all three of the following are also true of invited instructors: they continue learning in the field (e.g., they have attained or are working on attaining a higher degree in mathematics education); they are active in the field of mathematics education (e.g., they attend conferences, participate in local mathematics councils, or conduct professional development for peers); and they have a substantial amount of teaching experience (generally ranging from 8–25 years).

8. This determination is based on a U.S. federal program that provides free or reduced-price school lunches to children of low-income families. (See http://www.fns.usda.gov/cnd/Lunch/.)

9. The Modified Fennema-Sherman Mathematics Attitude Scale omits a 12th item that appeared in the original Male Domain subscale: "Mathematics is for men; arithmetic is for women." No explanation is provided for this omission; however, it may be that the distinction between mathematics and arithmetic is obscure to young people.

10. Combining racial/ethnic minorities into one group based on racial classification can be problematic in failing to acknowledge important differences among the groups. Further, it fails to distinguish race and ethnicity. Nevertheless, non-White groups tend to share lower societal status compared with Whites.

11. Parents had adequate insight into the program to answer this question by being provided with a detailed camp schedule, attending camp opening and closing sessions (optional events) that foreshadowed camp activities and demonstrated camp learning, and discussing the program with their daughters during the camp (via phone calls, email, physical mail, and/or visits) and in the several months after the camp had ended until the survey was administered.

REFERENCES

Alexander, K. L., Entwisle, D. R., & Olson, L. S. (2004). Schools, achievement, and inequality: A seasonal perspective. In G. D. Borman & M. Boulay (Eds.), *Summer learning: Research, policies, and programs* (pp. 25–51). Mahwah, NJ: Erlbaum.

Andreescu, T., Gallian, J. A., Kane, J. M., & Mertz, J. E. (2008). Cross-cultural analysis of students with exceptional talent in mathematical problem solving. *Notices of the AMS, 55*(10), 1248–1260.

Antunes, C., & Fontaine, A. M. (2007). Gender differences in the causal relation between adolescents' maths self-concept and scholastic performance. *Psycologica Belgica, 47*(1/2), 71–94.

Blue, J. & Gann, D. (2008). When do girls lose interest in math and science? *Science Scope, 32*(2), 44–47.

Blume, G. W., Galindo, E., & Walcott, C. (2007). Performance in measurement and geometry from the viewpoint of *Principles and Standards for School Mathematics*. In P. Kloosterman & F. K. Lester (Eds.), *Results and interpretations of the 2003* mathematics assessment of the National Assessment of Educational Progress (pp. 95–138). Reston, VA: National Council of Teachers of Mathematics.

Bouffard, S., Little, P., & Weiss, H. (2006). Building and evaluating out-of-school time connections. *The Evaluation Exchange, 12*(1/2), 2–6.

Brown, T. (2008). Helping girls envision a tech-savvy future. *AAUW Outlook, 102*(1), 12–14.

Bureau of Labor Statistics. (2008). *Highlights of women's earnings in 2007* (Report 1008). Washington, DC: U.S. Department of Labor. Retrieved April 27, 2009, from United States Department of Labor Web site: http://www.bls.gov/cps/cpswom2007.pdf

Burkam, D. T., Ready, D. D., Lee, V. E., & LoGerfo, L. F. (2004). Social-class differences in summer learning between kindergarten and first grade: Model specification and estimation. *Sociology of Education, 77*(1), 1–31.

Cavanagh, S. (2007). Science camp: Just for the girls. *Education Week, 26*(45), 26–28.

Chacon, P., & Soto-Johnson, H. (2003). Encouraging young women to stay in the mathematics pipeline: Mathematics camps for young women. *School Science and Mathematics, 103*(6), 274–284.

Chin, T., & Phillips, M. (2005). Season of inequality: Exploring the summer activity gap. *American Educator, 29*(2). Retrieved April 27, 2009, from American Federation of Teachers Web site: http://www.aft.org/pubs-reports/american_educator/issues/summer2005/chin.htm

Cooper, H. (2003). *Summer learning loss: The problem and some solutions* [ERIC Digest EDO-PS-03-5]. Champaign, IL: ERIC Clearinghouse on Elementary and Early Childhood Education.

Cooper, H., Nye, B., Charlton, K., Lindsay, J., & Greathouse, S. (1996). The effects of summer vacation on achievement test scores: A narrative and meta-analytic review. *Review of Educational Research, 66*(3), 227–268.

Doepken, D., Lawsky, E., & Padwa, L. (1993). *Modified Fennema-Sherman attitude scales.* Retrieved July 8, 2009, from Woodrow Wilson National Fellowship Foundation Web site: http://www.woodrow.org/teachers/math/gender/08scale.html

Duffett, A., & Johnson, J. (2004). *All work and no play? Listening to what KIDS and PARENTS really want from out-of-school time.* New York: Public Agenda. Retrieved May 6, 2009, from The Wallace Foundation Web site: http://www.wallacefoundation.org/SiteCollectionDocuments/WF/Knowledge%20Center/Attachments/PDF/All%20Work%20and%20No%20Play.pdf

Edwards, T., Kahn, S., & Brenton, L. (2001). Math Corps Summer Camp: An inner city intervention program. *Journal of Education for Students Placed at Risk, 6*(4), 411–426.

Fiebrink, R. A., & Alcott, T. R. (2003). *Designing a programming workshop for girls.* Unpublished paper. Retrieved April 27, 2009, from http://rfiebrink.tripod.com/designing.pdf

Frenzel, A. C., Pekrun, R., & Goetz, T. (2007). Girls and mathematics—A "hopeless" issue? A control-value approach to gender differences in emotions towards mathematics. *European Journal of Psychology of Education, 22*(4), 497–514.

Froschi, M., Sprung, B., Archer, E., & Fancsali, C. (2003). *Science, gender, and afterschool: A research-action agenda.* New York: Educational Equity Concepts & Academy for Educational Development.

Halpern, D. F., Aronson, J., Reimer, N., Simpkins, S., Star, J. R., & Wentzel, K. (2007). *Encouraging girls in math and science: IES Practice Guide.* Washington, DC: U.S. Department of Education, National Center for Education Research, Institute of Education Sciences.

Halpern, D. F., Benbow, C. P., Geary, D. C., Gur, R. C., Hyde, J. S., & Gernsbacher, M. A. (2007). The science of sex differences in science and mathematics. *Psychological Science in the Public Interest, 8*(1), 1–51.

Hughes, T. (2006-07). The advantages of single-sex education. *National Forum of Educational Administration and Supervision Journal, 23*(2), 5–14.

Hyde, J. S., Lindberg, S. M., Linn, M. C., Ellis, A. B., & Williams, C. C. (2008). Gender similarities characterize math performance. *Science, 321*(5888), 494–495.

Institute of Education Sciences. (2008). *Digest of education statistics 2007.* Washington, DC: U.S. Department of Education, National Center for Education Statistics. Retrieved April 27, 2009, from National Center for Education Statistics Web site: http://nces.ed.gov/programs/digest/d07/

Kortering, L. J., deBettencourt, L. U., & Braziel, P. M. (2005). Improving performance in high school algebra: What students with learning disabilities are saying. *Learning Disability Quarterly, 28*(3), 191–203.

Lauer, P. A., Akiba, M., Wilkerson, S. B., Apthorp, H. S., Snow, D., & Martin-Glenn, M. L. (2006). Out-of-school-time programs: A meta-analysis of effects for at-risk students. *Review of Educational Research, 76*(2), 275–313.

Lawhead, P., Loyd, R., Schep, M., Laws, M., & Price, K. (2005). Experiences in math, science and technology summer camps for young females. *ACM-SE 43: Proceedings of the 43rd annual Southeast regional conference* (Vol. 1). New York: Association for Computing Machinery.

Liu, O. L., Wilson, M., & Paek, I. (2008). A muldimensional Rasch analysis of gender differences in PISA mathematics. *Journal of Applied Measurement, 9*(1), 18–35.

Ma, X., & Cartwright, F. (2003). A longitudinal analysis of gender differences in affective outcomes in mathematics during middle and high school. *School Effectiveness and School Improvement, 14*(4), 413–439.

Ma, X., & Johnson, W. (2008). Mathematics as the critical filter: Curricular effects on gendered career choices. In H. M. G. Watt & J. S. Eccles (Eds.), *Gender and occupational outcomes: Longitudinal assessment of individual, social, and cultural influences* (pp. 55–83). Washington, DC: American Psychological Association.

Marks, G. N. (2008). Accounting for the gender gaps in student performance in reading and mathematics: Evidence from 31 countries. *Oxford Review of Education, 34*(1), 89–109.

McGraw, R., Lubienski, S. T., & Strutchens, M. E. (2006). A closer look at gender in NAEP mathematics achievement and affect data: Intersections with achievement, race/ethnicity, and socioeconomic status. *Journal for Research in Mathematics Education, 37*(2), 129–150.

Meelissen, M., & Luyten, H. (2008). The Dutch gender gap in mathematics: Small for achievement, substantial for beliefs and attitudes. *Studies in Educational Evaluation, 34*(2), 82–93.

Miller, B. M. (2003). *Critical hours: Afterschool programs and educational success.* Brookline, MA: Miller Midzik Research Associates.

Morrow, C. (2006). Effective mathematics learning environments for females. *IO-WME Newsletter, 20*(3), 16–28.

Morrow, C., & Schowengerdt, I. (2008). Stepping beyond high school mathematics: A case study of high school women. *ZDM, 40*(4), 693–708.

National Association for Year-Round Education (NAYRE) (2007). Statistical summaries of year-round education programs: 2006–2007. Retrieved May 8, 2009, from NAYRE Web site: http://www.nayre.org/STATISTICAL%20SUMMA-RIES%20OF%20YRE%202007.pdf

National Council of Teachers of Mathematics (NCTM). (2008). *Intervention resources: Creating or selecting intervention programs.* Retrieved April 27, 2009, from NCTM Web site: http://www.nctm.org/intervention.aspx

National Council of Teachers of Mathematics (NCTM). (2000). *Principles and standards for school mathematics.* Reston, VA: NCTM.

Nevada Department of Education. (2006). Nevada mathematics standards: Integrating content and process. Carson City, NV: Author. Retrieved April 27, 2009, from Nevada Department of Education Web site: http://www.doe.nv.gov/Standards_Mathematics_Standards.html

Organisation for Economic Co-operation and Development. (2007). *PISA 2006: Science competencies for tomorrow's world* [executive summary]. Paris: Author.

Phi Delta Kappa International. (2006). Effect of after-school programs. *Topics and Trends, 6*(3).

Phi Delta Kappa International. (2008). Effective summer programs. *Topics and Trends, 7*(10).

Pollock, L., McCoy, K., Carberry, S., Hundigopal, N., & You, X. (2004). Increasing high school girls' self confidence and awareness of CS through a positive summer experience. *ACM SIGCSE Bulletin, 36*(1), 185–189.

Reid, P. T., & Roberts, S. K. (2006). Gaining Options: A mathematics program for potentially talented at-risk adolescent girls. *Merrill-Palmer Quarterly, 52*(2), 288–304.

Roberts-Ohr, S. (2008). *Expanding Your Horizons Network: 2007 annual report.* Retrieved May 6, 2009, from Expanding Your Horizons Network Web site: http://www.expandingyourhorizons.org/about/reports/0607annual.pdf

Spielhagen, F. R. (2006). How tweens view single-sex classes. *Educational Leadership, 63*(7), 68–69, 71–72.

Stutler, S. L. (2005). Breaking down the barriers: Adventures in teaching single-sex algebra classes. In S. K. Johnsen & J. Kendrick (Eds.), *Math education for gifted students* (pp. 93–104). Waco, TX: Prufrock Press.

Stevens, T., Wang, K., Olivárez, A., & Hamman, D. (2007). Use of self-perspectives and their sources to predict the mathematics enrollment intentions of girls and boys. *Sex Roles, 56*(5/6), 351–363.

Wiest, L. R. (2004). Impact of a summer mathematics and technology program for middle school girls. *Journal of Women and Minorities in Science and Engineering, 10*(4), 317–339.

Wiest, L. (2008a). Conducting a mathematics camp for girls and other mathematics enthusiasts. *The Australian Mathematics Teacher, 64*(4), 17–24.

Wiest, L. R. (2008b). *Tool technologies as powerful aids to girls' mathematics learning.* Unpublished manuscript.

Wimer, C., & Gunther, R. (2006). *Summer success: Challenges and strategies in creating quality academically focused summer programs* [no. 9]. Cambridge, MA: Harvard Graduate School of Education, Harvard Family Research Project.

Wimer, C., Hull, B., & Bouffard, S. M. (2006). *Harnessing technology in out-of-school time settings* [no. 7]. Cambridge, MA: Harvard Graduate School of Education, Harvard Family Research Project.

Wolbach, M. (2007). Strength in numbers: Sustaining girls' interest in math and science. *AAUW Outlook, 101*(2), 15–18.

Wong, C. M. (2006). *An evaluation of the Expanding Your Horizons science and mathematics conference for girls at four sites.* Unpublished master's thesis, San Francisco State University.

AUTHOR ACKNOWLEDGMENT

The author would like to thank the following major funding sources ($10,000 or more) for the Girls Math and Technology Program during the 2001–2007 time period: Lemelson Education and Assistance Program (LEAP) through the Nevada Women's Fund; University of Nevada, Reno Regents Award Program; International Game Technology (IGT); E.L. Cord Foundation; Mathematical Association of America/Tensor Foundation.

CHAPTER 5

FREEDOM TO CHOOSE?

Girls, Mathematics and the Gendered Construction of Mathematical Identity

Fiona Walls
James Cook University

INTRODUCTION

This chapter is based on data gathered through a longitudinal ethnographic study in which the mathematics education histories of ten New Zealand children—four girls and six boys—were tracked from the beginning of their third year of elementary schooling as seven-year-olds, into their upper secondary schooling and occupational decision-making, as eighteen-year-olds. The experiences of the children are examined for the ways in which *male* and *female* are produced as sexuate occupational identities through the learning of mathematics—a process I call mathematical genderfication—and how this process is implicated in the ways in which girls and boys engage with mathematics, whether they choose to study mathematics during their upper secondary schooling and the kinds of occupation or further study they elect to take up beyond school. The chapter casts further light on the complex social processes that produce gendered patterns of math-

International Perspectives on Gender and Mathematics Education, pages 87–110
Copyright © 2010 by Information Age Publishing
All rights of reproduction in any form reserved.

ematical participation. Key findings of the study are presented that link:
(1) the children's interests outside of school; (2) the children's changing
vocational aspirations through their schooling; (3) the schooling, qualifications, and mathematical experiences of the parents; and (4) the children's
evolving affective engagement with mathematics over time and the children's participation in mathematics over their schooling careers.

Beginning in 1998, the study set out to investigate a phenomenon recognized in the TIMSS research findings:

> While a majority of students have positive attitudes to learning mathemat-
> ics... beginning from a fairly young age, there is an increasing proportion
> of students having lost interest in the subject, with a concomitant decline in
> their achievement. This effect is considerably greater for girls than for boys.
> (Garden, 1997, p. 252)

The correlation between sex, enjoyment, and achievement in mathematics continued to be reported in many countries. Thompson and Fleming
(2004) in Australia for example, stated that:

> Students' self-confidence in mathematics had a clear positive relationship
> with mathematics achievement. Males had higher self-confidence in learning
> mathematics than females.... At both year Levels [4 and 8] males enjoy learn-
> ing mathematics more than females. (p. 9)

Their report also showed a dramatic overall decline from Year 4 to Year 8
in children's enjoyment and confidence with mathematics, a trend echoed
elsewhere.

The TIMSS results demonstrated that mathematics as a school subject is
productive—that is, it constructs children as social and mathematical entities. As these data show, a significant proportion of children experience
discomfort, alienation, and disengagement from mathematics from very
early in their schooling, and this gathers momentum as a growing number
of students switch off mathematics. Girls are more likely to be produced
as disaffected and marginalized learners in this process. The connections
between losing interest, loss of confidence, and diminishing achievement
are now well-recognized, as is the gendered nature of the connection (see
Mendick, 2006).

Despite sustained efforts in many countries over recent years to address
sex-based disparities in mathematical participation, marked differences
continue to be observed (Mendick, 2006). As women's right to equal participation in society is increasingly accepted as an ethical principle enshrined
in legislation, it is presumed that girls now can (and do) do anything. Girls'
overall achievement in mathematics has drawn almost level with that of boys
in many countries, but the continuing alienation of girls from mathematics

in higher proportions than boys, and women's significantly lower participation rates in mathematics-based careers despite their relevant qualifications are therefore perplexing.

Popular explanations (e.g., Lubinski & Benbow, 2006; McArdle, 2008) hold that masculinity and mathematics are somehow naturally aligned; given women's (presumed) agency in a world that no longer discriminates against them, women with qualifications in mathematics are seen as exercising their freedom to choose by pursuing occupations that involve working with people or animate objects, while equally highly qualified males opt in far greater numbers for careers that engage them with tools in less 'organic' situations. This pattern of preference is attributed to distinctive masculine/ feminine characteristics.

By contrast, Mendick (2006) has reasoned that differentiated self-sorting through subject choice is a social construction. Her research involving upper secondary students of mathematics in the UK demonstrated that mathematics is a domain that continues to be strongly associated with masculinity as a social power relationship rather than a genetically inevitable one. Gilbert and Gilbert (1998) noted the ways in which subjects vested with high masculine status and power such as mathematics and physics can create a personal sense of failure for boys who do not succeed within these male domains of achievement. Studies such as these suggest that it is the power dimensions of the masculinity/mathematics connection that must be addressed if genuine freedom to choose for both girls and boys is to be achieved.

Researching the Constructive Process of Genderfication in Mathematics

If masculine and feminine identities are wrought in the social processes of teaching and learning of mathematics as Mendick contends, sociological studies that focus on individual children's experiences of learning mathematics are likely to yield useful data in our understanding of this construction. A number of studies already exist (e.g., Bartholomew, 2005; Boaler, 2002; Mendick, 2006) that make valuable contributions to our understanding of gender, pupil experience, and subject choice in mathematics, but these are located in a specific phase in students' lives.

The study reported in this chapter set out to follow a small group of children for a significantly longer period of their mathematical learning, beginning from early schooling. Originally designed to document the children's learning from the ages of 7 to 10 years (from the beginning of their third year at school to the end of their fifth), the study resumed towards the end of Year 11 when the children were 15 years old and facing their first national examinations. The children's engagement with mathematics was then tracked through to the end of their schooling and into the workforce or tertiary study. The study group consisted of four girls and six boys randomly selected from the elementary schools in a large metropolitan area. The children belonged to markedly differing social networks within a large urban region of New Zealand. Their schools differed by type, size, and overall socioeconomic status. Diverse ethnicity, home background, and schooling experience were found within the group. One of the girls identified herself as indigenous (New Zealand Maori), and the father of one of the boys is an Asian immigrant. Six of the children (two of the girls) experienced living with only one parent at a time, or with step-parents. One of the boys was raised by his mother as solo parent. One of the girls spent her final year of schooling on a funded exchange program in a Swedish school. Two of the boys spent several years of their schooling in other countries including Switzerland, the UK, and Australia. One spent most of his secondary schooling in a large Australian city.

The study was designed to gather data on a wide range of variables in children's learning of mathematics, including the gendered disaffection/ achievement connection. Information was sought about the children's lives and interests in general, their vocational aspirations, experiences of learning mathematics, and beliefs and feelings about mathematics and mathematical competence. Data were gathered by various means. The children were visited three times a year, from the beginning of their third year at school until the end of their fifth, during which time the children were observed in their classrooms and interviewed using a flexible semi-structured conversational approach. The teachers and parents were also interviewed, and the children's workbooks, textbooks, school reports, and assessment records examined for insights into the children's mathematical learning environments within their school and family contexts. The children were contacted again as they were nearing the end of their eleventh year at school, and informal interviews were conducted with children and families over the following two years until the children had either reached the end of their schooling, taken up further study, or joined the workforce. The children were able to provide compelling descriptions of how they learned mathematics at this stage of their schooling. By chance, some of the children attended the same secondary schools and their descriptions provided corroborating evidence. Additional data were obtained during this phase of study from school reports, external examination results, mathematics exercise books and textbooks.

The children's stories demonstrate the gendered nature of learning mathematics and how this plays out through a child's schooling career. Significant differences by gender were apparent from early in their schooling but became much more pronounced when the children needed to choose subjects to study at upper secondary school, courses of study at tertiary level, or paid employment. Learning mathematics thus came to be viewed as much more than the acquisition of a set of cognitive skills and processes we call "mathematical."

Mathematical Genderfication as Occupational Identity Work

Studies of occupation offer us useful theoretical frameworks for investigating career choice and the part that mathematics might play in this process. According to Kielhofner (2008), for example, the activities in which an individual engages for both leisure and work can be regarded as her/his *occupational identity*. Occupational identification can be viewed as an ongoing process of construction as the *real* and *ideal* "self," which are produced in unfolding occupational narrative responding to such influences as social interactions, schooling experiences, and access to opportunities. Gottfredson (2005) argued that such narratives are gendered. She suggested that occupational choice involves compromise between what is ideal and what is possible within socially circumscribed occupational space in which masculine/feminine and high/low social value form intersecting axes of opposition, explained as follows:

> Regardless of their own social origin . . . newborns will develop essentially the same view of occupations by adolescence. Like adults, they will distinguish occupations primarily along two dimensions—their masculinity-femininity and their overall social desirability (prestige level). They will also share common stereotypes about the personalities of different kinds of workers—accountants vs. artists, engineers vs. teachers, and so on. Despite their similar perceptions, their occupational aspirations will nonetheless reproduce most of the class and gender differences of the parent generation: girls will aspire mostly to "women's" work, boys to "men's" work, and lower class youngsters to lower level jobs than their higher social class peers. (p. 72)

For the children in this study, choices about continuing to learn mathematics into upper secondary school, engaging in tertiary study after leaving school or joining the paid workforce, could be seen as part of wider identity work rooted in early childhood. Their choices of playthings, their organized after-school activities as primary school students, their interactions with family and friends, and their reported feelings about learning mathematics revealed a growing awareness of what constitutes women's and men's work and the social valuation of occupations as reflected in their evolving occupational aspirations. These were linked to the qualifications, occupations, and educational aspirations of their parents, and parental experiences of learning mathematics, as outlined in the following sections of this chapter.

The Role of Children's Play in Mathematical Genderfication

Children's play is of interest to sociologists because it is believed to contribute to children's growing processes of identification. During the first three years of the study the children were asked about their favorite toys (see Table 5.1). Clear distinctions can be seen between girls' and boys' preferences, demonstrating the early appearance of a masculine/feminine divide in the children's occupational identity narratives. Development of models for thinking, reasoning, and working mathematically on the one hand, and social positioning on the other, were thus gendered through the children's play from a very young age. It could be argued that the boys' greater engagement with toys such as Lego, transformers, computerized games, and remote controlled vehicles exposed them to the kinds of logical and spatial thinking that were required in mathematical problems in their classrooms. Girls' toys were mostly non-mechanical and engaged girls in caring and grooming. The boys' toys included occupational role models of males such as soldiers and superheroes and scale models of real objects they might one day use such as cars and skateboards. The girls' toys were less

TABLE 5.1 Favorite Toys

	7 Years	8 Years	9 years
Fleur	Stuffed toy rabbit, Tickle me Elmo	Stuffed toy panda bear	Stuffed toy panda bear
Georgina	Teddy bear, Barbie Doll	Furbie, Pokemon toys	Hair tie toys
Jessica	Telly Tubby	Dolls	Computer games
Rochelle	Computer, Gigapet	Furbie	
Dominic	Beast Ball (transformer toy), toy soldiers	Tech decks (finger skateboards)	Soldiers and war toys, computer war games
Jared	Spiderman Suit	Pokemon toys, baseball set	Pokemon toys, Game Boy
Liam	Remote control car	Army men, toy cars	Scooter, Playstation
Mitchell	Remote control fire engine	Pokemon toys	
Peter	Toy soldiers	Star Wars toys, Game boy	
Toby	Water guns, Lego	Not into toys any more, sports gear	Playstation

Note: Girls' names shaded

obviously presenting them with views of themselves as occupationally active, powerful, and socially admirable adults. Occupational identity work and its gendered and mathematical dimensions were thus inscribed in complex ways through the children's engagement with their toys.

Table 5.2 shows the kinds of organized activities in which children were regularly engaged outside school. Particularly noticeable is the boys' involvement in competitive sport from an earlier age and in greater proportions than for the girls, who tended to be more engaged in non-competitive activities such as dance and music lessons. This was significant for the ways in which classroom pedagogies of mathematics in this study were found to draw heavily on recognized elements of sporting culture such as team games, individual competition, and speed tests. The boys in the study who took part in organized sports from a young age were the most comfortable with the competitive nature of much of the classroom mathematics they experienced, while those children who did not, both boys and girls, found this to be severely alienating.

Documentation of the differentiated/differentiating modes by and through which the children learned mathematics within the complex cultural spaces of school and home revealed how everyday actions served to delineate and strengthen masculine domination in this subject area, as the following excerpt from an interview with 9 year-old Dominic showed when he was asked whether he thought boys or girls were better at mathematics:

TABLE 5.2 Children's Organized Out-of-School Activities

	7–8 years	9–10 years	15–16 years	17–18 years
Fleur		Tap and jazz dance	Dance	
Georgina	Swimming	Netball		
Jessica	Ballet	Violin, flipper ball	Speech and drama, Water polo	Water polo
Rochelle		Netball	Netball	
Dominic	Soccer	Soccer, cricket	Soccer	Soccer
Jared	Tball	TBall		BMX cycling, indoor soccer
Liam	Golf	Golf, cricket	Rugby League	Rugby League
Mitchell	Scouts			
Peter	Cricket, soccer, swimming	Cricket, soccer	Soccer	Soccer
Toby	Taekwondo, cricket, soccer	Cricket, soccer, piano	Soccer	Roller hockey, Soccer

Note: Girls' names shaded

> **Dominic:** I don't really know but we play this game called Maths Challenge... there's 2 boys standing up and 2 girls, and [the teacher] will choose someone, and they'll say, like, who they want to challenge and if the person who's standing up gets it wrong, then the person who called it will go up [to replace the person who got it wrong]. Out of all the games that we've played... we've [the boys] won seven and the girls have only got two.

Although Maths Challenge was played in the mathematics session, it had little to do with learning mathematics. Structured as a contest between boys and girls, the game was productive of *mathematics* as a body of facts to be committed to memory, *doing mathematics* as responding as speedily as possible with right answers to externally imposed questions, *mathematical achievement* as winning competitive sport-like mathematics games where provision of the correct answer became a public and judged performance, and *mathematical ability* as gendered, since the teams were segregated by sex and boys won significantly more of these matches than girls. Throughout the initial three years of the study, timed basic facts tests and games such as these were found to be a predominant feature of everyday mathematics classroom practice. The boys in the study were generally far more positive about such activities than the girls, particularly those who played competitive sport outside school, since they readily embraced the "striving to win" motivational ethos of such activities, as the following interview excerpts show:

> **Researcher:** *How do you feel about the* Quick Twenty*?*
> **Liam:** Good.... 'Cause it's a competitive thing and I like to compete.
> **Researcher:** *What about* Buzz*? You said earlier that you didn't like that as much.*
> **Liam:** No, I can't compete with it. (8 years)
> **Toby:** We play this thing called *Around the World.* I like it, like when I beat somebody when they come to my desk. (8 years)
> **Toby:** Speed maths that's my favorite part. I like being ahead of other people. (9 years)

In her early years at school, Rochelle was the only girl in the study who performed confidently in these kinds of competitive mathematical activities.

> **Researcher:** *What about that game* Around the World *that I saw you playing?*
> **Rochelle:** It's fun... you get to beat people at saying the answers.
> **Researcher:** *Do you usually beat people?*
> **Rochelle:** Sometimes. I nearly got around the world. (9 years)

Jessica was typical of those who found such games alienating.

> **Jessica:** We do *Around the World*, things like times tables, adding and dividing and there's this boy, he goes around and he's, like, really, really good. He's made it, like, three-quarters of the way around.
> **Researcher:** *How do you feel when he comes around to stand by you?*
> **Jessica:** Not Good. I feel kinda nervous. Because there's the whole class there and stuff. (8 years)

It was significant perhaps that she noted the sex (boy) rather than the name of the child who won this game. In Georgina's class, mini 'tests' in the form of ten quick questions with children racing against the clock occurred daily at the beginning of each mathematics session, with the longer written tests interspersed at monthly intervals. Before our first interview Georgina volunteered:

> **Georgina:** I hate maths. 'Cause I hate it when we do tests. I only get three or four or five or something, 'cause it's really hard. (7 years)

Although Georgina felt bad about failing, she attributed her failure to the difficulty of the questions. Through the deeply alienating experience of such tests, Georgina developed an aversion not only to the tests, but to mathematics itself.

The Construction of Gendered Narratives of Occupational Aspiration

Supporting Gottfredson's (2005) claims about the differentiated process of occupational choice, Table 5.3 shows the children's strongly gendered career aspirations over the 11 years of the study. The levels of mathematics required for these occupations are indicated in the table as appropriate.

Table 5.3 shows the links between competitive sport and masculinity played out in the imagined careers of the boys for whom sport was a major part of everyday life.

The Role of Family in Mathematical Genderfication

Studies (e.g., Garden, 1997) show strong connections between parental levels of education, qualifications, and occupations and those chosen by their children. In Table 5.4, the qualifications, school mathematics, and

TABLE 5.3 Children's Vocational Aspirations and Current Employment

	7–8 yrs (Yr 3)	9–10 yrs (Yr 5)	15–16 yrs (Yrs 11–12)	17–18 yrs (Yr 13)
Fleur	Florist	Florist	Psychology or sociology degree	Wants: Psychology degree
Georgina	Teacher, hairdresser, animal rescue (likes her pets) masseuse (parents say has gentle hands)	Doctor or vet (parents say is good with people)	Teacher—Year 12 math required	Wants: Own business Current job: call center
Jessica	Violin teacher, bus driver (aunt is a bus driver)	No idea	Commerce	Wants: No idea Actual job: School sport coach
Rochelle	Dental nurse, teacher	Teacher	Nurse—Year 13 math required	Wants: Policewoman—Year 12 math required Current job: temporary secretarial work, basic numeracy required
Dominic	Army, air force	Pilot	Pilot—advanced Year 13 math with calculus required	Wants: Economics degree
Jared	Judge—make people go to jail (father in prison); not a baker, too messy (mother a cook)	Cartoonist	Food technology or degree in Sports and Fitness	Wants: policeman—Year 12 math required, builder Current job: shelving supermarket goods
Liam	Policeman, helicopter pilot, builder, mechanic	Professional cricket player	Professional sportsman/personal trainer	Wants: Professional sportsman/personal trainer/coach
Mitchell	Army	Fireman (like father)	Supermarket checkout	Current job: Kitchen hand Wants: Cook
Peter	No idea	No idea	Not sure	Wants: History or Geography degree
Toby	Air force, professional sportsman	Professional soccer player	Economics or Commerce	Wants: Commerce degree

Note: Girls' names shaded

TABLE 5.4   Parental Backgrounds and Occupations

	Mother's highest qualification	Father's highest qualification	Mother's occupation	Father's occupation
Fleur	No academic qualification— no mathematics	University entrance—Year 11 mathematics	Home maker/ Typist/ Retail assistant	Credit manager large business
Georgina	No academic qualification— no mathematics	School certificate—Year 11 mathematics	Small business employee	Manager, small business
Jessica	School Certificate—Year 11 mathematics	School certificate—Year 11 mathematics	Home maker/ Personal trainer	Self-employed small business
Rochelle	No academic qualification— no mathematics	No academic qualification	Personal assistant small business	Professional sportsman/ Tradesman
Dominic	Bed—Year 11 mathematics	MBA—Year 12 Mathematics	Human resources large business	Systems manager large business
Jared	University entrance	Unknown (no contact with son)	Manager small business	Sales rep (stepfather)
Liam	No academic qualification— no mathematics	Higher school certificate	Retail assistant	Ordering clerk medium-sized business
Mitchell	No academic qualification— no mathematics	No academic qualification	Secretarial work	Fireman
Peter	BSc—Year 12 Mathematics	PhD Science— Year 13 mathematics	CEO small NGO	Policy analyst government department
Toby	University entrance Year 11 mathematics	BCom—Year 13 mathematics	Self-employed small business	Adviser government department

Note: Girls' names shaded

careers of the children's parents are shown. The mothers in the study were less likely to have gained higher qualifications than their husbands, were engaged in different kinds of occupations than those of their husbands, and earned less than their husbands. When mapped against Table 5.3, we can see that the girls' educational histories, mathematical learning experiences, and occupational aspirations diverged very little from those of their mothers and likewise for the boys and their fathers.

The interactions between children and families, including how parents supported children in their learning of mathematics and parental explanations for the children's mathematical (in)capabilities, revealed how parents and siblings played a significant role in children's learning of mathematics. Without exception, the fathers in the sample group were judged by their families to be better at mathematics than the mothers, who generally found mathematics difficult at school. Fleur's mother, for example, did not study advanced mathematics at secondary school and her recounted experiences of learning mathematics show how her turning to the males in her life for help with mathematics became a recurring pattern that positioned her as female.

Fleur's mother: Maths was a real struggle for me all my life. [On one occasion] I sat on the floor for an hour and a half and cried because I couldn't do it and my big brother came along went through all these ways until I understood. Maths was always the subject that I hated because I couldn't get to grips with it . . . I have to say I'm dreading when I can't help Fleur. (*Turning to Fleur's father*) You're going to have to help her with maths. (Early Year 3)

Fleur's teachers reported that Fleur also cried when mathematics became difficult for her to understand. Fleur's mother enlisted her husband's support with their daughter's mathematics homework.

Fleur's mother: When they did the power of ten. On the first day of that she came home with her homework and couldn't do it, and when I looked at it, I couldn't work out how they'd taught her. I got her father to look at it when he came home. He went through it with her patiently for three-quarters of an hour. (Late Year 4)

In these narratives, Fleur's mother identified and positioned herself in relation to more able (male) others for whom mathematics seemed naturally to make sense, and who patiently assisted the less able (females) to understand. She expressed in a very real sense the alienating domain that mathematics and its teaching and learning represented for her, an androcentric discursive space from which she was excluded but for the proffered male hand which she gratefully, and in the same instant self-deprecatingly, accepted. Although Fleur stayed at school longer than her mother and planned to go to university, she dropped mathematics as soon as the subject became optional.

Jessica also reported seeking help with mathematics from her elder brother. Her mother did not see herself as good at mathematics, but not entirely incapable saying, "I wasn't, um, useless at it, I passed Year 11 School certificate." She had opted for shorthand typing instead of mathematics and dropped out of school in Year 12. When talking of Jessica as a 7 year-old, she saw her daughter as a particular 'type' of girl:

Jessica's mother: She wants to sing in the choir and she's going to be that sort of girl, you know, she may be more creative . . . I just get this feeling there's a lot of the creative side coming out of her.

Jessica stayed at school longer than her mother and continued to study Year 12 mathematics but failed her external examinations. Like her mother, she chose subjects that did not demand mathematics and science—what the children referred to as 'bum subjects'—such as drama, computer and media studies, and physical education.

Georgina's mother did not do well at school mathematics and it was her father who provided Georgina with help. The following excerpts from family conversations highlight the roles played by mother and father in Georgina's family.

Georgina's mother: I left [school] early in [Year 10] so I don't have any qualifications . . . I'm not very good at mathematics. Her father shows her how to do it. (Early Year 3)

Georgina: He [Dad] tells me what to do and then I write it down, and then I tell him what the answer is and he says if it's correct or not. (Early Year 4)

Georgina: In my maths book the first week there was these really hard working forms and my dad's, like, taught me how to do the first one and then I got to do it and I got all of them right without anyone helping me. (Early Year 5)

Georgina's father: I'm frustrated because Georgina doesn't seem to be getting it (basic facts). I could do it easily—I've got a scientific mind. (Mid Year 5)

Masculinity was aligned in these statements with mathematical capability and scientific mindedness. Despite Georgina's teachers' belief that there were no significant cognitive barriers to her learning of mathematics, Georgina was repeatedly placed in the lowest ability group for mathematics, and opted for practical mathematics in Year 11 rather than general mathematics like most of her peers.

Four of the six boys in the study continued to take mathematics to Year 13 level, although none of them chose the more difficult options that included calculus. Peter had not felt particularly confident with mathematics throughout his secondary schooling but when his parents engaged the help of a private tutor—a male university student—this made a significant difference to Peter's achievement. Peter's father, a scientist and confident user of advanced mathematics, provided a strong masculine role model for his son.

Peter's father: He's doing Stats this year, that's one thing I did a lot at university and with my PhD and I've had a look at some of the Stats that he has been doing this year and I know if he came and asked me it wouldn't take me very long to get up to speed.

Sibling comparisons played a part in producing gendered mathematical subjectivities as feminine and masculine occupational attributes were "recognized" by parents.

Liam's mother: He's a little smarty. We'll say to Chantelle [Liam's sister, older by two years] "What's such and such?" [a mathematical question] and he'll go like, (*clicks her fingers*) not a problem really. (Early Year 3)

Liam's father: He'd probably be quicker, if anything he might be quicker than Chantelle. Yeah, but they're different people. He's not going to be a reader or story teller like Chantelle is. (Early Year 3)

Classroom Mathematics and Genderfication

From the children's and teachers' descriptions of mathematics lessons during the eleven years of this study, it was found that a high proportion of mathematics time was spent on written tasks in the form of worksheets, textbook pages, or work from the board, with practical work, group work, and discussion diminishing the older they became, as shown in the following typical accounts from the Year 12 interviews:

Fleur: Every day it was the textbook . . . our class is like, for the first like, 20 minutes you just write down notes and then you'd have 20 minutes of doing the work and then you do it at home . . . 3rd and 4th form we did a bit more practical. 5th form was real textbook and notes.

Georgina: I get bored having the same. It just gets so repetitive and boring, [I would like] going outside and something and diagrams not just notes all the time.

Peter: We usually just do exercises and stuff and they tell us the formulas that we need to know and that doesn't change much throughout the year for different things . . . we've got like a quite a big text book and it just has all the exercises that we do in it and some, like, exam questions and stuff.

Toby: The teacher gives us notes . . . if it gets dragged on for a long time it just gets boring.

Rochelle: [The teacher said] 'Right do this page and when you're finished bring it up or go onto the next page,' and stuff like that.

Their descriptions speak of a subject that is universally disengaging for its lack of variation, lack of stimulation, lack of student choice, and lack of personal relevance. However, there were marked differences between the boys and girls in the ways in which the children responded to these predominant pedagogical practices of their mathematics classrooms, as their reported feelings about doing mathematics reveal in Table 5.5.

Negative feelings were found to be closely linked to achievement. Jared and Mitchell, the boys who reported the greatest alienation and disaffection, were placed in the lowest groups for mathematics.

As a school leaver Georgina applied for entrance into university to enroll in a teaching degree. On failing the literacy and numeracy entrance tests she enrolled instead in short polytechnic course in computing and secretarial skills. When asked to reflect on who it was that succeeded at mathematics in her experience, Georgina responded:

Georgina: I remember at [primary] school there were three of us girls that were really bad at maths, and we always got taken aside and shown how to do it step by step, and all the boys, there were about seven boys had no trouble whatsoever. I'm not to sure if I read it in the paper that boys are better than girls at maths, but I think it's a load of crap to be honest, I think girls are just as good as guys. (17 years)

TABLE 5.5   Children's Responses When Asked How They Felt About Mathematics

	Year 3	Year 4	Year 5	Year 12	Year 13
Fleur	Negative	Negative/OK	Negative/OK	Negative	No mathematics
Georgina	Very negative	Negative	Negative	OK	No school
Jessica	Negative	Negative/OK	Negative/OK	Negative	Very negative
Rochelle	Very positive	Positive	Positive/OK	Very negative	No school
Dominic	Positive	Positive	Positive	Negative/OK	OK/positive
Jared	OK/Negative	OK	OK	Negative	Very negative
Liam	Very positive	OK	OK	Negative	Negative
Mitchell	Negative	Negative	Very negative	Very negative	Very negative
Peter	Positive	OK	OK	Positive/OK	Positive/OK
Toby	Very positive	Positive	Positive	OK	OK

Note: Girls' names shaded

She also noted that in her polytechnic course which required no mathematical qualifications or entry test mathematics, there were only two males.

> **Georgina:** You don't see many females working as mechanics and technicians and engineers and things like that, there are very few females that take that on so I'm thinking that possibly males have more mechanically minded brains and they have more, it's easier for them to figure out equations and things, but then there are some things that females are better at doing, it all equals out in the long run, everything has an opposite, so you give and take things. (17 years)

Throughout her schooling, Georgina had experienced being positioned in the tail of underachieving girls in mathematics and as a school leaver, experienced genderfication by career choice. While she rejected the stereotypical view that boys are better at mathematics, she accepted that natural and therefore inevitable gender differences may exist when it comes to certain kinds of thinking, and that such distinctions are equitably opposed.

By the time Rochelle had reached upper secondary school, mathematics had become something she did not enjoy. When she moved from an all-girls school to a co-educational school in Year 12, the presence of boys in the class made a difference to her level of comfort:

> **Rochelle:** The boys, when there was a [mathematics] teacher, a female teacher, they would muck around a bit, whereas if there was a male teacher they wouldn't, and I don't know why that was. (17 years)

Rochelle perceived that for boys, a female mathematics teacher was different from a male and had a bearing on boys' behavior in class. Rochelle also noted that outside the classroom, boys and girls differentiated between the kinds of time they spent with each other:

> **Rochelle:** Girls wouldn't hang out with boys unless, I mean, some would, but a majority of them wouldn't, yeah ... I don't think boys want to, you know, play kicks or play basketball with girls, but you would see the odd girl playing basketball ... but not very often, unless they were like, family, or like cousins or something like that.
>
> **Researcher:** *I wonder why that is, why that happens?*
>
> **Rochelle:** Maybe having their space from girls, like you know, you just need, if you're a girl, you need your girls' ... you want your girl time, you know, and I think the same goes for boys, they

wanna hang out with their boys, 'cause even my boyfriend you know wants to hang out with his boys like heaps of time without me, but not because he wants to get up to something, just because he misses his boy time.

Rochelle's identification of what she terms "girl time" and "boy time" suggests that sex-segregated activities are self-evidently natural. Her description of how girls and boys operated in different ways at school including how they interacted with one another and with male or female teachers, spoke strongly of boys' and girls' experiences of life as gendered beings, supporting theories that gender is produced in our everyday enactments of "self."

Choosing to Study Mathematics

Table 5.6 shows how the children in this study participated in mathematics education over the final three years of their upper secondary schooling and/or into the workforce. By the final year of the study, none of the four girls was continuing to study advanced mathematics of any kind and two were already in fulltime employment. Georgina, the first of the group to gain tertiary qualifications, had undertaken a 20-week polytechnic computer studies course. The table shows the gendered process of subject selection for upper secondary school and planning careers beyond school, and the implications for the life chances of these ten children.

For all of the children, the choice about whether to continue studying mathematics, and if so, what kinds of mathematics, was tied to their perceptions of the occupational value of mathematics, as the girls' explanations for their choices about studying mathematics illustrated.

> **Fleur:** I'm not doing it next year [Year 12 mathematics], I'd rather do classics and history and geography... I wanna do psychology or sociology [at university], and Miss Highly said that you need maths in bio [tertiary study in biology] but the [other subjects] you don't have to.

Jessica's decision not to continue studying maths into her final year at secondary school was one of self-confirmation, as the following conversation showed:

> **Jessica:** I was really considering dropping [maths] before last year... I am pleased, [I continued taking it] because it meant that by the time I got to the end of it I really knew I didn't want to do it any more, I was like I'd really had enough now,

TABLE 5.6 Participation in Mathematics Subjects in Years 11–13

	Year 11 (mathematics compulsory)	Year 12 (mathematics not compulsory)	Year 13 (mathematics not compulsory)
Fleur	General mathematics	No mathematics	Remained at school. No mathematics
Georgina	Numeracy Mathematics	No mathematics	Left school, Polytechnic—Computer skills certificate (20 weeks' study); Employed fulltime secretarial work
Jessica	General mathematics	General mathematics—selected reluctantly	Remained at school. No mathematics
Rochelle	General mathematics	General mathematics (first 3 months of year); Left school, Employed—supermarket sales assistant	Employed— temporary secretary in government departments
Dominic	General mathematics	General mathematics	Remained at school. Mathematics with statistics
Jared	General mathematics	General mathematics	Left school, unemployed (6 months) Employed part time / Studying mathematics by distance
Liam	General mathematics	General mathematics	Remained at school. Mathematics with statistics
Mitchell	Numeracy mathematics	Numeracy mathematics	School/polytechnic transition to work program—kitchen hand/waiting.
Peter	General mathematics	General mathematics	Remained at school. Mathematics with Statistics
Toby	General mathematics	General mathematics	Remained at school. Mathematics with Statistics

Note: Girls' names shaded

you know at the end of Year 11, I was like, 'I'm so over maths, I don't want to do it any more,' at the end of Year 12. I was like, 'Hey I'm really done in maths.' I'm pleased I made that decision then rather than before. Now I'm like I'm totally sweet with the fact that it's gone... I found it quite easy to make the decision to drop it.

Researcher: *No one tried to persuade you to carry on?*

Jessica: No, my mum was kinda like, 'Oh, so no maths?' and I'm like 'No! I'm not doing it.' 'Ok then.' I think she would have loved it if I'd continued, but nevertheless she is not really that fussed, I don't think that it's really... I think it's more that she felt she had to encourage, because she doesn't really care a lot about maths that much, she's not doing anything with maths and I think it is more that she felt like it was one of those things that was expected... that everyone does maths, so she encouraged it for that reason, but I don't know that she really felt that strongly about me doing it or not. (Mid Year 13)

Linked to comments made by Jessica's mother in my first interview when Jessica was a 7-year-old, this statement revealed how choices about studying mathematics play out across generations, as feminine identification is strongly preserved.

Mother: [Mathematics is] really important, and more so for girls. Not because I'm being sexist, I just feel that, I know that the girls get to a certain level in college like we did, like I did, and it's almost expected that you should drop out, you know... and maybe it wasn't obvious, but it was like, you know, you don't have to do well at this, there are other things [for girls]. (Early Year 3)

Made more than ten years apart, these two statements illustrated mathematical genderfication as intergenerational. Jessica did not perceive mathematics as something that was likely to have value for her in the future, just as it had not for her mother. Just as her mother had, Jessica saw the choice of studying mathematics as governed more by societal expectation—something girls *can* do but don't have to—than personal necessity. This process works in ways that are hard to pin down, as Jessica's mother astutely observes when she says, "Maybe it isn't obvious."

For girls, failure in mathematics is still not seen as an issue in occupational identity, as Georgina explains:

Georgina: At college and at school I was always really struggling with maths and I was trying to work real hard... but [dropping maths] hasn't disadvantaged me in any way at all and I don't think [lack of maths] will ever stop me from doing anything.

The fact that Georgina's failure in the numeracy entry test for university had already stopped her from pursuing a range of career options was either deliberately or inadvertently discounted in her personal account of the usefulness or not of mathematics in her life. The boys in the study were more likely to see the occupational value in continuing to study mathematics as something that kept 'doors open' or enabled them to get where they wanted to go.

> **Toby:** It'll be a good thing to have (Year 13 mathematics credits). Obviously some stuff we do isn't relevant but it'd be a good thing to have for later in life and all that, yeah, it might keep some doors open.
>
> **Peter:** I wanted to keep my options open so I took Stats (in Year 13)
>
> **Liam:** Good if I get it (Year 13 maths credits) but if I don't I'll be gutted. I should be able to like, scrape through, I should get enough credits to get where I want to go.

For Liam, continuing to study mathematics into Year 13 was difficult because many of his male mates had already dropped out of mathematics to take subjects that would lead to some kind of trade. For Toby and Peter the choice was much clearer since they were following in their father's occupational footsteps towards university study.

Mathematical Discourse As Sexuated

The case stories in this study support Mendick's idea that *boys* and *girls* are subjects formed in mathematical discourse, and that individual agency to act or freedom to choose is shaped by what is socially (discursively) doable. Expanding on this idea, Irigaray (2002) argued that discourses themselves, including those of discursive domains such as the teaching and learning of mathematics, are sexuated. She described the ways in which women are presented in dominant everyday discourse as an invisibility, their "otherness" manifesting as a discursive exclusion she termed *monosexuality* or *sexual indifference*. In this view, girls' apparent rejection of mathematics can be explained as a process of cognizing mathematics as pre-gendered (male) territory, a mono-sexual and therefore exclusive/excluding domain, a domain that is indifferent to the female voice. Bourdieu (2001) likewise argued the existence of an omnipresent masculine domination he termed the *androcentric unconscious*, something so embedded in everyday thought and action as to be barely perceptible even when we are looking for it. An illustration of masculine domination at work was provided by the children in this study when asked where mathematics comes from. Their responses revealed an unquestioning acceptance of mathematics as a subject that has "been around forever," and its exclusively masculine origins and evolution.

Georgina: From the Romans isn't it? Wasn't there a guy called Archimedius [sic] or something? And like Da Vinci, he had to figure out numbers and equations to figure out how to make something fly.

Jessica: I have no idea... I really don't think I've ever been told.

Fleur: Ah, this might sound dumb, I'm not actually sure, but I think it's always been around, because math is around us all the time even if we don't realize it... So I think it's been around forever.

Peter: We don't really do the history of it we just do it, they just give us, tell us what to do and we don't really ask questions.

Dominic: To look historically I'd say the Renaissance and you know, sort of the development of science and that kind of thing, moving away from religion into science like, logic... Just look at famous mathematicians like Fibonacci from the Renaissance. The sequence is not the man. I'd say that would be the only famous mathematician that I know, it's just a name that cropped up.

Liam: I don't know. Maybe Roman numerals or something, that's the only thing I can think of.

Jared: People have been doing it for years, like, since the beginning of time.

Even within the mathematical discourses of diversity and inclusion, this acceptance/indifference can be discerned as a pervasive negligence, a systematic oversight, an exclusory deafness that fails to substantiate female/feminine/woman/girl mathematical narratives and contributions to the discipline of mathematics as distinct—not to be confused with singular or essential as Fuss (1990) describes—as sexually nuanced, and as situated in women's and girls' contextualized, gendered, everyday experiences in society. The female register is scripted within male-centric mathematical discourses only as echo, mimicry, and ventriloquism.

The (mono)sexuation of mathematics education discourses is particularly apparent when we examine the choice of plenary conference speakers, of pressing (gender blind) issues in mathematics education such as overall scores on international comparative tests on which to focus our collective attention, of theories by which we *examine, interrogate,* and *inform* our practice, or the method and form of what counts as acceptable/believable (male sanctioned) research, consistently revealing a marginalization, minimization, and dematerialization of the *female* in mathematics. The mathematics adopted worldwide as *the* mathematics children should learn, has a powerful patriarchal genealogy. Wertheim's examination of the cultural/historical construction of Western mathematics (Wertheim, 1997) shows how the intertwined discourses of *mathematics, physics,* and *God* constitute a *gender war* that consciously and explicitly constructs the feminine out of mathematics through a binary oppositional alignment of female with evil, ungodly, impure, irrational, and illogical. This is reinforced by Walkerdine (1998) whose research demonstrated how the *feminine* is *subtracted* from mathematics, as science, reason and the male mind are co-produced through psychological discourses of education.

The experiences of the children in this study provide evidence of the androcentric unconscious at work, manifesting in their orientations to life, vocational aspirations, and their learning of mathematics. By their final year at school, all of the boys continued to study mathematics but none of the girls. The girls had chosen careers that they believed did not require high levels of mathematical attainment. The girls became alienated from mathematics from a younger age and were more likely to attribute their lack of success to the "hardness" of mathematics, their failure to understand the teachers, and their own (natural) lack of ability, than were the boys, who were more likely to blame their personal lack of effort, the teachers' failure to teach, or teachers simply failing to see their abilities. Parents' beliefs about their children mirrored these causal attribution patterns. Both groups commented on the irrelevance and boring nature of mathematics lessons but the girls exhibited a much earlier rejection of mathematics' importance in their lives within and beyond school.

CONCLUSIONS

Case studies such as these do not provide us with definitive proof that mathematics somehow counts boys in and girls out, nor do they aim to. Their purpose is to highlight the processes by which individual children are constructed, and construct themselves as gendered beings in their learning of mathematics, and the implications of such construction. As this study shows, children live gendered lives, revealed in their preferences from favored toys to the choice of school subjects. Through their interactions with friendship groups and family, children are engaged in social performances that distinguish them as "boys" or "girls." In turn, children's experiences of learning mathematics, which occur not only at school but within other social settings such as family and peer-group, are both gendered and genderfying. In this study, genderfication can be seen to be ineluctably implicated in the processes of teaching and learning mathematics.

Lynch (1999) argued that, "Our society is profoundly androcentric in the sense that male norms and values are persistently privileged over female norms and values; things named as 'feminine' are treated as inferior, unworthy and subordinate. These injustices of recognition have deeply inegalitarian implications for women" (p. 138). If this is so, the pervasive global (self)exclusion of girls and women from upper secondary school mathematics, from mathematics subjects at universities, from the field of mathematics, and from equal and different standing in mathematics education policy formation, implementation, research, and debate, particularly in developing countries, points to the potency of connections between masculinity, mathematics, and social power, the often obscured social realities of which studies such as this help us to recognize. The study shows how, in order for them to engage successfully in mathematics education either as learners or teachers, girls and women are required to assume a female invisibility that affords them temporary status as honorary males trespassing in a male domain in which they become the masculated disciples, conduits and traffickers of a male-centric mathematical discourse. Social patterns of interaction including pedagogies of school ensure that the masculine domination of mathematics remains, thus boys "muck around more" for female mathematics teachers, girls opt out of mathematics because they say they find it "hard" and "boring" and they "don't really need it," and boys continue to study a subject they too say seems "boring" and "irrelevant" but nevertheless "keeps doors open."

Parents' and peers' actions to support decisions to drop or continue studying mathematics in upper secondary school endorse these differentiated and differentiating views. For the boys in this research group, the content of mathematics itself may have been unappealing, but they viewed its study as a necessity, an inevitability, and an unquestionably masculinizing, empowering signifier of their schooling. The psycho-social demand of studying mathematics as a distinctively masculinized and masculinizing subject became untenable for all of the research girls, their continuing study of mathematics rejected with relief rather than regret. The issue appears to lie not with girls' differing mathematical abilities or learning styles, nor with pedagogies of mathematics that do not cater to girls' cognitive needs, but with something much more deep-seated and therefore difficult to address— the entrenched alignment of masculinity and mathematics that manifests as a subject of and about those things that have a masculinizing meaning for boys and men and which a significant proportion of girls, faced with an apparent freedom to choose, reject for its failure to recognize them.

REFERENCES

Bartholomew, H. (2005). Top set identities and the marginalisation of girls. In M. Goos, C. Kanes, & R. Brown (Eds.), *Mathematics education and society: Proceedings of the 4th International Mathematics Education and Society Conference* (pp. 69–77). Gold Coast, Australia.

Boaler, J. (2002). *Experiencing school mathematics: Traditional and reform approaches to teaching and their impact on student learning.* Mahwah, NJ: Lawrence Erlbaum Associates.

Bourdieu, P. (2001). *Masculine domination.* Palo Alto, CA: Stanford University Press.

Fuss, D. (1990). *Essentially speaking.* New York: Routledge.

Garden, R. (Ed.). (1997). *Mathematics and science performance in the middle primary school: Results from New Zealand's participation in the Third International Mathematics and Science Study.* Wellington: Ministry of Education.

Gilbert, R., & Gilbert, P. (1998). *Masculinity goes to school.* St Leonards, NSW: Allen & Unwin.

Gottfredson, L. (2005). Using Gottfredson's theory of circumscription and compromise in career guidance and counseling. In S. D. Brown & R. W. Lent (Eds.), *Career development and counseling: Putting theory and research to work* (pp. 71–100). New York: Wiley.

Irigaray, L. (2002). *To speak is never neutral.* New York: Routledge.

Kielhofner, G. (2008). *Model of human occupation: Theory and application.* Fourth Edition. Baltimore, Maryland: Lippincott, Williams & Wilkins.

Lubinski, D. & Benbow C. J. (2006). Study of mathematically precocious youth after 35 years: Uncovering antecedents for math-science expertise. *Perspectives on Psychological Science. 1*(4), 317–345.

Lynch, K. (1999). *Equality in education.* Dublin: Gill and Macmillan.

McArdle, E. (2008, May 18). The freedom to say no. *Boston Globe*, p. D1.

Mendick, H. (2006). *Masculinities in mathematics.* Berkshire: Open University Press.

Thompson, S. & Fleming, N. (2004). *Summing it up: Mathematics achievement in Australian schools in TIMSS 2002. (TIMSS Australia Monograph no 6).* Camberwell, Victoria: Australian Council for Educational Research.

Walkerdine, V. (1998). *Counting girls out: Girls and mathematics.* London: Falmer.

Wertheim, M. (1997). *Pythagoras' trousers: God, physics and the gender wars.* London: Fourth Estate.

CHAPTER 6

GENDER MAINSTREAMING

Maintaining Attention
on Gender Equality

Colleen Vale
Victoria University

We have gender equality in mathematics education in Australia, don't we? This is the prevailing view among mathematics teachers, and, leading up to the turn of the century, the gender gaps in achievement, participation, and affect in Australia were closing and in some studies not significant (Forgasz, Leder, & Vale, 2000). However since the turn of the century, while this trend has continued to be evident in some studies, gender differences are persistent and even widening in affect, participation, and achievement for some grade levels and domains in mathematics (Vale & Bartholomew, 2008; Vale, Forgasz, & Horne, 2004). How could this happen?

Gender mainstreaming is a strategy for achieving gender equality that was adopted by the United Nations at the Fourth World Conference on Women in Beijing 1995. Involving policy and program development, implementation and monitoring at all levels of government, and in research, the strategy of gender mainstreaming requires that attention be given to the goal of gender equality (OSAGI, nd). Many countries, both developed

International Perspectives on Gender and Mathematics Education, pages 111–143

and late developing, have adopted gender mainstreaming as government policy, including Australia. However, the meaning of gender mainstreaming is contested, as is the meaning of gender equality (Walby, 2005).

Using Australia as a case study, the extent to which and the way in which attention is given to gender equality and the approaches adopted to promote gender equality in mathematics education will be explored in this chapter. The chapter begins with a discussion of the contested theories of gender mainstreaming and gender equality in the context of mathematics education. In the second section, the policies and programs developed to promote gender equality in Australia are analyzed drawing on the various gender theories concerning mainstreaming and equality. I will describe the rise and fall of feminism in Australian education policy and argue that performativity, together with the feminist backlash accompanied by a change to conservative government denied girls equality in mathematics education in Australia. The role of feminism in the demise of a gender agenda focused on girls will also be canvassed. The research studies and data gathered by the education authorities in Australia that support—and contest—the presumption of equality in mathematics education are presented and analyzed in the third section of this chapter. I will conclude with some thoughts on the gender mainstreaming for realizing gender equality in mathematics education.

GENDER MAINSTREAMING

The United Nations determined that gender mainstreaming be understood as a strategy or a tool for improving the status and lives of women around the world. However, researchers have argued that it is a concept with multiple meanings and hence open to various interpretations and means of implementation by governments (Walby, 2005). Not surprisingly the multiple perspectives and meanings of gender mainstreaming are consistent with the theoretical distinctions concerning gender equality.

Assimilation and Liberal Feminism

One view of gender mainstreaming is to see it as a process of integration or assimilation. In this view gender perspectives are included without challenging the priority or direction of existing policies; women and girls are assimilated into the mainstream and monitoring is confined to accountability measures. Consistent with this interpretation of gender mainstreaming, the norm is understood to be defined according to male attributes, and the goal of "equality" is pursued with the assumption that equality means "the same" and is monitored through measures for equal opportunity, equal treatment, and equal outcomes. These meanings of equality have been associated with a liberal feminist paradigm based on a deficit view of women and acceptance of continued gendered practices in some realms (Jungwirth, 2003; Kaiser & Rogers, 1995). Indicators of equality in this sense may be categorized and measured according to equality of access and progression, resources, and outcomes (Sherman & Poirier, 2007).

According to Lombardo (2005), the liberal feminist approach may be successful in bringing attention to gender inequality but is less likely to result in substantial change. Numerous examples of the failure of liberal feminist policy to effect change can be observed in a range of policy areas. For example, in Australia and elsewhere, women's suffrage has not resulted in equal or proportional representation of women in political positions; equal pay has not resulted in equality of incomes for men and women; and equal access to education has not resulted in agency or economic independence for all. With respect to mathematics education, Fennema (1995), reporting for the ICMI study on gender and mathematics education, argued that many studies of gender and mathematics indicated that equal access and equal treatment were not sufficient to achieve equal outcomes.

Transformation and Radical Feminism

An alternate perspective of gender mainstreaming embraces valuing and respect for women as different and equal. The Council of Europe (1998) defined gender equality as "equal visibility, empowerment and participation of both sexes in all spheres of public and private life . . . and accepting and valuing equally the differences between women and men and the diverse roles they play in society" (pp. 7–8).

This interpretation is empathetic with radical feminist theory, which emphasizes differences between men and women and the need to include women's perspective. Gender mainstreaming is a strategy for changing the mainstream, when equality means accepting and valuing difference between women and men (Lombardo, 2005; Walby, 2005). Agenda setting is the approach for changing the mainstream and involves transformation of policies and processes (Lombardo, 2005; Walby, 2005).

Many mathematics education researchers have emphasized the importance of recognizing differences and including girls and women. For example, Burton (1995) argued that mathematics curriculum was based on male perspectives of mathematics and needed to change to include the perspective of women. Earlier, Belencky, Clinchy, Goldberger, and Tarule (1986) had argued that women learn differently and therefore teaching needed to adapt to women's ways of knowing. Similarly, Boaler (1997) argued for different approaches to teaching and learning that would connect with and engage girls in mathematics.

Radical feminist theory was criticized for essentializing women (Jungwirth, 2003). According to post-modern theory, gender is socially constructed, and inequality arises through the unequal distribution of power in social interaction. Gender is situated, and gender identity shifts according to sphere of life (Jungwirth, 2003). Moreover, women are not the same due to many other social differences including socioeconomic status, ethnicity, faith, sexual orientation, and age. An interpretation of gender mainstreaming that includes equally valuing difference among women combines the goal of equality with diversity by merging mainstreaming of gender with mainstreaming of other categories of difference (Rees, 1998, cited by Henningsen, 2004). Equally valuing diversity involves "the process of creating and maintaining a positive environment where the differences of all persons are recognised, understood and valued, so that all can reach their full potential and maximise their contributions" (Henningsen, 2004, p. 2).

Combining the goal of gender equality and diversity requires deliberative democracy (Squires, 2005; Walby, 2005). Approaches to gender mainstreaming that give priority to taking action, that is, a commitment to "closing the gap" for all forms of inequality, align with meanings of equity and social justice (Gewitz, 1998; Secada, Fennema & Adajian, 1995; Walby, 2005).

Democracy and Post-Modern Feminism

Mathematics education researchers using a post-modern theoretical framework have shown how girls are disadvantaged in the mathematics classroom when their mathematics identity does not accord with gendered stereotypes (for example, Rodd & Bartholomew, 2006; Shannon, 2004; Vale, 2002; Walden & Walkerdine, 1985). These studies and others illustrate Fennema's (1995) argument that equal access and equal treatment are not sufficient for equal outcomes. They point to the need to transform and democratize curriculum and classrooms.

The twelve practices for teachers and curriculum designers listed in Goodell and Parker's (2001) "connected equitable mathematics classroom" focus on the quality of the curriculum and pedagogy and the expectations of teachers. According to Anthony and Walshaw (2007), equity in mathematics classrooms involves paying attention to different needs of students according to diverse forms of social inequality and using different and appropriate pedagogical approaches for these students. Teachers are therefore practicing gender sensitivity when they acknowledge gender inequality as one of the differences in their classrooms and target the diverse learning needs of girls (Jungwirth, 2003). Moreover, combining responses to diversity with engagement in mathematical thinking affords more opportunities for developing agency and empowerment through mathematical learning (Boaler, 2002; Boaler, 2003).

If gender mainstreaming is to be both transformative and democratic, then monitoring policy, programs, and outcomes for equality requires giving attention to the quality of the educational experience and curriculum that girls, including girls from diverse equity groups, have access to; the quality of resources including the knowledge and attitudes of teachers; and the significance of the outcomes measured with respect to capacity building, empowerment, and participation in society.

The following section of this chapter explores the ways in which and the extent to which gender is mainstreamed in education policy and mathematics curriculum, and is implemented and monitored by the Federal Government and the state governments in Australia since 1975.

GENDER IN EDUCATION POLICY AND MATHEMATICS IN AUSTRALIA

In this section Australian government education policy related to gender and mathematics will be analyzed to discern whether gender mainstreaming is evident and which perspective on gender mainstreaming and equality is applied.

First Wave Feminism 1975–1984

Australia has a respectable history with respect to the development of policy regarding gender and education. Attention began to be given to girls in Australian schools during International Women's Year in 1975, with the publication of *Girls School and Society* (Schools Commission, 1975). It heralded the commencement of the Australian government's involvement in schooling policy and curriculum, constitutionally the responsibility of the state governments in Australia. This report on the status of girls in Australian schools adopted a liberal feminist stance arguing that the invisibility of women and girls in the curriculum and sexism experienced by girls in Australian schools resulted from an oversight that still identified girls as the problem (Yates, 1993). This report therefore commenced the mainstreaming of gender in education in Australia where the vision was to integrate and assimilate girls to promote gender equality where equality meant sameness.

Second Wave Feminism 1984–1996

The second wave of feminism during the 1980s heralded the development and publication of a series of reform-based programs and resources, many funded under the auspices of Australian Government programs, some of which concerned gender and mathematics teaching and learning (for example, Vale, 1987) along with further government reports. *Girls and Tomorrow* (Schools Commission, 1984) indicated that there was sufficient evidence of girls' disadvantage to warrant a policy (Lingard, 2003). The first policy on gender in education in Australia, the *National Policy for the Education of Girls in Australian Schools* (Schools Commission, 1987) was endorsed by all the Ministers of Education at the state and federal level in 1986. This policy was also the first national policy of any type on education in Australia and hence represented the first attempt to unify schooling in Australia through a reform agenda (Lingard, 2003).

According to Yates (1993), the *National Policy* was "an attempt to draw together, systematize and bring a new accountability in relation to the strands of reform that had developed" (p. 170). The four objectives of this policy were raising awareness, supportive school environment, equal access to and participation in appropriate curriculum, and equitable resource allocation. These elements reflected emerging feminist perspectives regarding the social construction of gender and radical feminist perspectives on gender differences in learning (for example, Belenky et al., 1986; Becker, 1995) and gender-appropriate curriculum materials and teaching approaches: "In 1987, there is no longer such a clear assumption that 'liberal education,' with adjustments, is all that is required for progressive action. The policy places much more emphasis on the 'educational needs of girls'" (Yates, 1993, p. 170).

Moreover, the policy indicated that these needs may be different for girls of different ethnicity or socioeconomic background. The policy therefore implied an agenda setting interpretation of gender mainstreaming with the intention of changing the mainstream to accommodate the needs of girls and to improve their educational outcomes.

In the period after the publication of the *National Policy*, implementation began and included the publication of a national newsletter, *The Gen*, a series of national statements on curriculum for Australian schools, mandatory government reports monitoring the implementation of the policy, and the development of curriculum materials to support teachers. *A National Statement on Mathematics for Australian Schools* (AEC, 1990) proposed that "access to and success in school mathematics should be independent of gender, social class or ethnicity" (Australian Education Council, 1990, p. 8). One example of a Federal Government-funded project for the development of appropriate mathematics education materials for girls was *Investigating Change*

(Barnes, 1991), a series of booklets for the teaching of calculus in senior secondary mathematics. Key features of this publication, and other similar materials developed during this period, were girl-friendly pedagogical approaches such as group work and discussion and an emphasis on relating mathematics to the real world. These materials were also consistent with the concurrent development of problem solving and modeling and constructivist approaches to learning in mathematics education. At the state level, the impact of the policy was evident in the changes to curriculum. For example in Victoria, the curriculum documents for primary and junior secondary mathematics included sections on pedagogical practices for teachers, and at the senior secondary level assessment practices were changed to include school assessed investigations and problem-solving tasks.

In 1992 the Ministers of Education in Australia endorsed the *National Action Plan for the Education of Girls 1993–1997* (Australian Education Council, 1993). The *Plan* was described as "a guide to mainstream policy making at the national, system, school and community levels in order to achieve the ongoing objectives of he *National Policy for the Education of Girls*, and as a practical manual to assist educators to achieve these objectives in their day to day work" (MCEETYA, nd).

This policy direction clearly foreshadowed the forthcoming gender mainstreaming strategy of the United Nations for giving attention to the goal of gender equality and invoking policy development, implementation and monitoring at all levels of the government and decision-making. The plan included eight priorities for action:

- Examining the construction of gender
- Eliminating sex-based harassment
- Improving the educational outcomes of girls who benefit least from schooling
- Addressing the needs of girls at risk
- Reforming the curriculum
- Improving teaching practice
- Broadening work education
- Changing school organization and management practice

Extending the dual emphasis on the feminism of equality and the feminism of difference of the *National Policy*, the *National Action Plan* also more explicitly acknowledged other forms of social disadvantage crippling girls' opportunities for improvement. A gender inclusive curriculum was required—that is, in mathematics, a curriculum that would be humanizing and value cultural, social, and personal connections as well as value different methods of solution and forms of proof and different ways of thinking (Burton, 1995). Changing curriculum was deemed to be insufficient. Classroom-based research indicated the need to change and improve teachers' attitudes and practices (for example, Leder, 1993; Forgasz & Leder, 1996).

Feminist Backlash 1996–2007

In 1996, the year following the Fourth World Conference on Women, there was a change of government. The conservative parties gained power at the national level in Australia. At this time evidence of girls' increased participation and achievement in non-traditional areas, such as mathematics and science, was beginning to emerge, especially for middle and upper class girls in private and/or single-sex schools (Teese, Davies, Charlton, & Polesel, 1995).

In 1996, Collins, Batten, Ainly, and Getty (1996) completed an evaluation of the *National Policy* and the *National Action Plan*. Perhaps indicative of what was to emerge in the new conservative policies, the evaluation was designed to "provide data on gender matters in relation to boys as well as girls" (Collins et al., 1996, p. ix), and the authors reported:

> This study offers considerable evidence that attention to gender issues by systems, schools and individual teachers does make a difference to the gender experiences of students in schools....There is nevertheless, still much to be done at all levels for both sexes for their fuller development as individuals and for their present and future lives together. (p. xiv)

This report was the first evidence of a shift away from a gender policy focused on improving the outcomes for girls.

Ironically, then, in the year after becoming a signatory to the UN policy on gender mainstreaming, the new report, *Gender Equity: A Framework for Australian Schools* (MCEETYA, 1996), was focused on improving the outcomes for girls and boys in Australian schools. Ailwood and Lingard (2001) argued that this document represented the "endgame" of gender mainstreaming with a focus on girls' schooling and the goal of gender equality. "The policy signals a substantial shift in focus—from girls and boys in relation to girls, to both girls and boys—within a framework of presumptive equality" (Ailwood & Lingard, 2001, p. 9). Ailwood (2003) referred to this "sidelining" of many issues regarding girls' education that still warranted attention as the "downstreaming" of gender. Feminists working within the bureaucracy found themselves in the position of taking a defensive role in policy formation rather than the agenda setting role previously afforded in policy development (Lingard, 2003). Furthermore, government abrogated its responsibility for implementation and accountability of gender policy by shifting these responsibilities to schools as part of the outcome-focused and performance-based devolution of management to schools occurring concurrently in this period (Ailwood, 2003; Ailwood & Lingard, 2001).

On the surface, the five strategic directions did not seem substantially different from those that preceded *Gender Equity*. They included understanding the process of gender construction; curriculum, teaching and learning; violence and school culture; post-school pathways; and supporting change. However, absent from these strategic directions was the advice for change implied by the terms "eliminating," "improving," "addressing," "reforming," and "broadening" of the *Action Plan*. The guidelines for action had been replaced with indicators for accountability.

Furthermore, the discourse of *Gender Equity* consistently referred to girls and boys being equally but differently disadvantaged in schooling; girls as a category were essentialized with little acknowledgement of which girls remained seriously disadvantaged; and girls' educational needs were seen to be in the past (Ailwood, 2003). While both boys and girls were encouraged to undertake work experience and vocational training in non-traditional areas, emphasis was given to placing higher value on the arts and humanities for both boys and girls; no reference was made to the value of mathematics and science for either girls or boys. This may have occurred because some feminists had questioned the "valorization" of mathematics and science in previous policies (for example, Kenway, Willis, Blackmore, & Rennie, 1997). While using the language of equity, the discourse of the framework was far removed from social justice with a commitment to closing the gap; equality was presumed.

In the period between the publication of *Gender Equity* and the turn of the century, the recuperative masculinists challenged the feminists, gaining attention in the media and staffrooms of Australian schools (Lingard, 2003). Instead of applauding the achievements of some females, the media fuelled the feminist backlash also emerging in other parts of the world (Skelton, 1998) by consistently reporting the poorer performance of boys in grade 12 subjects including English, and in Victoria, Further Mathematics (the least demanding mathematics subject) (Forgasz, et al., 2000). However, it was not the underachievement of working class boys in the least demanding mathematics subject (Teese, 2000) that turned gender policy on its head. Rather the media failed to effectively communicate interaction of gender and other social and economic factors of disadvantage cogently explained in the research conducted by the Australian government's department of education (Collins, Kenway, & McLeod, 2000). Instead the media essentialized the category of all boys as under-achievers, whose failure, the media claimed, was explained by boys' propensity to take "high risk" subjects to secure better post-school options and the feminizing of the curriculum; the media also promoted the best-selliing books prescribing the worst aspects of hegemonic masculinity (Forgasz et al., 2000; Lingard, 2003; Vale et al.,

2004). In June, 2000, the Federal Minister for Education reacted to the community concern about the poor performance of boys in year 12. He announced a Parliamentary Inquiry into the education of boys (Parliament of Commonwealth of Australia, 2000).

Collins et al. (2000) had recommended that policy and programs focus on *which* girls and *which* boys were most disadvantaged in school and with respect to their life chances post-school; however, the terms of reference for the inquiry were to focus on the factors affecting the education of boys in Australian schools, especially their literacy and socialization skills, and to identify strategies that schools had adopted to support boys' learning needs that may be more broadly implemented in schools (Parliament of the Commonwealth of Australia, 2000). The final report from this inquiry did identify teaching strategies appropriate for boys in some fields of study but it also stated that "it is important to remember that while improvements to education outcomes for some groups of girls are real they have eluded many other girls" (Parliament of the Commonwealth of Australia, 2002, p.18). But by then the damage was done. The media, politicians and academics had begun to talk of the "post-feminist era" and social researchers had reported the evacuation of gender mainstreaming in other major areas of policy related to women, especially in relation to work and domestic duties. Summers (2003) called it "the end of equality."

In the following section I will assess what damage and how much damage was done, if any, to equality in mathematics education in Australia. I will use the results reported in large international studies and analyze the accountability measures imposed on schools by Australian governments (i.e., state-wide and national tests in mathematics and teacher reports of student achievement) and compare these with other studies conducted by researchers in the field of mathematics education.

GENDER EQUALITY IN MATHEMATICS EDUCATION IN AUSTRALIA, 1995–2007: FAME OR FICTION?

In the previous sections I have argued that monitoring gender equality requires that attention be given not only to measures of access, resources, and outcomes, but also to the quality of the curriculum, teaching, and resources and the significance of the outcomes measured with respect to capacity building, empowerment, and participation in society. In the previous section I reported on key initiatives related to curriculum, teaching, and learning materials. In this section I will present various mathematics education outcomes since the signing of the UN policy in 1995, and discuss the significance of the results for gender mainstreaming policy and gender equality. Outcomes at the primary, junior secondary, and post-compulsory level for achievement, participation, and attitudes will be discussed in turn. The data on student achievement outcomes have been sourced from international studies (TIMSS and PISA), from national and state tests (National Benchmarks incorporating Western Australia Benchmarks, Victoria AIM tests and NAPLAN), and from teacher reports of student achievement (Victoria teacher judgments). Outcomes for the states of Western Australia and Victoria were selected because disaggregated results from state-wide tests by gender were readily available. Furthermore, Western Australia might be considered as a bellwether state, as results in this state seem to predict national results (McGaw, 2008), and some large-scale studies in Victoria such as the Early Years Numeracy Project (Horne, 2003; Horne, 2004) and the Middle Years Numeracy Project (Siemon, Virgona, & Cornielle, 2001) provide comparable studies.

International Studies

At the national level, since the mid-1990s, Australia has been a participant in two series of mathematics achievement studies: TIMSS (Third International Mathematics and Science Study conducted by the IEA) and PISA (Programme for International Student Assessment conducted by the OECD in 57 countries). TIMSS uses a multiple choice written test to assess the mathematics knowledge of 9-year-old (typically in grade 4 in Australia) and 13-year-old students (typically in grade 8) covering six content areas in mathematics and was conducted in 1995, 2002, and 2007. PISA assesses the mathematical literacy of 15-year-old students (typically in grade 10) and was conducted in 2000, 2003, and 2006. The multiple-choice, pen and paper test includes four sub-scales for mathematics literacy and measures students' "ability to apply their knowledge and skills to real-life problems and situations... [and] whether students, faced with problem situations that might occur in real life, are able to analyse, reason and communicate their ideas effectively" (Thomson, Cresswell, & Bortolli, 2004, p. vii).

The instruments used in these two international studies assess students' knowledge of routine and to some extent non-routine procedures for different kinds of mathematics problems and so, to some extent, measure the capacity of students to use mathematics for further learning and in other learning or life and work contexts. However, they are limited in their capacity to assess students' strategic thinking and reasoning for extended or authentic problems—skills and knowledge that are required for further learning or life and work contexts. Moreover, neither of these instruments is able to assess the mathematical literacy required for students in the digital age, which involves the facility to use digital technology in a range of contexts where mathematical thinking and problem solving are needed to solve problems and interpret information (Keitel, Kotzmann, & Skovsmose, 1993). Each of these international programs of assessment has also gathered data from students on affective factors and teachers' practices.

State-Wide Accountability Measures

At the state level in Australia, the governments and education authorities have monitored the performance of systems and schools using pen and paper mathematics tests for students in grades 3, 5, and 7 arising from the implementation of outcome-based curriculum and the concurrent shift to accountability and performance of devolved school management. Each state designed its own test that was aligned to the National Numeracy Benchmarks, that is, nationally agreed minimum accepted standards for students at each of these grade levels. "They represent minimum standards of performance below which students will have difficulty progressing satisfactorily at school" (MCEETYA, 2007, p. 2). The data recorded in the National Benchmarks reports for 2000–2007 are a compilation of the benchmarks reported for each state. The national benchmark results for the period 2000–2007 are recorded in Tables 4.A, 4.B, and 4.C in the appendix to this chapter. Results of inferential tests of significance, if conducted, were not published and could not be carried out on these data by the author. The mean scores for numeracy of boys and girls in Western Australia (see Table 6.D of the appendix) and Victoria (AIM[1] tests) are discussed in this chapter. The published results also do not include comparison of means.

In 2008 the Federal Government implemented the National Assessment Program in Literacy and Numeracy (NAPLAN) for students in grades 3, 5, 7, and 9. The second stage report of NAPLAN (MCEETYA, 2008a) included results for national benchmark by gender along with mean scores by gender.

Education authorities within each state are responsible for the summative assessment and certification of students in their final year of secondary schooling (grade 12). There is more than one mathematics subject at this level in each state, and with the exception of Queensland, the assessment involves a mixture of school-assessed tasks and external examinations. Queensland uses a system of moderated school-based assessment. Depending on the subject, there is some diversity in the kinds of assessment tasks, though there remain few examples of authentic investigations and modeling projects that existed in the early 1990s due to advocacy for gender inclusiveness among mathematics reforms at the time (Leder, Brew, & Rowley, 1999; Vale, 1993). All states except New South Wales advise students to use digital technologies, and the regulations regarding their use in particular assessment tasks vary according to state and task.

Teacher Judgments

Since 2001, following Federal government funding obligations, state governments have required teachers to base their student achievement results on standards-based assessment criteria described in mathematics curricula in each state. Mean scores of teacher judgments for Victoria are analyzed in this chapter for primary grades (P–6)[2] in the domains of number and measurement, and for junior secondary grades (7–10) in the domains of algebra and chance and data for the period 2002–2005 (DEECD, nd; see Appendix Table E for mean scores). Results of comparisons of means, if conducted, were not available; hence only differences in mean scores can be identified and discussed. I have included discussion of teacher judgment results because they provide some insight into classroom-based assessment of performance. Involving subjective characteristics, these data are likely to reflect changing emphasis in curricula, new approaches to teaching and learning, and/or teachers' response to the gender equality debate.

Outcomes for Primary School Girls and Boys

Achievement Outcomes

The results of studies showing the performance of primary school girls and boys from 1995 to 2007 are summarized in Table 6.1. That is, the gender recording the higher mean score for international studies (TIMSS and PISA), Western Australian Benchmarks, teacher judgments of school-based assessment for Victoria, and the EYNRP and MYNRP conducted in Victoria is reported. Results of comparisons of means (tests for significance) could be reported only for international and research-based studies. Contrary to the prevailing rhetoric during the period regarding boys' under-achievement, the findings reported in Table 6.1 show that boys tend to record higher mean scores for all age groups, including National Benchmarks in Western Australia and Victoria. Significant differences favoring males were reported in grades P–2 by the Early Years Numeracy Report (EYNP) (Horne, 2003, 2004). Only in the upper grades (5 and 6) were there equivalent scores in teacher judgments or a significant difference favoring females as reported by MYNP (Siemon et al., 2001).

For the three years in which the TIMSS was conducted, the mean scores for 9-year-old girls and boys were similar, since the higher mean score recorded for boys was not significantly different from that of girls' (Lokan, Ford, & Greenwood, 1997; Thomson & Fleming, 2004; Thomson, Wernert, Underwood, & Nicholas, 2008; Zammit, Routitsky, & Greenwood, 2002). Analysis of the distribution of scores showed that males were slightly more

TABLE 6.1   Gender Differences in Outcomes for Primary Students

	Aust. TIMSS (means)	Australia benchmarks (%)		W.A. benchmarks (means)		Victoria: Teacher judgments (means)						Vic. EYNRP (means)	Vic MYNRP (means)
						Gr 3		Gr 4		Gr 5			
	9-y-o	Gr 3	Gr 5	Gr 3	Gr 5	Nu	Me	Nu	Me	Nu	Me	Gr 2	Gr 5 & 6
1995	M												
1996				M									
1997				—									
1998				F	M								
1999				F	M								
2000		F	F	M	M								F*
2001		F	F	M	M							M*	
2002	M	F	F	M	M	M	E	M	M	E	E		
2003		F	F	M	M	M	E	M	M	E	F		
2004		F	F	M	M	M	M	M	M	E	E		
2005		F	F	M	M	M	M	M	M	E	E		
2006		F	F	M	M								
2007	M	F	F	M	M								

Note: M = males higher score, F = female higher score, E = equal score, * significant difference, Nu = number strand, Me = measurement strand

highly represented among the highest and lowest achievers in mathematics at this age in each of the TIMSS studies.

Higher proportions of girls than boys in grade 3 achieved the National Benchmark from 2000 to 2007 (MCEETYA, 2007), indicating that a higher proportion of grade 3 boys are at risk than girls, and this proportion has increased during the eight year period (see Appendix Table 6.A). At the grade 5 level, there are also slightly higher proportions of boys at risk, though this difference has fluctuated over the period in question (see Appendix Table 6.B). These findings are consistent with the distribution of scores for TIMSS where boys are more highly represented among the lowest achievers. What is noticeable is the increase in proportion of girls and boys who are at risk in grade 5 compared to grade 3. These results show that a national testing program, in the absence of a focus on teaching and learning or a program for intervention, has not been able to reduce the proportion of boys or girls at risk in the primary years.

The National Benchmarks results suggest that boys are the under-performing gender in mathematics; however, these data are nowhere near adequate for monitoring gender mainstreaming. They shed no light on achievement at expected levels of performance or at higher than expected levels. Other state-wide measures (from Western Australia and Victoria) show a completely different picture that is more consistent with the international studies. In Western Australia, with the exception of two years at grade 3, the mean scores for boys were higher than for girls in each of the years in the period 1996–2007; only at grade 3 is there a trend showing a narrowing of the gap. In Victoria, in 2007, means scores for the AIM tests also showed higher mean scores for boys at grade 3 (mean$_M$ = 2.28; mean$_F$ = 2.15) and at grade 5 (mean$_M$ = 3.16; mean$_F$ = 3.00).[3] Without comparison of means it is not possible to discern whether any of these differences are significant. At best they reflect international trends in the primary grades where differences favor males but are not significant (Thomson et al., 2008). The recently published NAPLAN results for 2008 suggest, however, that gender differences in the primary years are significant (MCEETYA, 2008a). Boys in grades 3 and 5 achieved higher mean scores than girls in all Australian states with the greatest difference recorded in Victoria for both grades. Since confidence intervals did not overlap, these differences are statistically significant. While differences in the proportion of girls and boys achieving at or above the benchmark are negligible, there are higher proportions of boys than girls achieving in the top two bands.

Other studies of primary student achievement during this period have focused on particular areas of mathematics or the implementation of new curriculum and approaches to teaching and learning. Significant gender differences in achievement emerged after three years of schooling for students participating in the Early Years Numeracy Research Project (Horne, 2003, 2004). With entry performance as a covariate, boys moved significantly ahead of girls in the number domain, whereas there was no difference in measurement or space. Horne (2003) suggested that these findings differed from other studies because of the nature of the assessment, which was completed using an individual interview. In this program, and others like it in other states and countries, the child's mental strategies are a key indicator of performance. Derived strategies rather than counting strategies are evidence of conceptual and procedural development for number operations.

Alternately, in the Middle Years Numeracy Research Project, Siemon et al. (2001) found significant gender differences favoring girls. Their study of students from grade 5 to grade 9 focused on multiplicative thinking and proportional reasoning that pervades the number, algebra, and measurement curricula for these grade levels. The items required interpretation of problems in context. The results of this study are contrary to mean score comparisons throughout the period, and the authors speculated that the significant difference favoring females was due to the literacy requirements of the items in the instrument and girls' advantage in literacy.

The gender recording the higher mean score for teachers' judgments in Victoria for the content areas of number and measurement for grade 3, 4 and 5 students are shown in Table 6.1 (see Appendix Table 6.E for mean scores). The mean scores are not surprising since 2.5 is the expected standard for grade 3 students, 3.0 for grade 4 students and 3.5 for grade 5 students according to the curriculum that teachers were following. At grade 3 and 4, both girls and boys are below the expected standard, whereas at grade 5 for both content areas and all years both girls and boys are above the expected standard. These data show teachers have judged that boys were at a higher standard in grades 3 and 4 for all years, a result consistent with the EYNP in Victoria (Horne, 2003, 2004), international studies, and WA Benchmarks. On the other hand, grade 5 teachers in Victoria have judged boys and girls to be at the same standard, except in 2003 when girls were higher than boys for measurement. While teachers in upper primary may not have implemented a "girl-friendly" mathematics curriculum that gives advantage to students performing at expected standards for literacy, classroom-based assessment at these grade levels afford recognition of girls' mathematical performance not evident in state-wide tests or international studies.

These findings show that girls and boys are performing similarly in mathematics in primary school. If there were equality of achievement outcomes, one would expect mean scores to be higher for females more frequently in the range of assessments reported here.

Affective Outcomes

The TIMSS 2002–2003 also gathered data on affective factors that were compared with the previous test results (Thomson & Fleming, 2004). Even though primary-aged girls and boys were similar in achievement, 9-year-old boys scored significantly higher levels of self-confidence in mathematics learning than 9-year-old girls (see Table 6.3), and self-confidence was strongly correlated with achievement. Furthermore, a higher proportion of boys than girls enjoyed mathematics. Thomson and Fleming (2004) also found that enjoyment of mathematics had increased significantly for 9-year-old Australian students since the previous study.

Despite the general absence of significant differences in achievement in the primary grades, boys are overwhelmingly more comfortable with mathematics than girls. It would seem that neither democratic nor transformative practices associated with gender mainstreaming have strongly influenced curricula or teaching and learning in the primary years. Rather, the media and policy attention given achieving boys may have encouraged teachers in the early years to provide positive learning experiences for boys, perhaps at the expense of girls.

Outcomes for Junior Secondary Girls and Boys

Achievement Outcomes

As for the primary grades, the gender recording the higher mean score or proportion for the various studies conducted during the period is recorded in Table 6.2. (See original data in the Appendix.) Until PISA 2006 (Thomson & De Bortolli, 2008) and TIMSS 2007 (Thomson et al., 2008), gender similarities characterized the findings for junior secondary students from the two different international studies occurring during the period 1995–2003. These most recent results signal a change away from the trend toward closing the gap established over the previous decade and the re-emergence of gender differences in achievement in secondary mathematics in Australia.

For the three TIMSS conducted with 13-year-old students from 1995–2002, no significant differences in the mean scores were found; nor were there any significant differences for the six content areas (Lokan, Ford, & Greenwood, 1996; Thomson & Fleming, 2004; Zammit, Routitsky, & Greenwood, 2002). Afrassa and Keeves (1999) noted that there had been a change in the direction of mean scores since the FIMSS, from boys scoring higher to girls scoring higher. PISA 2000 and PISA 2003 also recorded no significant gender differences in mean scores for mathematical literacy, though the mean score was higher for males for each study (Lokan, Greenwood, & Cresswell, 2001; Thomson, Cresswell, & De Bortolli, 2004). The authors of the PISA 2003 report noted that, in comparison to other OECD countries,

TABLE 6.2   Gender Differences in Outcomes for Junior Secondary Students

	Aust TIMSS (means) 13-y-o	Aust PISA (means) 15-y-o	Aust. B'marks (%) Gr 7	W.A. B'marks (means) Gr 7	Victoria: Teacher judgments (means)						Vic. MYNRP (means) Gr 7&8	Aust. LSAY (means) Gr 9
					Gr 8		Gr 9		Gr 10			
					Al	Ch	Al	Ch	Al	Ch		
1995	F											M*
1996				M								M*
1997				—								M*
1998				M								M*
1999	E			—								
2000		M	F	M							F*	
2001			F	M								
2002	E		F	M	F	F	F	F	F	F		
2003		M	F	M	F	F	F	F	F	F		
2004			F	M	F	F	F	F	F	F		
2005			F	M	F	F	F	F	F	F		
2006		M*	F	M								
2007	M*		E	M								

Note: M = males higher score, F = female higher score, E = equal score, * significant difference, Al = algebra strand, Ch = chance and data strand

"Australia seems to have been able to contain the widening of gender disparity with age in mathematics" (Thomson et al., 2004, p. 32).

However, the distribution of scores for males and females was not similar. For PISA 2003, males were more highly represented at the most advanced level of mathematical literacy in five of the eight states and territories, with substantial differences recorded in New South Wales, Victoria, and Tasmania. Only in the ACT were there a higher proportion of females than males at the most advanced level. For the lowest level of proficiency, females were more highly represented than males in three states, including the Northern Territory, which recorded a higher mean score for females than males. Only in the ACT were there a much higher proportion of males than females at the lowest level. Perhaps the higher proportion of girls performing at the lowest level was an indication of what was to emerge three years later.

In the previous decade and during the 1990s, males record significantly higher means than females in the Longitudinal Study of Australian Youth (Rothman, 2002) from 1975 to 1998, which assessed mathematics knowledge of grade 9 students. The only divergent view came from the Middle Years Numeracy Research Project, which was conducted for students in grades 5 to 8 and registered a statistically significant mean difference favoring girls (Siemon et al., 2001). As noted above, the items in this test revealed girls' relative competence with respect to interpreting text in mathematics problems. Up until 2006 and 2007, the contradictory findings in these studies suggested that the results may depend on the nature of the assessment tasks rather than any particular trends over time.

Given the findings of previous international studies, PISA 2006 and TIMSS 2007 came as a shock to the education community. Reversing the findings and trends of the previous international studies at this level, Australia recorded a significant difference favoring males for 15-year-olds (Thomson & De Bortolli, 2008). This difference was higher than the OECD average. McGaw (2008) argued that the change in results—such that for the first time in a decade gender differences in average scores favored males and were significant—was due to the fall in performance of girls in two states: Western Australia and the Northern Territory. Mean scores were higher for males than females in all states and territories and significantly so for Victoria, Western Australia, Tasmania, and Queensland. Males were more highly represented in the most advanced level of mathematical literacy for each state and territory, and contrary to the distributions in previous years, females were more highly represented in the lowest level in half the states and territories. The changing results for Western Australia mirrored the changes for Australia as a whole.

Contrary to the international trend in the TIMSS 2007 study of 13-year-old students and previous results for Australian students, a significant difference in the mean score favoring males was found for Australian students

(Thomson et al., 2008). Only four other countries, all late developing countries from Africa or South America, had larger gender differences favoring males than Australia. Mean scores favored males in five states: Tasmania, Western Australia, Queensland, ACT, and New South Wales and were significant in Queensland. So what might the data collected by the state governments reveal that would explain this change in the direction of 15-year-olds in 2006 and 13-year-olds in 2007?

The numeracy benchmarks show that at grade 7, as for grades 3 and 5, more boys are at risk than girls (Table 6.2), a pattern consistent with the distribution of scores in TIMSS 2007 but at variance with PISA 2006. At grade 7 the proportion of students at risk rises sharply and consistently throughout the period (Appendix, Tables 4.A, 4.B, 4.C). Almost 20% of students are at risk, compared with 10% at grade 5 and 7% at grade 3.

In Western Australia, grade 7 boys' mean scores on the state numeracy test were higher than grade 7 girls' for every year that data were collected (see Table 6.1 and Appendix Table 6.D). These differences were generally greater for grade 7 students than the other grade levels, and were largest for the cohort of students likely to have participated in PISA 2006. In Victoria, in 2007, mean scores for the AIM tests also show higher mean scores for boys at grade 7 (mean$_M$ = 4.24; mean$_F$ = 4.18) and grade 9 (mean$_M$ = 4.97; mean$_F$ = 4.86).

The return to significant gender differences in mean scores was confirmed by the results of 2008 NAPLAN (MCEETYA, 2008a). Boys in grades 7 (mean$_M$ = 552.3) and 9 (mean$_M$ = 586.5) achieved significantly higher mean scores than girls (mean$_F$ = 537.2 and mean$_F$ = 577.6, respectively) in Australia. Gender differences favoring males were recorded in all Australian states. As for the primary grades, while differences in the proportions of girls and boys achieving at or above the benchmark are negligible, there are higher proportions of boys than girls achieving in the top two bands.

Mean teacher judgment scores for secondary students in Victoria favored females in grades 8, 9, and 10 from 2002 to 2006 (Table 6.1). The results are for the algebra and chance and data domains (see Appendix Table 6.F for mean scores). These findings would appear to be in contrast with the results of the international studies, especially in 2006 and 2007, with consistently higher mean scores recorded for females. According to the judgment of teachers, girls perform better on average on classroom-based assessment tasks. It is not clear why this pattern of difference favoring girls is so consistent. The finding is, however, consistent with a view uncovered in research regarding the maths as a male domain, where both boys and girls are identifying girls as being better at mathematics (Forgasz, Leder, & Barkatsas, 1998). As noted earlier, teacher judgements are subjective, and these findings make one wonder whether teachers are using girl-friendly classroom assessment tasks (Leder et al., 1999) or being influenced by gendered patterns of behavior in classrooms (Vale et al., 2004), or the publicity of under-achieving boys, whereby teachers' judgements are a consequence of lowered expectations.

Far from leading the world on gender equity in junior secondary mathematics achievement, Australia now trails all the developed countries. The considerable effort afforded the development of gender-inclusive teaching and learning materials made prior to the turn of the century have clearly dissipated. With the attention given to boys' underachievement and priority given boys' education in the national agenda, single-sex classes in mathematics, an intervention originally designed to target the needs of girls, enabled teachers to identify and tailor their approaches to suit the needs of boys (Forgasz et al., 2000; Vale et al., 2004).

Affective Outcomes

The affective domain for junior secondary students has been researched quite rigorously (Table 6.3). Through a series of studies, Forgasz and colleagues have established a shift in gendered perceptions of mathematics. Students in Australia generally regard mathematics as a gender-neutral domain (Forgasz et al., 1998; Forgasz, Leder, & Klosterman, 2004). However,

TABLE 6.3   Gender Differences in Affect in Australia

2004–2007	Affective scale	Gender differences
TIMSS (9-yr-olds)	self-confidence	M*
	enjoyment	M
TIMSS (13-yr-olds)	self-confidence	M*
	enjoyment	M
PISA (15-yr-olds)	enjoyment & interest	M*
	self-efficacy	M*
	self-concept	M*
	anxiety	F*
Watt et al. (Gr 10)	self-perception	M*
	expectation of success	M*
	intrinsic value of maths	M*
Vale & Leder, 2004 (Gr 8 & 9)	Attitudes to computer-based mathematics (ACBM)	M*
Pierce et al., 2007 (Gr 8–10)	Mathematics and Technology Attitudes scale (MTAS)	M*
Forgasz, 2004 (Gr 7–10)	Attitudes to computers for learning mathematics (ACM)	M*

Note: M = males higher mean score, F = females higher mean score;
* indicates significant difference.

studies of secondary students' self-perception show consistent gender differences favoring males.

Watt (2000) discerned that these attitudes had become more negative over time, and that boys had more positive self-concepts of mathematics ability, greater interest in mathematics, and higher expectations of success (Watt & Eccles, 2000). Both of the international studies reported on affect during the period. TIMSS 2002/3 found that 13-year-old males scored significantly higher levels of self-confidence in mathematics learning and that self-confidence was more strongly correlated with achievement than value of mathematics (Thomson & Fleming, 2004). Findings from the PISA study were similar. Fifteen-year-old males reported higher levels of enjoyment, interest, self-efficacy, and self-concept, and they held stronger beliefs in the value of mathematics than females (Thomson et al., 2004). Females were more anxious than males but reported more favorable relations with teachers and a stronger sense of belonging. Self-efficacy and self-concept had the strongest relationship with mathematical literacy among the affective variables measured in the PISA study. Males are also significantly more positive about the use of technology in secondary mathematics, according to the various instruments that have measured students' attitudes (Forgasz, 2004; Pierce, Stacey, & Barkatsas, 2007; Vale & Leder, 2004).

These findings show persistent gender differences in affective factors favoring boys. Failure to deal with these differences through gender mainstreaming initiatives since 1995 has had disastrous consequences for gender equality at the post-compulsory level of schooling.

Outcomes for Girls and Boys at the Post-Compulsory Level

Participation

It is when students have the option to choose to study mathematics that persistent gendered practices emerge. Data regarding participation in the various post-compulsory mathematics subjects for each state in Australia are generally available in the public domain. Forgasz (2006) reported on a study of enrollment in year 12 mathematics subjects of Australian students from 2000–2004. She focused her analysis on the mathematics subjects, across the different states in Australia, that are the minimum pre-requisites for continued mathematics study in tertiary education. The findings for Australia overall showed that for the five-year period, enrollments in these particular mathematics subjects had declined. There were more males than females enrolled in these subjects in each state, except the ACT (where there have been higher proportions of females achieving at the highest levels in international studies, as reported above). Declining enrollments by girls during this period contributed to the overall fall in enrollments in each of the states that recorded a decline in enrollments in this subject. These findings contradict the trend of closing the gender gap in participation observed during the 1990s (Forgasz et al., 2000).

Affective Outcomes

Even though mathematics is considered gender neutral by students, affective factors appear to hold the key to explaining gender differences in participation. Although prior success in mathematics was found to influence subject choice, students who rated the intrinsic value of mathematics and their self-perception higher took higher-level mathematics subjects, even when prior success was a covariate (Watt, 2006). Furthermore, while boys who saw mathematics as moderately useful were likely to aspire to mathematics-related careers, only girls who saw mathematics as highly useful were likely to do so (Watt, 2006). Thomson et al. (2004) argued that participation of females in tertiary mathematics and related disciplines could be increased if teachers attended to girls' attitudes, while Watt (2006) claimed that we need to find out why girls who perform similarly to boys regard themselves as lacking talent. Both of these views perceive girls to be the problem. Shannon (2004), on the other hand, argued that mathematics teaching was the problem:

> 'Fixing' mathematics so that girls are not alienated by its current pedagogical structure is, in my opinion, a much more important and critical goal to be addressed. Until girls *enjoy* mathematics, I believe that we will find it difficult to justify to them that it is socially relevant and a way of improving their quality of life. (p. 515)

Achievement Outcomes

The only international study conducted for students at the post-compulsory level found that Australian males were significantly ahead of females in mathematical literacy (Mullis, Martin, Fierros, Goldberg, & Stemler, 2000). Surprisingly, given the media attention on grade 12 results, only a few of the states publish results of grade 12 high-stakes assessment in the public domain or by gender. Analysis of grade 12 results for mathematics subjects showed that the nature of assessment does make a difference to gendered patterns of performance. Females outperformed males on school-assessed tasks and tasks involving extended problem solving (Cox, 1996). Since 2000 for senior secondary mathematics in Victoria, females received higher mean scores than males in the least demanding subject, except in 2006, while males have achieved higher mean scores in the subject that is the minimum requirement for further mathematics study in 2000 and for the years since 2004 (VCCA).

Various studies show that males may be advantaged by the use of technology in senior secondary mathematics, with males performing better on graphic calculator advantage items (Forster & Mueller, 2001) and higher proportions of male students in the highest three grade categories in CAS environment than in the graphics calculator environment (Forgasz, Griffith, & Tan, 2006).

At the post-compulsory level, boys are more positive about mathematics and likely to participate in mathematics required for further study in mathematics and related fields. They are also more likely to succeed at the highest levels. Clearly post-school options for girls are narrowing as a consequence of moving from an action plan to a performative framework in gender equity policy.

MAINTAINING ATTENTION ON GENDER EQUALITY

Since the Fourth World Conference on Women in 1995, Australia has adopted an assimilation approach to gender mainstreaming as evidenced by the Australian government report to the Commonwealth Secretary-General in 2001: "The Australian Government has maintained its strong commitment to promoting gender mainstreaming across all government agencies. In line with international best practice, the Australian Government has pursued a strategy of integrating women's issues into mainstream policy making and practice across all government agencies" (Department of Prime Minister and Cabinet, 2001, p. 2).

Given that conservative parties were in power from 1996 it is hardly surprising that they would adopt a liberal feminist approach that would not threaten the mainstream. Typically, however, girls and education are invisible and rarely mentioned in government reports on gender mainstreaming (Office for Women, 2001, 2003, 2007), and when education is the focus it is generally vocational and higher education. With some exceptions, almost no mention is made of differences among women.

With respect to girls and mathematics, the international and national testing over the past decade indicates that there are more similarities than differences in average scores and in teacher judgments at all levels of schooling, though distributions are not the same, with boys more likely to achieve at the highest levels. However this trend toward gender equality appears to be reversing. The most serious damage of the "downstreaming" of gender policy since the election of the conservative government in 1996 is the falling participation of girls in mathematics subjects that are the minimum requirements for further mathematics study, and the failure to address the persistent lower levels of self-perception of girls compared to boys. The most recent Australian results from the PISA and TIMSS studies are cause for concern and warrant immediate attention.

In this chapter I have shown how the interests and needs of women and girls in education policy were thwarted by a feminist backlash. Collins et al. (2000) argued that girls had to stay on longer at school and in education to reach comparable levels of economic independence with boys. While supporting women's right to choose, these evaluators of gender policy failed to press the importance of women's and girls' right to enjoy, participate, and succeed in mathematics at all levels of schooling. Reform must address this consistent problem in the quality of teaching and learning. McGaw (2004) argued that there is a need for governments and systems to act in order to "level up" performance for socially disadvantaged students through policy and program development, implementation and monitoring. Such attention would highlight and act to eliminate the persistently poorer mathematics outcomes for girls, especially Indigenous girls, girls from a low socioeconomic background, or girls who are recent arrivals.

In 2008 Australia began to develop a national curriculum for mathematics. With the return of a labor government toward the end of 2007, social justice is back on the agenda and the recently released draft *National Declaration on Educational Goals for Young Australians* (MCEETYA, 2008b) includes the promotion of equity. In this manifesto there is an urgency to provide "equality of opportunities to access and participate in high-quality schooling" (p. 6), to improve learning outcomes, especially for indigenous students, and to create a "socially cohesive society that respects and appreciates diversity" (p. 6). The discourse of this document is action and reform, but gender is not given the attention it deserves. The national curriculum is important, but action to improve the quality of teaching and learning is paramount for gender equality in mathematics education.

NOTES

1. Achievement Improvement Monitor, Victorian Curriculum and Assessment Authority.
2. Preparatory grade, the first year of schooling in Victoria.
3. Results provided by the Victorian Curriculum and Assessment Authority in private communication.

APPENDIX

TABLE 6.A   Percentage of Grade 3 Students Achieving the Benchmarks by Gender and Subgroup, Australia, 2000–2007

	All students	Males	Females	Indigenous students	LBOTE students
2000	92.7 ± 2.0	92.7 ± 2.1	92.8 ± 2.1	73.7 ± 7.1	90.3 ± 2.7
2001	93.9 ± 1.2	93.7 ± 1.3	94.3 ± 1.3	80.2 ± 3.9	92.5 ± 1.5
2002	92.8 ± 1.3	92.5 ± 1.4	93.1 ± 1.5	77.6 ± 3.6	91.3 ± 1.4
2003	94.2 ± 1.1	93.8 ± 1.1	94.7 ± 1.2	80.5 ± 3.7	93.3 ± 1.1
2004	93.7 ± 1.2	93.3 ± 1.2	94.1 ± 1.3	79.2 ± 4.1	92.3 ± 1.2
2005	94.1 ± 1.1	93.5 ± 1.1	94.7 ± 1.1	80.4 ± 3.8	94.0 ± 1.2
2006	93.0 ± 1.4	92.5 ± 1.4	93.5 ± 1.4	76.2 ± 4.3	92.6 ± 1.3
2007	93.2 ± 1.4	92.8 ± 1.4	93.7 ± 1.5	78.8 ± 4.1	92.1 ± 1.4

Source: Adapted from: MCEETYA, 2007, Table D3

TABLE 6.B   Percentage of Grade 5 Students Achieving the Benchmarks by Gender and Subgroup, Australia, 2000–2007

	All students	Males	Females	Indigenous students	LBOTE students
2000	89.6 ± 1.7	89.4 ± 1.7	89.8 ± 1.8	62.8 ± 4.5	87.1 ± 2.1
2001	89.6 ± 1.3	89.5 ± 1.4	89.8 ± 1.5	63.2 ± 3.7	87.9 ± 1.6
2002	90.0 ± 1.3	89.9 ± 1.4	90.2 ± 1.5	65.6 ± 3.7	87.9 ± 1.5
2003	90.8 ± 1.2	90.3 ± 1.3	91.4 ± 1.3	67.6 ± 3.9	89.3 ± 1.4
2004	91.2 ± 1.2	91.0 ± 1.2	91.5 ± 1.3	69.4 ± 3.9	89.3 ± 1.4
2005	90.8 ± 1.3	90.5 ± 1.3	91.2 ± 1.4	66.5 ± 3.9	90.0 ± 1.4
2006	90.3 ± 1.3	90.0 ± 1.3	90.6 ± 1.4	66.0 ± 3.8	90.0 ± 1.4
2007	89.0 ± 1.3	89.1 ± 1.3	89.1 ± 1.4	65.5 ± 3.4	87.6 ± 1.4

Source: Adapted from MCEETYA, 2007, Table D6

TABLE 6.C   Percentage of Grade 7 Students Achieving the Benchmarks by Gender and Subgroup, Australia, 2000–2007

	All students	Males	Females	Indigenous students	LBOTE students
2001	82.0 ± 0.9	81.7 ± 1.0	81.9 ± 1.1	48.6 ± 2.8	77.8 ± 1.4
2002	83.5 ± 0.9	83.3 ± 0.9	83.8 ± 1.0	51.9 ± 3.0	79.2 ± 1.2
2003	81.3 ± 0.8	81.0 ± 0.9	81.6 ± 0.9	49.3 ± 2.9	76.6 ± 1.2
2004	82.1 ± 0.8	81.9 ± 0.9	82.3 ± 0.9	51.9 ± 2.8	77.9 ± 1.3
2005	81.8 ± 0.9	81.6 ± 0.9	82.0 ± 1.0	48.8 ± 2.9	78.8 ± 1.3
2006	79.7 ± 1.1	79.7 ± 1.1	79.7 ± 1.3	47.5 ± 2.9	76.3 ± 1.5
2007	80.2 ± 0.9	80.0 ± 0.9	80.6 ± 1.0	46.0 ± 2.7	78.0 ± 1.2

Source: Adapted from MCEETYA, 2007, Table D9

TABLE 6.D   Means Scores for Numeracy for Boys and Girls, Grade 3, 5, and 7, 1996–2007, Western Australia

	Grade 3		Grade 5		Grade 7	
	Males	Females	Males	Females	Males	Females
1996	344	341	—	—	469	466
1998	353	354	—	—	486	479
1999	334	337	392	388	—	—
2000	329	326	397	391	476	474
2001	329	324	396	393	471	463
2002	332	321	403	396	487	476
2003	336	329	410	402	483	473
2004	338	331	403	399	472	472
2005	341	337	408	401	477	464
2006	339	339	407	398	477	466
2007	341	339	412	399	476	468

Source: Adapted from Department of Education and Training, 2007.

TABLE 6.E   Mean Teacher Judgments for Number and Measurement, Grades 3, 4, and 5, Victoria, 2002–2006

	Year 3				Year 4				Year 5			
	Measurement		Number		Measurement		Number		Measurement		Number	
	M	F	M	F	M	F	M	F	M	F	M	F
2002	2.58	2.58	2.59	2.57	2.93	2.92	2.93	2.92	3.53	3.53	3.54	3.54
2003	2.59	2.59	2.60	2.58	2.95	2.93	2.96	2.93	3.54	3.55	3.55	3.55
2004	2.60	2.59	2.61	2.59	2.96	2.95	2.97	2.95	3.55	3.55	3.56	3.56
2005	2.60	2.59	2.61	2.59	2.96	2.95	2.98	2.96	3.56	3.56	3.56	3.56

Source: Adapted from DEECD, n.d.

TABLE 6.F   Mean Teacher Judgments for Number and Measurement, Grades 8, 9, and 10, Victoria, 2002–2006

	Year 8				Year 9				Year 10			
	Chance & Data		Algebra		Chance & Data		Algebra		Chance & Data		Algebra	
	M	F	M	F	M	F	M	F	M	F	M	F
2002	4.75	4.80	4.75	4.80	5.42	5.47	5.42	5.47	5.69	5.73	5.67	5.72
2003	4.75	4.79	4.75	4.79	5.41	5.47	5.39	5.44	5.70	5.74	5.68	5.72
2004	4.75	4.79	4.76	4.80	5.41	5.47	5.40	5.46	5.70	5.75	5.69	5.73
2005	4.75	4.79	4.76	4.80	5.40	5.47	5.41	5.47	5.72	5.76	5.70	5.74

Source: Adapted from DEECD, n.d.

REFERENCES

Afrassa, T. M., & Keeves, J. P. (1999). Differences between the sexes in mathematics achievement in Australia. In J. M. Truran & K. M. Truran (Eds.), *Making the differences, Proceedings of the 22nd annual conference of MERGA, Adelaide* (pp. 485–493). Sydney: MERGA.

Ailwood, J., & Lingard, B. (2001). The endgame for national girls' schooling policies in Australia? *Australian Journal of Education, 45*(1), 9–22.

Ailwood, J. (2003). A national approach to gender equity policy in Australia: Another ending, another opening? *International Journal of Inclusive Education, 7*(1), 19–32.

Anthony, B., & Walshaw, M. (2007). *Effective pedagogy in mathematics/Pangarau: Best evidence synthesis in mathematics.* Wellington, New Zealand: Ministry of Education.

Australian Education Council. (1990). *A National statement on mathematics for Australian schools.* Melbourne, Australia: Curriculum Corporation.

Australian Education Council. (1993). *National action plan for the education of girls 1993–1997.* Carlton, Australia: Curriculum Corporation.

Barnes, M. (1991). *Investigating change: An introduction to calculus for Australian schools.* Carlton, Australia: Curriculum Corporation.

Becker, J. R. (1995). Women's ways of knowing in mathematics. In P. Rogers & G. Kaiser (Eds.), *Equity in mathematics education: Influences of feminism and culture* (pp. 163–174). London: The Falmer Press.

Belencky, M. F., Clinchy, B. M., Goldberger, N. R., & Tarule, J. M. (1986). *Women's ways of knowing: The development of self, voice and mind.* New York: Basic Books.

Boaler, J. (1997). Reclaiming school mathematics: The girls fight back. *Gender and Education, 9*(3), 285–306.

Boaler, J. (2002). Learning from teaching: Exploring the relationship between reform curriculum and equity. *Journal of Research in Mathematics Education, 33*(4), 239–258.

Boaler, J. (2003). Studying and capturing the complexity of practice—The case of the 'dance of agency.' In N. A. Pateman, B. J. Dougherty, & J. Zilliox (Eds.), *Proceedings of the 2003 joint meeting of PME and PMENA* (Vol.1, pp.1-3–1-16). Honolulu, HI: Center for Research and Development Group, University of Hawai'i.

Burton, L. (1995). Moving towards a feminist epistemology of mathematics. In P. Rogers & G. Kaiser (Eds.), *Equity in mathematics education: Influences of feminism and culture* (pp. 209–226). London: Falmer Press.

Collins, C., Batten, M., Ainley, J., & Getty, C. (1996). *Gender and school education.* Canberra, Australia: Australian Council for Educational Research, Commonwealth of Australia.

Collins, C., Kenway, J., & McLeod, J. (2000). *Factors influencing the educational performance of males and females in school and their initial destinations after leaving school.* Canberra, Australia: Commonwealth of Australia.

Council of Europe. (1998). *Gender mainstreaming: Conceptual framework, methodology and presentation of good practice.* Council of Europe. EG-S-MS(8), 2nd rev.

Cox, P. J. (1996). Sex and CATs: Findings from a detailed statistical analysis. In H. Forgasz, T. Jones, G. Leder, J. Lynch, K. Maguire, & C. Pearn (Eds.), *Mathematics: Making connections. Proceedings of the 33rd annual conference of the MAV* (pp. 354–361). Melbourne, Australia: MERGA.

Department of Education and Early Childhood Development. (nd). *Teacher judgements against the CSF II.* Produced by Student Outcomes, DEECD. Retrieved September 19, 2008 from http://www.education.vic.gov.au/

Department of Education and Training. (2007). *Performance of Years 3, 5 and 7 students Western Australia: WALNA2007.* East Perth, Australia: Department of Education and Training.

Department of Prime Minister and Cabinet. (2001). *Australia's response to the Commonwealth Secretary-General's Report 2001 on the implementation of the 1995 Commonwealth Plan of Action on Gender and Development and Update (2000–2005).* Canberra, Australia: Department of the Prime Minister and Cabinet, Office of the Status of Women.

Fennema, E. (1995). Mathematics, gender and research. In B. Grevholm & G. Hanna (Eds.), *Gender and mathematics education. An ICMI study* (pp. 21–38). Lund, Sweden: Lund University Press.

Forgasz, H. J. (2004). Equity and computers for mathematics learning: Access and attitudes. In M. J. Johnsen Høines & A. B. Fuglestad (Eds.), *Proceedings of the 28th conference of the International Group for the Psychology of Mathematics Education [PME]* (2-399–2-406). Bergen, Norway: Bergen University College.

Forgasz, H. J. (2006). Australian year 12 "intermediate" level mathematics enrolments 2000–2004: Trends and patterns. In P. Grootenboer, R. Zevenbergen, & M. Chinnappan (Eds.), *Identities, cultures and learning spaces. Proceedings of the 29th annual conference of the Mathematics Education Research Group of Australasia* (pp. 211–220). Adelaide, Australia: MERGA.

Forgasz, H. J., Griffith, S., & Tan, H. (2006, December). Gender, equity, teachers, students and technology use in secondary mathematics classrooms. In C. Hoyles, J. Lagrange, L. H. Son & N. Sinclair (Eds.), *Proceedings of the seventeenth ICMI study: Technology revisited, Hanoi.* Hanoi, Vietnam: ICMI. [Available on CD-ROM.]

Forgasz, H. J., & Leder, G. C. (1996). Mathematics classroom, gender and affect. *Mathematics Education Research Journal, 8*(2), 153–173.

Forgasz, H. J., Leder, G. C., & Barkatsas, T. (1998). Revising the Fennema-Sherman 'Mathematics as a male domain' scale: Results from the first phase of the research. In G. Dimakos, C. Salaris, & A. Malefekas (Eds.), *Mathematics in New Times, 15th conference of the Mathematical Society of Greece* (pp. 389–400). Xios, Greece: HMS.

Forgasz, H. J., Leder, G. C., & Kloosterman, P. (2004). New perspectives on the gender stereotyping of mathematics. *Mathematical Thinking and Learning, 6*(4), 389–420.

Forgasz, H., Leder, G. C., & Vale, C. (2000). Gender and mathematics: Changing perspectives. In K. Owens & J. Mousley (Eds.), Mathematics education research in Australasia: 1996–1999 (pp. 305–340). Turramurra, Australia: MERGA.

Forster, P. A., & Mueller, U. (2001). Outcomes and implications of students' use of graphics calculators in the public examination of calculus. *International Journal of Mathematical Education in Science and Technology, 32*(1), 37–52.

Gewitz, S. (1998). Conceptualising social justice in education: Mapping the territory. *Journal of Educational Policy, 13*(4), 469–484.

Goodell, J.,& Parker, L. (2001). Creating a connected, equitable mathematics classroom: Facilitating gender equity. In B. Atweh, H. Forgasz, & B. Nebres. (Eds.), *Sociocultural research on mathematics education: An international perspective* (pp. 411–431). Mahwah, NJ: Lawrence Erlbaum Associates.

Henningsen, I. (2004, July). *Gender mainstreaming of research on adult mathematics education: Opportunities and challenges.* Paper presented at ICME10, the International Congress of Mathematics Education, Copenhagen, Denmark.

Horne, M. (2003) Gender differences in the early years in addition and subtraction. In N. A. Pateman, B. J. Dougherty, & J. Zilliox (Eds.), *Proceedings of the 2003 joint meeting of PME and PMENA,* Vol.3 (pp. 79–86). Honolulu, HI: Center for Research and Development Group, University of Hawai'i.

Horne, M. (2004). Early gender differences. In M. J. Johnsen Høines & A. B. Fuglestad (Eds.), *Proceedings of the 28th conference of the International Group for the Psychology of Mathematics Education*, Vol. 3 (pp. 65–72). Bergen, Norway: Bergen University College.

Jungwirth, H. (2003). What is a gender-sensitive mathematics classroom? In L. Burton (Ed.), *Which way social justice in mathematics education?* (pp. 3–26). Westport, CT: Praeger.

Kaiser, G., & Rogers, P. (1995). Introduction: Equity in mathematics education. In P. Rogers & G. Kaiser (Eds.), *Equity in mathematics education: Influences of feminism and culture* (pp. 1–10). London: Falmer Press.

Keitel, C., Kotzmann, E., & Skovsmose, O. (1993). Beyond the tunnel vision: Analysing the relationship between mathematics education, society, and technology. In C. Keitel & K. Ruthven (Eds.), *Learning from computers: Mathematics education and technology* (pp.242–279). Berlin: Springer.

Kenway, J., & Willis, S., Blackmore, J., & Rennie, L. (1997). *Answering back: Girls, boys and feminism in schools.* St Leonards, NSW: Allen and Unwin.

Leder, G. C. (1993). Teacher/student interactions in the mathematics classroom: A different perspective. In E. Fennema, & G. Leder (Eds.), *Mathematics and gender* (pp. 149–168). Brisbane, Australia: Queensland University Press.

Leder, G. C., Brew, C., & Rowley, G. (1999). Gender differences in mathematics achievement – Here today and gone tomorrow. In G. Kaiser, E. Luna, & I. Huntley (Eds.), *International comparisons in mathematics education* (pp. 213–224). Dordrecht, The Netherlands: Kluwer.

Lingard, B. (2003). Where to in gender policy in education after recuperative masculinity? *International Journal of Inclusive Education, 7*(1), 33–56.

Lokan, J., Ford, P., & Greenwood, L. (1996). *Mathematics and science on the line: Australian junior secondary students' performance in the Third International Mathematics and Science Study.* Melbourne: ACER.

Lokan, J., Ford, P., & Greenwood, L. (1997). *Maths & science on the line: Australian middle primary students' performance in the Third International Mathematics and Science Study.* Melbourne: ACER.

Lokan, J., Greenwood, L., & Cresswell, J. (2001). *15-Up and counting, reading, writing, reasoning… How literate are Australia's students? Programme for International Student Assessment (PISA).* Melbourne: ACER.

Lombardo, E. (2005). Integrating or setting the agenda? Gender mainstreaming in the European constitution-making process. *Social Politics: International Studies in Gender, State & Society 12*(3), 412–432.

MCEETYA. (nd). *National action plan for the education of girls 1993–1997.* Retrieved September 19, 2008 from http://www.mceecdya.edu.au/mceecdya

MCEETYA. (1996). *Gender equity: A framework for Australian schools.* Curriculum Corporation. MCEETYA. Retrieved September 19, 2008 from http://www.curriculum/mceetya

MCEETYA. (2007). *National report on schooling in Australia (preliminary paper): National benchmark results reading, writing and numeracy Years 3, 5 and 7.* MCEETYA. Retrieved September 19, 2008 from http://www.curriculum/mceetya

MCEETYA. (2008a). *National assessment program: Literacy and numeracy (Achievement in reading, writing, language conventions and numeracy).* MCEETYA.

Retrieved November 3, 2009 from http://www.naplan.edu.au/verve/_resour
ces/2ndStageNationalReport_18Dec_v2.pdf

MCEETYA. (2008b). *National declaration on educational goals for young Australians—
Draft*. MCEETYA. Retrieved September 19 2008 from http://www.curriculum.
edu.au/verve/_resources/Draft_National_Declaration_on_Educational_
Goals_for_Young_Australians.pdf

McGaw, B. (2004, June). *Australian mathematics learning in an international context*. Paper presented at 27th annual conference of Mathematics Education Research
Group of Australasia, Townsville, Australia.

McGaw, B. (2008, February). *International benchmarking of Australian schools*. Paper
presented to the Annual General Meeting of Victorian Council of School Or-
ganisations, Melbourne, Australia.

Mullis, I. V. S., Martin, M. O., Fierros, E. G., Goldberg, A. L., & Stemler, S. E. (2000).
*Gender differences in achievement. IEA's Third International Mathematics and Science
Study (TIMSS)*. Chestnut Hill, MA: TIMSS International Study Center, Boston
College.

Office for Women. (2001). *Australia's response to the Commonwealth Secretary-General's
report 2001 on the implementation of the 1995 Commonwealth Plan of Action on Gender and Development and Update (2000–2005)*. Department of Prime Minister
and Cabinet, Canberra. Retrieved October 14, 2008 from http://www.ofw.
facs.gov.au/publications

Office for Women. (2003). *Women in Australia*. Canberra: Department of Prime Minister and Cabinet, Canberra. Retrieved June 29, 2009 http://www.fahcsia.gov.
au/sa/women/pubs/govtint/

Office for Women. (2007). *Women in Australia*. Canberra: Australian Government.
Retrieved October 14, 2008 from http://www.ofw.facs.gov.au/publications

OSAGI. (n.d.). *Gender mainstreaming*. Office of the Special Advisor on Gender Issues
and Advancement of Women (OSAGI), United Nations. Retrieved December
19, 2007 from http://www.un.org/womenwatch/osagi/gendermainstream-
ing.htm

Parliament of the Commonwealth of Australia. (June 2000). *Inquiry into the education of boys. Media release from the Parliament of Australia, House of Representatives
Standing Committee on Employment, Education and Workplace Relations*, 19.

Parliament of the Commonwealth of Australia. (2002). *Boys: Getting it right (Report on
the inquiry into the education of boys)*. House of Representatives Standing Committee on Education and Training. Canberra: Commonwealth of Australia.

Pierce, R., Stacey, K., & Barkatsas, A. (2007). A scale for monitoring students' attitudes to learning mathematics with technology. *Computers and Education, 48*,
285–300.

Rodd, M., & Bartholomew, H. (2006). Invisible and special: Young women's experiences as undergraduate mathematics students. *Gender and Education, 18*(1),
35–50.

Rothman, S. (2002). *Achievement in literacy and numeracy by 14 year olds, 1975–1998
(LSAY Research Report No. 29)*. Hawthorn, Australia: ACER.

Schools Commission. (1975). *Girls, school and society. Report of a study group to the
Schools Commission*. Canberra: AGPS

Schools Commission. (1984). *Girls and tomorrow*. Canberra: AGPS.

Schools Commission. (1987). *National policy for the education of girls in Australian schools.* Canberra: AGPS.

Secada, W.G., Fennema, E., & Adajian, L.B. (1995). *New directions for equity in mathematics education.* Cambridge: Cambridge University Press.

Shannon, F. (2004). Classics counts over calculus: A case study. In I. Putt, R. Faragher & M. McLean (Eds.), *Mathematics education for the third millennium: Towards 2010. Proceedings of the 27th annual conference of MERGA* (pp. 509–516). Pymble, Australia: MERGA.

Sherman, J. D., & Poirier, J. M. (2007). *Educational equity and public policy: Comparing results from 16 countries (UIS Working Paper No. 6).* Quebec, Canada: UNESCO Institute for Statistics.

Siemon, D., Virgona, J., & Corneille, K. (2001). *The middle years numeracy research project: 5–9, final report* (A project commissioned by the Department of Education, Employment and Training, Victoria, Catholic Education Commission of Victoria and Association of Independent Schools of Victoria. RMIT University). Retrieved December19, 2003 from http://www.sofweb.vic.edu.au/mys/research/MYNRP/index.htm

Skelton, C. (1998). Feminism and research into masculinities and schooling. *Gender and Education, 10*(2), 217–227.

Squires, J. (2005). Is mainstreaming transformative? Theorizing mainstreaming in the context of diversity and deliberation social politics. *International Studies in Gender, State & Society, 12*(3), 366–388.

Summers, A. (2003). *The end of equality: Work, babies, and women's choices in 21st century Australia.* Milsons Point, Australia: Random House.

Teese, R. (2000). *Academic success and social power: Examinations and inequality.* Carlton South, Australia: Melbourne University Press.

Teese, R., Davies, M., Charlton, M., & Polesel, J. (1995). *Who wins at school? Boys and girls in Australian secondary education.* Canberra: Commonwealth of Australia.

Thomson, S., & Fleming, N. (2004). *Summing it up: Mathematics achievement in Australian schools in TIMSS 2002 (TIMSS Australia Monograph No 6).* Camberwell, Australia: ACER.

Thomson, S., & De Bortolli, L. (2008). *Exploring scientific literacy: How Australia measures up (The PISA 2006 survey of students' scientific, reading and mathematical literacy skills).* Hawthorn, Australia: ACER.

Thomson, S., Cresswell, J., & De Bortolli, L. (2004). *Facing the future: A focus on mathematical literacy among Australian 15-year-old students in PISA 2003.* Melbourne: ACER.

Thomson, S., Wernert, N., Underwood, C., & Nicholas, M. (2008). *TIMSS 07: Taking a closer look at mathematics and science in Australia.* Melbourne: ACER.

Vale, C. (1987). *Recreating mathematics and science for girls. Schools resource program: The participation of girls in maths and science, Participation and equity program.* Melbourne: Ministry of Education.

Vale, C (1993). Sex differences in achievement in year 12 mathematics and non-routine problem solving. In B. Atweh, C. Kanes, M. Carss, & G. Booker (Eds.), *Contexts in mathematics education, Proceedings of the 16th annual conference of the Mathematics Education Research Group of Australasia [MERGA]*(pp. 563–568). Brisbane, Australia: MERGA.

Vale, C. (2002). Girls back off mathematics again: The views and experiences of girls in computer based mathematics. *Mathematics Education Research Journal, 14*(3), 52–68.

Vale, C., & Bartholomew, H. (2008). Gender and mathematics: Theoretical frameworks and findings. In H. Forgasz, A. Barkatsas, A. Bishop, B. Clarke, S. Keast, W-T. Seah, & P. Sullivan (Eds.), *Research in mathematics education in Australasia 2004–2007* (pp. 271–290). Rotterdam, The Netherlands: Sense Publishers.

Vale, C., Forgasz, H., & Horne, M. (2004). Gender and mathematics: Back to the future? In B. Perry, C. Diezmann & G. Anthony (Eds.), *Review of research in mathematics education in Australasia 2000–2003* (pp. 75–102). Sydney: MERGA.

Vale, C., & Leder, G. (2004). Student views of computer based mathematics in the middle years: Does gender make a difference? *Educational Studies in Mathematics, 56*(3), 287–312.

Walby, S. (2005). Gender mainstreaming: Productive tensions in theory and practice. *Social Politics: International Studies in Gender, State & Society, 12*(3), 321–343.

Walden, R., & Walkerdine, V. (1985). *Girls and mathematics: From primary to secondary schooling, Bedford Way Papers 24.* London: Heinemann.

Watt, H. M. G. (2000). Measuring attitudinal change in mathematics and English over the 1st year of junior high school: A multidimensional analysis. *Journal of Experimental Education, 68*(4), 331–361.

Watt, H. M. G. (2006). The role of motivation in gendered educational and occupational trajectories related to maths. *Educational Research and Evaluation, 12*(4), 305–322.

Watt, H. M. G., & Eccles, J. S. (2000, December). *An international comparison of students' maths- and English-related perceptions through high school using hierarchical linear modelling.* Paper presented at the combined annual meeting of the Australian Association of Research in Education and the New Zealand Association for Research in Education, Melbourne, Australia. Retrieved December 19, 2003 from http://www.aare.edu.au/99pap/wat99215.htm.

Yates, L. (1993). Feminism and Australian state policy: Some questions for the 1990's. In M. Arnot & K. Weiler (Eds.), *Feminism and social justice in education: International perspectives* (pp. 167–185). London: Routledge.

Zammit, S., Routitsky, A., & Greenwood, R. (2002). *Mathematics and science achievement of junior secondary school students in Australia (TIMSS Australia monograph, no. 4).* Melbourne: ACER.

SECTION II

NATIONAL FOCUS

CHAPTER 7

STUDIES IN MEXICO ON GENDER AND MATHEMATICS

Sonia Ursini, Martha P. Ramírez, and Claudia Rodríguez
Departamento de Matemática Educativa,
Cinvestav, IPN, México

María Trigueros and Ma. Dolores Lozano
Departamento de Matemáticas, ITAM, México

INTRODUCTION

In Mexico, interest in conducting research on the condition of women emerged in the late 1970s. Early studies dealt with fields such as health, education, science, and work. Problems related to violence, equity, and discrimination also attracted researchers' attention.

In the area of mathematics education, research on gender, although relatively recent, has produced interesting findings. In some studies, gender differences in relation to results attained on standardized tests designed to evaluate performance in mathematics have been analyzed, while the focus of other studies has been on more specific topics, such as gender differences in spatial visualization; the differential relationships mathematics teachers might establish with female and male students at various educational levels; the different attitudes of girls and boys towards mathematics

International Perspectives on Gender and Mathematics Education, pages 147–172
Copyright © 2010 by Information Age Publishing
147

in general and, more specifically, towards the use of technology as an aid in teaching and learning mathematics; and how these aspects change during schooling. With respect to mathematics learning outcomes, the differences found between boys and girls, in practice, always favor males, a situation aggravated by biases that result from societal socioeconomic differentiation.

With respect to Mexico's educational policy, prior to 2000, issues such as gender and equity in basic education were identified and handled by pre-service teacher education institutions including the General Direction of Professional Development for Teachers, the National Pedagogical University, the National Autonomous University of Mexico, the Metropolitan Autonomous University, and the *Universidad Michoacana de San Nicolás Hidalgo*. In 2000, a more systematic line of research in the field of gender and mathematics education emerged at the Centre for Research and Advanced Studies of the National Polytechnic Institute (Cinvestav, IPN).

Given the particular socio-cultural and economic conditions of Mexico as a multi-ethnic, developing country, disseminating information on the research projects conducted there and their results may be of special interest to nations in similar circumstances, and to the international community more broadly in terms of organizing research and support networks that can help resolve problems of sexism and other forms of discrimination. In this chapter an overview of some of the research conducted in Mexico is presented.

THE GENDER PERSPECTIVE

Research in this chapter uses the gender perspective as an analytical tool that allows the examination of the ways in which men and women are socially and culturally constituted, resulting in gender differences.

Initially, attempts at investigating gender differences related to the teaching and learning of mathematics included the clarification of the term "gender." While in English this word is limited in meaning and is applied only to the sexes, in the Spanish language this term has several distinct and additional referents. For example, it is used to refer to the classes, species, types, or categories in which people, animals, or things are classified; and in grammar it indicates the feminine, masculine, or neuter value of nouns, adjectives, articles, and pronouns. A strong achievement of Mexican studies, when using a gender perspective, has been to clearly establish the fact that gender differences are symbolized in sexual differences, and that they influence both men and women. Lamas (1977) shows sound theoretical foundations by setting up three viewpoints from which gender can be considered: *gender assignment*, from the moment of birth and as a function of the baby's genitals; *gender identity*, established at the same time as language acquisition when boys come to identify themselves as males and girls as females, and that then permeates all of their vital experiences; and, third, *gender roles*, which are assigned socially and culturally in the form of stereotypes that set the appropriate and acceptable behavior of women and men.

A more refined conceptualization has been implemented in more recent investigations (e.g., Rodríguez & Ursini, 2008; Ursini, in press), which considers gender as an ideological system (Flores, 2000), that is, a social regulation system of values, norms, attitudes, beliefs, and practices that determines the distinct ways in which men and women model their perceptions of the world and their roles in society in a given historical moment. It structures a cognitive organization, different for females and males, that coordinates their social practices and makes them relate to mathematics in different ways.

THE RESEARCH STUDIES

The research studies presented in this chapter have been organized into two main categories: gender differences in results on standardized tests; and gender differences in specific areas related to mathematics. Our aim is to show how research on gender and mathematics in Mexico has evolved from both theoretical and methodological perspectives. We conclude with a discussion of the results from the gender perspectives adopted.

Gender Differences on Standardized Mathematics Tests

In this section, two studies are presented in which results on standardized tests on mathematics performance were analyzed with the purpose of exploring gender differences. The first study focused on possible gender differences in career choice, while in the second, differences in secondary school students' mathematics achievements were analyzed.

Study 1
Bosch and Trigueros (1996) analyzed data from the Mexican educational system and the results attained by students on high school and university entrance examinations. Through a detailed analysis of official information, gender inequity in access to higher education was identified and was found to be primarily due to dropout rates after the completion of basic, compulsory levels of education. It was also found that gender inequity in access to higher education was clearly related to the degree of economic development in different geographical areas of the country. However, there were no significant differences in mathematics performance up to the end of

the secondary school (12–15 years old). These results were similar to those reported in other studies where no gender differences were found for students younger than 12 (e.g., Brusselmans-Dehairs, Henry, Beller, & Gafni, 1997; Friedman, 1989; Mullis, Martin, Fierros, Goldberg, & Stemier, 2000).

In the same study, Bosch and Trigueros (1996) also examined data from a large database gathered in an earlier project (Gil-Anton, 1994) intended to analyze general university lecturers' changes in career choice, in research preferences and in teaching preferences across time. Bosch and Trigueros aimed to show changes in the gender of university lecturers' profiles of the population of university lecturers and students, over a 10-year period. Upon analyzing the distribution of female students and lecturers by subject area, these researchers found that although population numbers studying sciences and mathematics had doubled in ten years, the percentage of female students had not changed in that time. In relation to university lecturers, they found that the percentage of female mathematics lecturers had also increased in the previous ten years, but there were still more male lecturers, and the working conditions (research and promotion opportunities, salary) for females were less rewarding than for men.

To complement these data, Bosch and Trigueros (1996) interviewed 18 female students in their final year of high school who were interested in the field of mathematics, and 12 female mathematics lecturers working at different universities. The findings revealed that despite their high school performance levels in mathematics (usually the same as their male counterparts), the female students rarely considered careers in the field. Bosch and Trigueros hypothesized that this was mainly due to social beliefs and conceptions about women's roles in society. Studying mathematics and being successful in this field were still strongly related to the male sphere, thus inhibiting the choice of mathematics-related careers by women. Those who did choose mathematics-related careers tended to be good students and high achievers, as was the case for the 12 female mathematics university teachers interviewed. However, the female lecturers' research output was lower than that of their male colleagues. The female lecturers explained that this was mainly because their multiple roles in family and work precluded them from devoting more time to research.

Study 2

González (2003) examined the mathematics performance of 18,751 female and male secondary school students (13–15 years old) on different mathematics tests:

- 9,600 (15 years old) answered the National Examination for Admittance to High-school Education (EXANI 1) in different academic years (1996–1999, 2000, 2001)

- 7,759 (13–14 years old) took the Test of Mathematical Knowledge and Ability (PCHM) in 2000
- 1,392 (13–14 years old) sat the Test of Performance in Mathematics (PRM) in 2001

The marks assigned to these students by their mathematics teachers were also examined.

The data were analyzed using d-tests (for interval measures) and χ^2 (for ordinals, $p < 0.05$) to examine if there were statistically significant gender differences in mean scores (the d-test is the difference between women's and men's mean scores, divided by the standard deviation—see Cohen, 1992). From a total of 309 items that were found in the database, for 88% there were no statistically significant gender differences ($p > 0.01$); for 9% of the items men scored higher, while women did better on 3% of the items. Only in geometry was it found that men, on average, outperformed women on a greater number of questions (14% compared to 2%). There was no significant gender difference on questions related to the rest of the mathematical content.

On these performance tests, average differences were small and, as already reported by other researchers (e.g., Doolittle & Cleary, 1987; Harris & Carlton, 1993) results sometimes favored boys, while on other occasion girls fared slightly better. These results led González (2003) to suggest that the gender differences found might be related to the instrument used.

In aptitude tests (EXANI 1, 1996–2001), women did not show higher scores on any of the questions. When considering all the items (192) in the database, 43% showed significant differences favoring male students. In particular, on questions related to geometric sequences (62%) and ratio (50%), males scored higher than females. Gender differences in the results of the mathematical aptitude tests were consistent with the earlier findings of Eccles, Adler, and Meece (1984).

González (2003) also found that boys performed significantly better than girls on items requiring spatial visualization. However, upon analyzing the marks assigned to these same students by their mathematics teachers, it was found that on average, girls performed equal to or sometimes better than boys, as was noted by Kimball (1995). Although González (2003) did not know the criteria used by these mathematics teachers to assign marks to students, she stressed that these findings were in contrast to some other studies showing that mathematics teachers generally tended to have lower expectations of success for girls than for boys (Eccles, Adler, & Meece, 1984; Fennema & Leder, 1990). González (2003) concluded that the data sets analyzed suggest that, while on average girls performed just as well as—or on occasion better than—boys, the items included in the aptitude test used to evaluate their capability to continue studying at a higher level seemed to constrain their educational opportunities.

Summary

The strengths of the two studies described above are in the detailed analyses of large and reliable databases. González (2003), for example, had access to a database that was not publicly available or easily accessible. The statistical analyses carried out in these studies offered an overview of the gender differences in mathematics outcomes for Mexican students aged 13 to 15 (González, 2003), and for high school students and female university mathematics lecturers (Bosch & Trigueros, 1996). Qualitative data, aimed at formulating plausible explanations for the results obtained, complemented the results of the latter study. However, one of the limitations of the findings from the database analyzed by Bosch and Trigueros (1996) is the lack of comparative findings from studies conducted elsewhere.

González's (2003) study did not include a qualitative approach. Doing so would have allowed researchers to formulate hypotheses about possible factors to account for the results obtained. It is likely that socio-cultural and socioeconomic issues were important; however, they were not mentioned in the study, and it would have be interesting to investigate these aspects. Steinthorsdottir and Sriraman (2008), for example, used a qualitative approach to show that the economic conditions and socio-cultural aspects found in Iceland had influenced girls' and boys' achievements in PISA 2003—Iceland was the only country among the nations involved in PISA 2003 in which girls performed significantly better than boys.

GENDER DIFFERENCES IN SPECIFIC AREAS RELATED TO MATHEMATICS

In this section, research studies in which gender differences in specific areas related to mathematics were examined are discussed, including spatial visualization, attitudes towards mathematics, male and female teachers' views on gender, and social representations of mathematics.

Gender Differences in Spatial Visualization

To explore the spatial visualization capabilities of male and female secondary school students and possible gender differences between them, Rivera (2003) administered a multiple choice 20-item questionnaire to

231 participants (106 girls, 125 boys; 13–15 years old) from two schools in Mexico City. This study had institutional support, allowing the researcher to gain access to the two schools in the same area in Mexico City. This allowed control of socio-economic and cultural variables that could have had an effect on gender differences. All the tested students were from lower and middle socioeconomic backgrounds. The percentages of correct and incorrect responses and the number of questions left unanswered by male and female students were calculated and compared. Additionally, five girls and five boys were interviewed to identify the solution strategies they had used.

The results showed that both boys and girls had serious difficulties in carrying out transformations, whether simple (2-dimesional [2-D] to 3-dimensional [3-D]), medium–difficult (3-D to 2-D), somewhat more complex (3-D to 3-D), or rotations (in 3-D). These common difficulties might be due to the fact that at school they worked only with two-dimensional representations. While no significant gender differences were found in students' ability to solve the items of the questionnaire, there were some differences in the kinds of wrong answers they gave to some questions. For example, in a problem where a drawing of a three dimensional object was presented and students were asked to draw the three hidden lines, most girls answered by altering the original object by adding a new one to it, as it is shown in Figure 7.1. In contrast, most male students tried to draw the three hidden lines, without altering the original shape (Figure 7.2).

Rivera (2003) commented that these kind of answers suggested that girls and boys might perceive and reason differently when facing the same problem, but this needs further investigation.

Another important point was that girls left more problems unanswered than boys did, a tendency that clearly appeared when answering the three

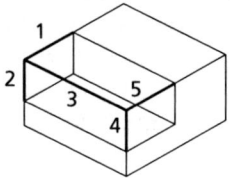 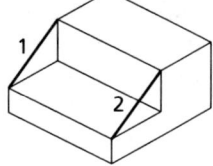

Figure 7.1

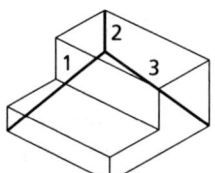 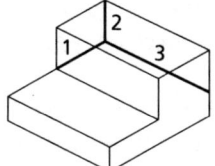 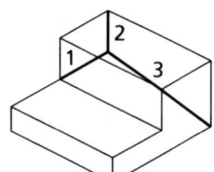

Figure 7.2

items involving 2-D to 3-D transformations. For all the items in this category, the percentage of girls who did not answer was higher than boys (61.3% vs. 55.2%; 64.1% vs. 60.8%; 15% vs. 12%).

A significant gender difference also emerged in the formats students who participated in this study used to support their answers. While boys justified their responses verbally, relying on the drawing, and strongly based their explanations on two-dimensional representations, girls tended to reply only verbally. As in earlier studies (Atweh & Cooper, 1995; Bishop, 1980; Fennema, 2000; Fennema & Sherman, 1977), it was stressed that the differences detected seemed to be a consequence of the formal and informal education girls and boys had received; that is, they were socio-cultural constructions. These gender differences might be due to everyday experiences and practices (e.g., certain games and activities engaged in at home and outside that might be considered "exclusively" apt for girls or for boys). They might stem from social beliefs and expectations inculcated from an early age by family, school, and the environments where children grew up (e.g., assigning attributes like strength, bravery, audacity, intelligence, and quickness to boys and, in contrast, those of sensibility, docility, obedience, tranquility, and carefulness to girls). Due to their socio-cultural origin, Rivera stated that these differences may be susceptible to change.

An interesting aspect of Rivera's (2003) study was the attempt to link gender stereotypes with girls' and boys' different perceptions and ways of reasoning when facing the same mathematical problem. To highlight this, attention was paid to gender differences in the wrong answers students gave and in the way they justified their answers. The participants in the study might be considered representative of students aged 13 to 15 in Mexico City; however, they represent only one part of the Mexican population. It is therefore not possible to generalize these results. Mexico is a multicultural country with different ethnic groups, different socio-cultural traditions, and groups with different socioeconomic conditions. Girls and boys therefore can have very different life experiences depending on the group they belong to. It would be interesting to explore in greater detail how different socio-cultural environments influence students' learning of mathematics and, in particular, their visual-spatial abilities.

GENDER DIFFERENCES IN ATTITUDES TOWARDS MATHEMATICS

In the last six years there have been several studies in Mexico concerning students' attitudes towards mathematics, computer-based mathematics, and the use of technology. One of the first studies was conducted by Ursini, Sánchez, and Orendain (2004). This study included the development of a scale called AMMEC (Attitude to Mathematics and Mathematics Taught with Computers). It is a 29-item, 5-point Likert scale organized in three sub-scales: 11 items that test students' attitudes towards mathematics (AM), 11 items that test students' attitudes towards computer-based math (AMC), and 7 items that test students' self-confidence in math (CM). The reliability of the AMMEC scale (split-halves correlation, .71; Cronbach's Alpha, .79) can be considered as good (De Vellis, 1991) or acceptable (Kerlinger & Lee, 2002), depending on the criteria used. Three factors, pertaining to each one of the three sub-scales, were detected using factor analysis. Reliability analysis provided satisfactory Cronbach's alpha for each sub-scale (AM, .81; AMC, .77; CM, .68), thus indicating good or acceptable internal consistency. The face validity of the scale was affirmed by four experts. Factor analysis using Principal Components Analysis and varimax rotation showed that the scale factors explained 36% of the total observed variance. As a result, only items with a factorial load between 0.38 and 0.78 were selected. The scale was administered to a group of 439 students. Their teachers were also asked to evaluate the students on the same topics. The AMMEC and the teachers' scores correlated strongly ($r = 0.98$). More details on the procedures followed to determine AMMEC scale validity and reliability can be found in Ursini, Sanchez, and Orendain (2004).

Though the results of administering this scale to 228 girls and 211 boys aged 12 to 15 years old showed no significant gender difference in their attitudes towards mathematics and computer-based mathematics, they did reveal a significant gender difference favoring boys in terms of self-confidence in this discipline.

In a later study (Ursini, Ramirez, & Sánchez, 2007), data derived from the administration of the AMMEC scale to two groups of students (458 who used Excel or Cabri-Géomètre and 221 who did not use technology) were compared. The results revealed that attitudes towards mathematics were more positive among the technology-users. However, even after using computers for two years, a significant percentage of students maintained a negative or neutral attitude towards mathematics. On average, girls were more likely to feel that way. These researchers found that the great majority of students, whether they used computers or not, felt positively towards computer-based mathematics (more boys than girls), though a substantial percentage remained neutral (on average, girls were more likely to feel that way).

Recently, a comparative longitudinal study of changes in girls' and boys' attitudes towards mathematics, towards computer-based mathematics, and self-confidence in mathematics was conducted (Ursini & Sánchez, 2008). A total of 430 students (236 girls, 194 boys) who used technology for mathematics and 109 (57 girls, 52 boys) who did not were monitored for three years. At the beginning of the study, all participants were in grade 7 (about 13 years of age). Using the AMMEC scale, there were very few changes over the three-year period in the gender differences in students' attitudes and self-confidence. Significant gender differences favoring boys were seen in attitudes towards mathematics in grades 8 and 9, but only for the technology-using group. Boys scored significantly higher than girls in attitudes towards computer-based mathematics, but only in grade 7 in the technology-using group. An intra-gender analysis showed that, in grade 8, girls using technology had significantly higher scores in attitudes towards computer-based mathematics than girls not using it. The use of technology did not have a positive impact on students' self-confidence. Whether students used computers or not, both boys and girls showed a decrease in self-confidence in mathematics from grades 7 to 9. These results were similar to those found by other researchers who reported that self-concepts of ability in mathematics declined through adolescence (Eccles et al., 1989; Jacobs, Lanza, Osgood, Eccles, & Wigfield, 2002; Watt, 2004). This can be explained through the *attitude decline hypothesis* (Osborne, Driver, & Simon, 1998; Ramsden, 1998), in which it is proposed that the development of negative attitudes towards science and mathematics is observable in students aged 11 or 12. In both groups, at grade 9, the scores obtained showed that boys and girls had low self-confidence. These results were enriched through interviews with 12 girls and 13 boys conducted at the end of the study. The analysis of the arguments that the students presented to explain and justify their attitudes towards mathematics, towards computer-based mathematics, and their self-confidence in working in mathematics provided evidence of important gender differences in the ways in which boys and girls construct their attitudes. These contrasts indicated that their constructions reflected the gender stereotypes present in Mexican society where passivity, discipline, ability, and obedience are characteristics linked to women while activity, aggressive behavior, indiscipline, and talent are associated to men (Flores, 2000).

While working with Australian students, Vale and Leder (2004) did not find gender differences in students' perceptions of their performance on mathematical tasks on the computer. In contrast to this, Ursini and Sánchez (2008) found that Mexican male and female students had different conceptions of the use of the computer when learning mathematics, and that those differences were closely related to the socio-cultural construction of gender. While Mexican male students had a pragmatic view of the use of the computer—as an instrument that made the solution of mathematical tasks easier and quicker—female students tended to see it as a resource that allowed them to construct mathematical knowledge, a device that motivated and helped them and allowed them to learn and verify their answers. Like Vale and Leder (2004), the findings of this study showed that the use of computers for teaching and learning mathematics depended on many factors, including economic ones. For example, it is important to stress that, in Mexico in 2006, only 19.4 % of families had a computer at home (INEGI, 2006).

Regarding self-confidence in relation to mathematics, consistent with the findings of Meyer and Kohler (1990), Ursini and Sánchez (2008) found that male students attributed their poor performance in mathematics to external factors (lack of support, motivation, and teachers), while they considered that successful outcomes were related to internal factors (intelligence, knowledge, ability, and interest). In comparison, female students attributed their low marks to issues such as lack of intelligence or knowledge, and high achievement to factors such as effort, dedication, and obedience.

The statistical analyses conducted in the studies reported above consistently indicated few gender differences in Mexican students' attitudes towards mathematics and towards computer-based mathematics. Significant gender differences were found in relation to students' self-confidence to work in mathematics. Even though students' scores exhibited low self-confidence, male students consistently scored higher than female students.

When qualitative approaches were used, a different picture emerged. The influence of local cultures was highlighted in relation to attitudes. Using interviews, Ursini and Sánchez (2008) found that boys and girls constructed their attitudes differently. It was observed that language became a powerful tool through which identity and gender roles were internalized. Students made use of specific vocabulary that belonged to their cultural context and which described mathematics with specific words associated to either males and/or females. Thus, for example, while boys associated math with such words as "like" and "ease" of learning, girls emphasized how "difficult" they found math. Boys used the words "well" and "easy" when asked about how they thought they had done on their most recent math exam, whereas girls tended to say "badly" and "difficult." Another noticeable finding was that boys with a negative attitude to mathematics tended to attribute their feelings to external factors, such as math teachers; while girls, though they mentioned the influence of teachers, also cited internal factors, such as difficulties in understanding math and a lack of intelligence. In contrast, internal factors were put forth by boys with positive attitudes to math, who referred to their ability to understand this subject and their intelligence. Girls with a positive attitude, in contrast, ascribed it to external factors such as teachers who offered clear explanations and support.

Discourse analysis (Ibañez, 1992; Pourtois & Desmet, 1992) revealed the influence of local culture on the way students construct their perceptions and ideas around mathematics, and their attitudes towards it.

MALE AND FEMALE TEACHERS' BELIEFS ABOUT GENDER

In Mexico, as in many other countries, the past decade has seen a strong trend towards the integration of technology into secondary school classrooms as a means of facilitating the teaching and learning of mathematics. Since 1997, the Mexican Ministry of Education has sponsored a national project called EMAT (Teaching Mathematics with Technology). The aim of the EMAT project was to incorporate computational technologies into the mathematical curriculum of secondary schools (children aged 12 to 15 years old) throughout the country (Ursini, 2006). Using a constructivist approach (Ursini & Rojano, 2000), it was expected that the EMAT materials (worksheets presenting curriculum-based activities to be solved using computer software, mainly Excel, Cabri-Géomètre, and TI calculator) would enrich and improve the current teaching and learning of the standard mathematics curriculum of secondary schools (Sacristán, Ursini, Trigueros, & Gil, 2006). Since 2000—and parallel to the implementation of the EMAT project—several studies were carried out to assess the impact of EMAT (e.g., Sacristán, 2005; Sacristán & Ursini, 2001; Ursini, Ramirez & Sánchez, 2007). Two of these studies are discussed in this section.

The first study was conducted by Ursini, Sánchez, Orendain, and Butto (2004), who investigated teachers' perceptions of the impact of technology on the behavior of girls and boys in mathematics class. They analyzed information provided by 24 teachers about 1,113 students (13–15 years). They probed mathematics teachers' views on how their students' behavior changed after 1, 2, or 3 years of computer use at school. They asked the teachers to consider the following parameters: participation, ability to analyze a problem and interpret worksheets, initiative, need for assistance, dedication, defense of their ideas, creativity, and preferences for individual or group work. For each of the parameters the teachers were required to assign a mark (1, when the considered aspect never appeared; 2, when it appeared sometimes; 3, when it appeared very often) to each student. Data were analyzed using Kruskal-Wallis' proof and χ^2. Results showed that teachers believed that the use of technology in mathematics classrooms modified most of the parameters mentioned above for all students but, depending on the particular aspect examined, the resulting changes were different for girls and boys. For example, after three years of participating in the project, the vast majority of students—with no difference by gender—had developed a strong ability to analyze the problems presented to them and to interpret worksheets, showed greater initiative than their classmates who had been involved in the project for less time, were more dedicated in their work, were better able to defend their ideas, and had a more creative attitude towards the problems presented. However, while the researchers observed that when students began to use technology, the great majority of boys preferred to work individually and only slowly became more favorably inclined towards group work, the vast majority of girls always preferred group work. In contrast, while teachers did not note any significant effect of the duration of technology use on participation and on requests for assistance by boys, they did observe this among girls. According to the teachers, girls who had participated in the project for three years were more participative and requested more help than those who had used technology for less time.

The data also showed that the differences in the aspects tested between boys and girls who had been using technology for three years were fewer than those seen among the students who had used it for just one or two years. This suggests that the use made of technology (propitiating group work and group discussion and guiding activities with worksheets) helped to modify certain cultural patterns of conduct that contributed to reinforcing gender differences, and thus fostered greater equality. In addition, Ursini, Sánchez, Orendain, and Butto (2004) pointed out that the marks assigned by the teachers might have been influenced by their own gender stereotypical beliefs. In effect, during their interviews with teachers, the researchers confirmed that both male and female teachers related many of the behaviors evaluated to characteristics they considered to be natural features of one gender or the other. These are some examples of teachers' comments:

> Girls are, you know, much more dedicated; much more earnest than boys... boys are not so persistent, they feel they do not need to study a lot. (female teacher)

> Girls interact a lot, they listen to each other, they reflect and defend their ideas...boys are passive, they are not flexible, they think they are never wrong. (female teacher)

> Girls ask for help much more than boys, they need help... I always try to help girls. (male teacher)

> Boys, you know, they do not pay much attention, it is difficult for them to be quiet, but finally they do well, they are good students. (male teacher)

In another study, Trigueros and Lozano (2008) focused on gender differences in the way in which teachers working in EMAT assessed their students' learning. Differences in the approaches adopted by a male and a female teacher evaluating their students' performance were identified from case studies of two teachers. The methodological approach used was "research in design," introduced by Brown (1992) and Collins (1992) and involved multiple iterative cycles in which the materials produced by participants provided the data for analysis. A task in which the two teachers were asked to assess students' learning in several cycles was designed. Finally, interviews with the teachers and students, complemented by observations in the classroom, comprised the data-gathering processes related to the teachers' activities and attitudes.

The results obtained in these two case studies showed that while involved in the EMAT project both teachers modified their ideas about mathematics learning in important ways. Interesting contrasts were, however, revealed in their respective beliefs and opinions. The female teacher's conceptions of mathematics learning became increasingly complex. As she reflected on her own work in the classroom and on students' behaviors, she began to incorporate several elements that took on greater importance as she thought about how students learn. She emphasized that various aspects—including students' attitudes and beliefs, language, understanding, independence, and autonomy—were significant components of learning, and thus she began to take all of these elements into account when designing new assessment practices. As far as this teacher was concerned, the role of computers in student learning was related to exploration or investigation. This was important because exploration (investigation) allowed students to become more independent. She also considered creativity and imagination to be important elements to be taken into account when discussing students' learning. Her emphasis on students coming to enjoy learning experiences provided additional evidence of the multifaceted approach to learning that this teacher developed.

The male teacher's views on learning also changed, though he maintained a more pragmatic view that focused on his students' ability to assimilate and understand mathematical concepts. Learning processes took on new importance for him as he came to see himself more and more as having the main responsibility for the outcome of those processes. The role of the computer was, he felt, to provide feedback for students and its use was also important because of its relevance to their futures. He was mainly concerned with training students so that they would be able to use mathematics in different contexts, and with their ability to move forward in their education.

Trigueros and Lozano (2008) also discussed how these differences revealed teachers' ideas concerning the role of technology in the classroom, and about science and mathematics learning more generally. As in a previous larger study (Trigueros & Carmona, 2006), it was found that female teachers tended to adopt a much wider perspective when thinking about students' learning—one that took into account emotional, social, and cognitive aspects. Male teachers, on the other hand, tended to focus primarily on the assimilation of mathematical concepts and procedures and applications of this learning in real life. Though it was not possible to generalize on the basis of such a small sample, general tendencies differentiating female and male teachers were detected, thus suggesting lines for future research on the EMAT project in which a larger number of teachers could be included.

Gender inequities in the interactions teachers establish in the mathematics class were studied by Ramirez (2006), who focused on elementary school teachers, and Espinosa (2007), who analyzed the way university teachers interact with their male and female students.

Ramirez (2006) interviewed 14 female teachers attending students aged 9 to 11 years. She decided to work only with female teachers because, as stated by researchers such as Elizondo (1999), Clarricoates (1993), and Fainholc (1997), the teaching profession in elementary school is viewed as a "natural" feminine domain. Additionally, in Mexico this school level is perceived as an extension of the domestic environment and primary teachers are viewed as the "*mater admirabilis*" (Elizondo, 1999).

Results from the interviews showed that teachers' self-confidence in relation to mathematics was low. They justified it by the "difficult nature" of mathematics and to the poor teaching they received during their school and professional training years. In relation to their ideas about boys and girls as mathematics students, they considered boys as "talented" and "good at learning and developing mathematics," while girls were thought of as "being able to do." This distinction is subtle yet powerful. Talent was associated with natural or genetic disposition, while "being able to do something" was considered a result of strong effort and constant practice. In the same manner, teachers associated talent, like aggressive behavior, as having a genetic component. All the teachers agreed on the fact that they considered mathematics important for the future professional development of their students and for enhancing possibilities for accessing better educational levels and higher social status. Twelve out of 14 teachers considered mathematics as a predominantly male arena.

Three teachers were selected to deeper investigate their practices through non-participant lessons' observations. Criteria for selection were based on their ideas on gender and its relationship to mathematics. Two of them viewed mathematics as a male domain and they distinguished between talent (associating it with boys) and ability (associating it with girls). One of these two teachers explained that she interacted differently with boys and girls during mathematics lessons in order to help girls who, she assumed, lacked talent. The third selected teacher perceived mathematics as a neutral domain, and she considered she was not making any difference between girls and boys when teaching mathematics.

The purpose of the classroom observations was to register how teachers interacted with boys and girls when posing mathematics questions of high and low cognitive demand (following the classification proposed by Wimer, Ridenour, Thomas, & Place, 2001) to the whole group or to individuals.

The analysis of the data showed that for whole-group questions, different patterns of interaction were found in relation to male and female students, thus revealing teachers' ideas on gender and its relationship to mathematics. In general, boys appeared to be more enthusiastic and eager to answer questions than girls, and teachers selected boys more frequently than girls to answer questions.

For those questions that were posed to a specific student in the group, the three teachers showed clear gender-related differences regarding the quality and the number of questions posed, favoring boys in all cases. Even though one of the teachers reported that she thought she favored girls given their "genetic deficiencies," observation records showed that all three teachers systematically posed higher level questions to boys while they chose girls to answer questions of lower cognitive demand. It is interesting to note here that results from a questionnaire answered by the students revealed that they did not perceive their teacher behaving differently when interacting with boys or girls.

These results suggest that the differences observed in teachers' ways of interacting with boys and girls might contribute in promoting the development of a passive-dependent type of personality for girls and an active-independent one for boys. This could be one of the variables that slowly, systematically, and progressively contribute to the construction of gender differences related to mathematics. More details of this study are available in English in Ramirez and Ursini (2008).

To study university teachers' perceptions of mathematics and the way they interacted with their male and female students, Espinosa (2007) interviewed seven calculus teachers from four different universities: three women and four men. Analysis of the interview data transcripts showed that all seven teachers considered the male students to be more proficient in mathematics than the females. Even those who felt that women could be successful in mathematics associated that possibility with greater effort and discipline on their part, a finding that indicates their perception of mathematics as a predominantly male domain. While observing the university teachers' mathematics classes, the researcher detected that female students tended to be passive during teachers' explanations, and when asked to solve exercises they tended to delay by looking at their notes, apparently waiting for one of their male classmates to answer first. They would then use the male student's response to confirm their own solutions. It was observed as well that all the university teachers observed tended to pose more questions to their male students. When they questioned a female student and she hesitated before answering, they did not wait long before turning to a male student to request his answer. To give but one example, Espinosa (2007) commented that in one class a questioned female student began by justifying the procedure she used before giving the result she had obtained. The result was not the one the university teacher expected. Instead of providing her some feedback the teacher turned to a male student he considered talented and asked him for the answer. Espinosa (2007) stressed that female students made no complaint about this lack of attention from the teachers, apparently considering it something normal. With respect to the analysis of the marks obtained by male and female students in a mathematics examination, no gender differences were found.

The interviews showed that the university teachers participating in the study were not aware of the differences they were making between male and female students. All of them believed they were giving the same amount of attention to all their students, but they agreed that women had a passive attitude towards mathematics and less mathematical ability, though they were orderly and tidy and showed a tendency to discuss amongst themselves while solving problems. The female students seemed to have good relationships with their male classmates and confidently asked them to provide explanations when they were not sure how to proceed. Only one student indicated that she occasionally felt uncomfortable because her male classmates tended to respond aggressively and made her feel inferior.

The studies described above highlight the importance of social aspects in the learning of mathematics. They also show how beliefs, values, and attitudes can be constructed and perpetuated through social interaction. In addition, the research reports emphasized the role of the teacher in the implementation and reproduction of gendered conceptions. These gendered conceptions probably contributed to the teachers' stereotyped views of girls' and boys' behaviors and were based in beliefs that female and male students have different mathematical abilities and that learning mathematics is more important for boys than girls. Similar results have been reported by many other researchers (Fennema & Hart, 1994; Gray, 1996; Grevholm & Hanna, 1995; Helwig, Anderson, & Tindal, 2001; Leder, 1992; Leder, Forgasz, & Solar, 1996; Moreno, 1996; Tiedemann, 2002).

In general, these research studies indicated that teachers' classroom behaviors reflected dominant male socio-cultural constructs and were more likely to stimulate the mathematical learning of male than female students. Additionally, the reports drew attention to the fact that both teachers and students viewed mathematics as a male domain. These findings differ from those reported by Forgasz (2001), for example, who pointed out that this view of mathematics was no longer as prevalent as it had been. The Mexican data highlight the importance of taking into account the socio-cultural and historical context of the people who are being studied. Most of the reported studies were carried out in Mexico City, but we believe that it would be very important to extend these kinds of studies to other areas of the country with different socio-economic and cultural conditions.

GENDER DIFFERENCES IN SOCIAL REPRESENTATIONS
OF MATHEMATICS

Gender difference in teachers' social representation (SR) (Moscovici, 1961, 1976, 1984) of the use of technology was studied by Rodriguez and Ursini (2008). These researchers departed from the SR theory developed by Moscovici (1961, 1976, 1984) where social representation was defined as the elaboration of a social object by a given community. The concept of social representation denoted a form of specific knowledge, the knowledge of "common sense," whose content demonstrated the operation of generative and functional processes marked socially. In its widest sense it denoted forms of practical thought oriented toward communication, understanding, and the control of social material and ideal environments (Moscovici 1961). Rodriguez and Ursini (2008) studied the SR male and female teachers had of *Enciclomedia*, a technological resource available to primary school teachers to support the teaching of mathematics and other school subjects. A total of 30 primary school teachers (22 women, 8 men) participated in the study. Each one of them had a minimum of three years' teaching experience. The following methods were used to detect teachers' SR: Abric's (1994) associative card; semi-structured interviews; and videotapes of 10 mathematics classes (one class from each of five male and five female teachers) in which *Enciclomedia* was used. The data were examined using traditional content analysis (Bardín, 1996) and the software program ALCESTE 4.9 (Reinert, 1986, cited in De Alba, 2005). Social Representations Theory helped the researchers explore and interpret the different ways that male and female teachers perceived the teaching of mathematics with *Enciclomedia*.

The results showed that the SRs of male and female teachers were indeed different, and revealed that female teachers viewed *Enciclomedia* as a means of adding an affective dimension to mathematics instruction (i.e., to make it more attractive, dynamic, and motivating), while male teachers, in contrast, imbued it with a more utilitarian sense because they considered *Enciclomedia* to be just one more tool to access and make their teaching easier.

The analysis of teachers' interviews revealed three central topics related to their SR: social image, professional development and needs, and teaching practices. With respect to the first, female teachers felt that their image as professionals depended on their effective use of *Enciclomedia*. In contrast, male teachers did not seem to consider this important, as they felt they would have the approval of the school community whether or not they used this multimedia teaching aid effectively. In relation to professional development and needs, female teachers argued that it was important and necessary for them to receive training in the use of this program, not only on how to use it, but also in terms of its pedagogical applications. Unlike their female counterparts, male teachers claimed that the training was sufficient for them to be able to work with *Enciclomedia*. With respect to their views on the introduction of technology for teaching, the great majority of the participants had a favorable attitude towards making greater use of technology in the classroom. Their discourse revolved around the question of social desirability, though some resisted incorporating such methods in their classrooms. The female teachers stated that they would have preferred access to *Enciclomedia* outside school so that they could explore it more and plan their classes more efficiently. The males, however, did not consider such exploration necessary. Finally, it was found that teaching practices adopted by the female teachers were characterized by greater collegiality or what Lagarde (1992) called sorority (female solidarity). Male teachers, meanwhile, rarely expected support from their peers, and rarely offered it. When comparing what was said and what was observed in practice, it was found that only one of the ten teachers observed, a male, based his teaching primarily on the problem-solving model.

In a further study, Ursini (in press) analyzed the social representations (Moscovici, 1961, 1976, 1984) of the mathematics of 12 female and 13 male grade 9 students. Discourse analysis revealed a markedly homogeneous—and apparently quite deeply rooted—perception of gender differences related to the intellectual, cognitive, and behavioral abilities that determine success or failure in mathematics. Although all participants agreed that learning mathematics required students to be interested in mathematics and to pay attention in class, as well as having intelligence, dedication, discipline, and hard work, these elements were differentially attributed to men and women. Both boys and girls linked females' success in mathematics to hard work, attentiveness, order, obedience, and the willingness to follow instructions; while achievement among males was more often related to intelligence and interest in the subject.

In general, female students manifested low confidence in their intellectual capacity. Those with low achievement in mathematics considered that it was due to their own lack of ability and intelligence, while those who had greater success assumed that it was due to effort, hard work, dedication, and obedience. The influence of external factors was deemed only marginal. In contrast, male students tended to trust their intellectual ability to learn math. They related poor performance to external factors and to a lack of hard work and dedication, but never to any lack of their own capacity. Not surprisingly, they related success to their own innate intelligence.

The SR that female students had of mathematics revealed an association with three fundamental requirements: external support, teachers' satisfactory explanations of mathematics' topics, and paying attention in class. The first two requirements suggest a reliance on external dimensions beyond their control and, therefore, severely limit the possibilities of them taking any steps that might increase their success in mathematics. The female students most often associated success in this subject with effort, compliance, discipline, and motivation, though they also mentioned enjoyment, interest, understanding, and intelligence. Most of them classified mathematics as "difficult" and said that they "don't do well" in this subject because they earned, at most, average marks. These data indicate a general tendency towards low self-confidence and a self-image characterized by low intelligence and insufficient ability to achieve success in math.

The boys, meanwhile, had an SR of mathematics that was very distinct from that of girls. They associated mathematics with such terms as understanding, enjoyment, and intelligence. They often said "I do pretty well." These elements determined the general representation expressed by the boys interviewed, with the "I do pretty well" being added as a marker of self-confidence about mathematics that was inculcated in males through social norms promoting the idea that they "naturally" understood mathematics better than females did. In their discourse, boys also stressed social valorizations of mathematics and the conditions that made it possible for males to succeed. Unlike girls, however, boys considered being successful in mathematics more a question of "having a knack" for the discipline than of being motivated or studying hard. They affirmed that "having intelligence" made mathematics somewhat easier and that, therefore, it was not so necessary to pay attention, push oneself, or complete all the work assigned by teachers. Finally, boys mentioned factors such as dedication, discipline, effort, and studying, but saw them as mechanisms that might help girls to attain good results in mathematics, based on the assumption that, generally speaking, females had more difficulty succeeding in mathematics.

Research studies by Rodríguez and Ursini (2008) and Ursini (in press) used a methodological approach that included several research methods that strengthened the validity of the results. For example, different techniques for data collection were employed (free association and interviews). Triangulation was used when analyzing data and different approximations to representations (structural and procedural). In this way, the theoretical approach used in these investigations was soundly sustained by appropriate methods.

We would like to draw attention to the studies reported in this section because they belong to a novel area of research in mathematics education. Previous studies have looked at classroom factors that contributed to gender differences in mathematics learning outcomes, and examined both students' and teachers' beliefs and behaviors (e.g., Forgasz, 1995; Forgasz & Leder, 1996). We have found only two reports that use Moscovici's (1961, 1976, 1984) theory of social representations applied to the investigation of mathematics teaching and learning: De Abreu and Cline (1998) and Flores (2007); only Flores (2007) included the gender perspective. The latter research study was carried out in Chile where the SRs of primary and secondary school teachers in relation to students' learning were analyzed, and gender differences highlighted. The author stated that the beliefs and expectations teachers held influenced their interactions with students in the classroom, limiting females' mathematical learning while enhancing that of males.

DISCUSSION AND CONCLUSIONS

This review of research on gender and mathematics conducted in Mexico shows not only the high level of interest in the topic that has emerged in recent years, but also how this research community has gradually been refining its gender perspective. This evolution is evidenced by the distinct ways in which the focus on gender has been used in the studies examined in this chapter. It seems clear that studies have taken one of two basic directions: descriptive analyses and studies probing the socio-cultural elements underlying certain generic constructions that perpetuate gender differences in relation to mathematics and its learning. The former types of research study focus essentially on gender differentiation and posit this as the fundamental cause of all observed differences. The starting point of the second type, on the other hand, is the explicit assumption that social norms, and not biological differences, give rise to inequalities between the sexes.

These different approaches are somewhat reminiscent of the evolution of the concept of gender itself. Research reports show development in theoretical approaches on gender that can now be understood as a socio-cultural construct organizing social relations and social practices among men and women and the way in which people interact with objects and cognitive constructions. On the basis of this conceptualization, gender is used as a category that allows us to analyze and explain social relations among men and women. Though it does not ignore biological differences, this posture strongly questions whether these differences are the causes of variations in the behavior and achievements of men and women, and places emphasis on the socio-cultural construction of "the feminine" and "the masculine."

In conclusion, we would emphasize the importance of the fact that these exploratory studies conducted in Mexico have used distinct methodological approaches. On the one hand, there have been studies of an eminently quantitative nature, some of them complemented by interviews. This type of study provided a first panorama as to the gender differences that existed with respect to different issues concerning mathematics. Although these data are necessary at the beginning of any exploration of gender differences in a specific space, the need soon emerges to inquire more deeply into the causes of such findings, and this has led some researchers to recognize that it is necessary to complement such empirical studies with inquiries of a more socio-cultural nature designed to produce data on the issues that underlie the emergence of gender differences in mathematics.

Finally, we would emphasize that one aspect identified in the findings of several of the studies cited herein is the predominance of an androcentric culture that permeates the perceptions of mathematics among girls and boys and that establishes this field as a male domain. It is interesting to note that this vision is present among both teachers and students and at different educational levels. Another theme arising from this body of research is the close relationship between the male vision held by mathematics teachers and its influence on pedagogical practices. Despite the advances achieved in research on the mathematics–gender binomial in Mexico, we are just starting to work in this direction. It is essential to undertake further research that will take into account the nation's diverse socioeconomic and cultural contexts. This would allow a better understanding of how such cognitive and representational systems are socially constructed, and then enable the discovery of the elements that will make it possible to design strategies that will contribute positively to diminishing gender differences.

REFERENCES

Abric, J.C. (1994). *Prácticas sociales y representaciones.* México City: Coyoacán.

Atweh, B., & Cooper, T. (1995). The construction of gender, social class and mathematics in the classroom. *Educational Studies in Mathematics, 28*(3), 293–310.

Bardín, L. (1996). *El análisis de contenido.* Madrid: Akal Universitaria.

Bishop, A. (1980). Spatial abilities and mathematics education. *Educational Studies in Mathematics, 11*(3), 257–269.

Bosch, C. & Trigueros, M. (1996). Gender and mathematics in Mexico. In G. Hanna (Ed.), *Towards gender equity in mathematics education. An ICMI Study* (pp. 277–284). London: Kluwer.

Brown, A. L. (1992). Design experiments: Theoretical and methodological challenges in creating complex interventions in classroom settings. *The Journal of the Learning Sciences, 2*(2), 141–178.

Brusselmans-Dehairs, C., Henry, G. F., Beller, M., & Gafni, N. (1997). *Gender differences in learning achievement: Evidence from cross-national surveys,* Paris: Organisation for Economic Co-operation and Development.

Clarricoates, K. (1993). Todo en el mismo día. In D. Spender & E. Sarah (Eds.), *Aprender a perder. Sexismo y educación* (pp. 95–109). Barcelona: Paidós.

Cohen, J. (1992). Statistical power analysis. Current directions. *Psychological Science, 1*(3), 98–101.

Collins, A. (1992). Toward a design science of education. In E. Legemann, & L. Shulman (Eds.), *Issues in education research: Problems and possibilities* (pp. 15–22). San Francisco: Jossey-Bass.

De Abreu, G., & Cline, T. (1998). Studying social representations of mathematics learning in multiethnic primary schools: Work in progress. *Papers on Social Representations, 7*(1/2), 1–20.

De Alba, M. (2005). Programa de análisis de textos Alceste (análisis de lexemas coocurrentes en los enunciados simples de un texto). *Cuadernillo de trabajo.* México City: Universidad Autónoma Metropolitana.

De Vellis, R. F. (1991). *Scale development. Theory and applications.* Thousand Oaks, CA: Sage.

Doolittle A. E., & Cleary, T. A. (1987). Gender-based differential item performance in mathematics achievement items. *Journal of Educational Measurement, 24*(2), 57–166.

Eccles, J. S., Adler, T., & Meece, J. L. (1984). Sex differences and achievement: A test of alternate theories. *Journal of Personality and Social Psychology, 46*(1), 26–43.

Eccles, J. S., Wigfield, A., Flanagan, C. A., Miller, C., Reuman, D. A., & Yee, D. (1989). Self-concepts, domain values and self-esteem: relations and changes at early adolescence. *Journal of Personality, 57*(2), 283–310.

Elizondo, A. (1999). Maternaje y Educación Preescolar. La desprofesionalización de la función docente. *Básica. Revista de la escuela y del maestro. 6*(27), 5–9.

Espinosa, C. (2007). *Estudio de las interacciones en el aula desde una perspectiva de género.* Unpublished Masters Thesis, Cinvestav-IPN, Mexico City, Mexico.

Fainholc, B. (1997). *Hacia una escuela no sexista.* Buenos Aires: AIQUE.

Fennema, E. (2000). *Gender and mathematics: What is known and what do I wish was known?* Paper presented at the fifth annual forum of the National Institute for Science Education, Detroit, Michigan. Retrieved June 6, 2009, from http://www.iwitts.com/html/fennema.pdf

Fennema, E., & Hart, L. (1994). Gender and the JRME. *Journal for Research in Mathematics Education, 25*(6), 648–659.

Fennema, E., & Leder, G. (1990). *Mathematics and gender.* New York: Columbia University.

Fennema, E., & Sherman, J.A. (1977). Sex-related differences in mathematics achievement, spatial visualization and affective factors. *American Educational Research Journal, 14*(3), 51–71.

Flores, F. (2000). *Psicología social y género. El sexo como objeto de representación social.* Mexico City: McGraw-Hill.

Flores, R. (2007). Representaciones de género de profesores y profesoras de matemática, y su incidencia en los resultados académicos de alumnos y alumnas. *Revista Iberoamericana de Educación, 43*(1), 103–118.

Forgasz, H. (1995). Gender and the relationship between affective beliefs and perceptions of grade 7 mathematics classroom learning environments. *Educational Studies in Mathematics, 28*(3), 219–239.

Forgasz, H. (2001). Australian and US preservice teacher's perceptions of the gender stereotyping of mathematics. In M. Van den Heuvel-Panhuizen (Ed.), *Proceedings of the 25th conference of the International Group of the Psychology of Mathematics Education* (pp. 433–440). The Netherlands: Utrecht University.

Forgasz, H. J., & Leder, G. C. (1996). Mathematics classrooms, gender and affect. *Mathematics Education Research Journal, 8*(2), 153–173.

Friedman, L. (1989). Mathematics and the gender gap: A meta-analysis of recent studies on sex differences in mathematical task. *Review of Educational Research, 59*(2), 185–213.

Gil-Antón, M. (1994). *Los rasgos de la diversidad: Un estudio sobre los académicos mexicanos.* Mexico City: Universidad Autónoma Metropolitana-Azcapotzalco.

González, R. M. (2003). Diferencias de género en el desempeño matemático. *Educación Matemática, 15*(2), 129–161.

Gray, M. (1996). Gender and mathematics: mythology and misogyny. In G. Hanna (Ed.), *Towards gender equity in mathematics education. An ICMI Study* (pp. 27–38). London: Kluwer.

Grevholm, B., & Hanna, G. (Eds.). (1995). *Gender and mathematics education. An ICMI study.* Lund, Sweden: Lund University Press.

Harris, A. M., & Carlton, S. T. (1993). Patterns of gender differences on mathematics items on the scholastic aptitude test. *Applied Measurement in Education, 6*(2), 137–151.

Helwig, R., Anderson, L., & Tindal, G. (2001). Influence of elementary student gender of teachers' perceptions of mathematics achievement. *The Journal of Educational Research, 95*(2), 93–103.

Ibañez, T. (1992). *Aproximaciones a la psicología social.* Barcelona: Sendai.

INEGI (Instituto Nacional de Estadística, Geografía e Informática). (2006). Retrieved November 12, 2006, from http://www.inegi.gob.mx/est/contenidos/español/sistemas/cem06/statal/coa/m030/index.htm.

Jacobs, J. E., Lanza, S., Osgood, D. W., Eccles, J. S., & Wigfield, A. (2002). Gender and domain differences across grades one through twelve. *Child Development, 73*(2), 509–527.

Kerlinger, F. N., & Lee, H. B. (2002). *Investigación del comportamiento. Métodos de investigación en ciencias sociales* (4th ed.). Mexico City: McGraw Hill.

Kimball, M. M. (1995). *Feminist visions of gender similarities and differences.* Binghamton, NY: Haworth Press.

Lagarde, M. (1992). Enemistad y sororidad: hacia una nueva cultura feminista. Isis Intenacional. *Ediciones de las Mujeres, 17*(3), 33–81.

Lamas, M. (1977). La antropología feminista y la categoría género. In M. Lamas (Ed.), *El género: La construcción cultural de la diferencia sexual* (pp. 97–125). Mexico City: PUEG/UNAM.

Leder, G. C. (1992). Mathematics and gender: Changing perspectives. In D. A. Grouws (Ed.), *Handbook of research in mathematics education* (pp. 597–622). New York: Macmillan.

Leder, G. C., Forgaz, H. J., & Solar, C. (1996). Research and intervention programs in mathematics education: A gender issue. In A. Bishop, K. Clements, C. Keitel, J. Kilpatrick & C. Laborde (Eds.), *International handbook of mathematics education, Vol. 2* (pp. 932–961). Utrecht, The Netherlands: Kluwer.

Meyer, M. R., & Koehler, M. S. (1990). Internal influences on gender differences in mathematics. In E. Fennema & G. C. Leder (Eds.), *Mathematics and gender* (pp. 60–95). New York: Teachers College Press.

Moreno, M. (1996). *Como se enseña a ser niña; el sexismo en la escuela.* Barcelona: Icaria.

Moscovici, S. (1961). *La psychanalyse, son image et son public.* Paris: Presses Universitaires de France.

Moscovici, S. (1976). *Social influence and social change.* London: Academic Press.

Moscovici, S. (1984). The phenomenon of social representation. In R. Farr & S. Moscovici (Eds.), *Social representation* (pp. 3–69). Cambridge: Cambridge University Press.

Mullis, I. V., Martin, M. O., Fierros, E. G., Goldberg, A. L., & Stemier, S. E. (2000). *Gender differences in achievement. IEA's third international mathematics and science study (TIMSS).* Chestnut Hill, MA: International Association for the Evaluation of Educational Achievement.

Osborne, J., Driver, R. & Simon, S. (1998). Attitudes to Science: Issues and concerns. *School Science Review, 79,* 27–33.

Pourtois, J., & Desmet, H. (1992). *Epistemología e instrumentación en ciencias humanas.* Barcelona: Herder.

Ramírez, M. P. (2006). *Influencia de la visión de género de las docentes en las interacciones que establecen con el alumnado en la clase de matemáticas.* Unpublished Masters Thesis, Cinvestav-IPN, Mexico City, Mexico.

Ramirez, M. P., & Ursini, S. (2008). *Influence of the female teachers' gender vision on the type of interactions they establish with boys and girls in the mathematics classroom.* Paper presented at the 11th International Congress of Mathematics Education (ICME 11), Topic Study Group 32: Gender and mathematics education, Monterrey, Mexico. Retrieved June 6, 2009, from http://tsg.icme11.org/document/get/165

Ramsden, P. (1998). Managing the effective university. *Higher Education Research & Development, 17*(3), 347–370.

Rivera, M. (2003). *Diferencia de género en la visualización espacial: un estudio exploratorio con estudiantes de 2º de secundaria.* Unpublished Masters Thesis, Cinvestav-IPN, Mexico City, Mexico.

Rodríguez, C., & Ursini, S. (2008). *Social representation and gender in the teaching of mathematics with multimedia devices.* Paper presented at the 11th International Congress of Mathematics Education (ICME 11), Topic Study Group 32: Gender and mathematics education, Monterrey, Mexico. Retrieved June 6, 2009, from http://tsg.icme11.org/document/get/166

Sacristán, A. I. (2005). Teacher's difficulties in adapting to the use of new technologies in mathematics classrooms and the influence on students' learning and attitudes. In G. M. Lloyd, J. L. Wilkins & S. L. Behm (Eds.), *Proceedings of the 27th annual meeting of the North American chapter of the International Group for the Psychology of Mathematics Education* (CD-ROM). Roanoke, VA.

Sacristán, A. I., & Ursini, S. (2001). Incorporating new technologies to the Mexican school culture: the EMAT project and its Logo extension. In G. Futschek (Ed.), *Eurologo 2001: A turtle odyssey. Proceedings of the 8th European logo conference* (pp. 237–244). Linz, Austria: Österreichische Computer Gesellschaft.

Sacristán A. I., Ursini S., Trigueros M., & Gil N. (2006). Computational technologies in Mexican classrooms: the challenge of changing a school culture. In Dagiene, V. & Mitterneir, R. (Eds.), *Selected papers of the 2nd International Conference Informatics in Secondary Schools: Evolution and Perspectives* (pp. 95–110), Vilnius, Lithuania.

Steinthorsdottir, O. B., & Sriraman, B. (2008). Exploring gender factors related to PISA 2003 results in Iceland: A youth interview study. *ZDM. The International Journal on Mathematics Education, 40*(4), 591–600.

Tiedemann, J. (2002). Teachers' gender stereotypes as determinants of teacher perceptions in elementary school mathematics. *Educational Studies in Mathematics, 50*(1), 49–62.

Trigueros, M., & Carmona, G. (2006). Nuevas perspectivas de evaluación. In T. Rojano (Ed.), *Enseñanza de la Física y la Matemática con Tecnología: Modelos de transformación de las prácticas y la interacción social en el aula* (pp. 231–241). SEP. Mexico.

Trigueros, M., & Lozano, M. D. (2008). *Teachers' assessment practices in mathematics courses. Does gender make a difference?* Paper presented at the 11th International Congress of Mathematics Education (ICME 11), Topic Study Group 32: Gender and mathematics education, Monterrey, Mexico. Retrieved June 6, 2009, from http://tsg.icme11.org/document/get/168

Ursini, S. (in press). Diferencias de género en la representación social de las matemáticas: Un estudio con estudiantes de secundaria. In N. Blazquez & F. Flores (Eds.), *Herramientas metodológicas y epistemología de género* (pp. 79–104). Mexico City: CIICH/UNAM.

Ursini, S. (2006). Enseñanza de las Matemáticas con Tecnología (EMAT). In T. Rojano (Ed.), *Enseñanza de la Física y la Matemática con Tecnología: Modelos de transformación de las prácticas y la interacción social en el aula* (pp. 25–41). SEP. Mexico.

Ursini, S., Ramirez, M. P., & Sánchez, G. (2007). Using technology in the mathematics class: How this affects students' achievement and attitudes. *Proceedings of the 8th ICTMT (Integration of ICT into Learning Processes)*. Czech Republic: University of Hradec Králové. [CD-ROM]

Ursini, S., & Sánchez, G. (2008). Gender, technology and attitudes towards mathematics: A comparative longitudinal study with Mexican students. *ZDM. The International Journal on Mathematics Education, 40*(4), 559–577.

Ursini, S., Sánchez, G., & Orendain, M. (2004). Validación y confiabilidad de una escala de actitudes hacia las matemáticas y hacia las matemáticas enseñada con computadora. *Educación Matemática, 16*(3), 59–78.

Ursini, S., Sánchez, G., Orendain, M., & Butto, C. (2004). El uso de la tecnología en el aula de matemáticas: Diferencias de género desde la perspectiva de los docentes. *Enseñanza de las Ciencias, 22*(3), 409–424.

Ursini, S., & Rojano, T. (2000). *Guía para Integrar los Talleres de Capacitación EMAT.* México: SEP-ILCE.

Vale, C. M., & Leder, G. C. (2004). Student views of computer-based mathematics in the middle years: does gender make a difference? *Educational Studies in Mathematics, 56*(2/3), 287–312.

Watt, H. M. G. (2004). Development of adolescents' self-perceptions, values, and task perceptions according to gender and domain in 7th-through 11th-grade Australian students. *Child Development, 75*(5), 1556–1574.

Wimer, J. W., Ridenour, C. S., Thomas, K., & Place, A. W. (2001). Higher order teacher questioning of boys and girls in elementary mathematics classrooms. *Journal of Education Research, 95*(2), 84–92.

CHAPTER 8

GENDER DIFFERENCES WHEN WORKING WITH ALGEBRAIC VARIABLES

A Study with Mexican Secondary School Students

Carolina Rubí Real Ortega and Sonia Ursini
Departamento de Matemática Educativa Cinvestav-IPN,
México

INTRODUCTION

Over the past forty years, a considerable number of studies seeking to determine a relationship between gender and mathematics learning have been conducted in various countries. In the 1970s, studies such as those carried out by Maccoby and Jacklin (1974), Fennema and Sherman (1976), and Harris (1987) declared the existence of significant differences in favor of male students in mathematics operations. These various studies hypothesized that beliefs and self-confidence when working with mathematics might influence student learning, in addition to the fact that males demon-

International Perspectives on Gender and Mathematics Education, pages 173–202
Copyright © 2010 by Information Age Publishing
173

strated an increased ability to apply visual-spatial skills when compared to females. Students' mathematics performance, participation, achievement, and attitudes have been analyzed since then to detect possible gender differences (Atweh & Cooper, 1995; Burton, 1990; Eudave, 1994; Fennema, 1990; Forgasz & Leder, 2000; Hanna, 1989; Jungwirth, 1991; Leder, 1992, 1996, 2001; Leder & Fennema, 1990; Leder & Forgasz, 1992; Ursini, Ramirez, & Sanchez, 2007; Ursini & Sanchez, 2008). In general, there is agreement in that the differences found, usually favoring males, are linked to the female condition in our societies. It is stressed that several causes contribute to the construction of gender differences: stereotypes related to females and mathematics built upon sexual differences; teachers' stereotyped view of girls' and boys' capabilities to learn mathematics that encourage male's learning; context influences on students' beliefs about mathematics that promote their choice or avoidance of professions or careers involving mathematics; and self-confidence that has a bearing on the levels of participation and mathematics achievement.

Studies devoted to highlighting gender differences have been conducted at different school levels. In general, no significant gender differences between boys and girls have been reported at the elementary school level (Ercikan, McCreith, & Lapointe, 2005; Fennema, 1974; Fierros, 1999; Johnson, 2000; Leahey & Guo, 2001; Leder, 1974; Zhang & Manon, 2000); while differences favoring male students seem to emerge with time (Campbell, 1995; Mullis & Stemler, 2002). Nevertheless, the results have been contradictory, and they seem to strongly depend on the socio-cultural and economic characteristics of the population studied (Forgasz, 2002; Francis & Skelton, 2005; Ursini & Sanchez, 2008). Although many researchers report gender differences favoring men in relation to performance, participation, and mathematical achievement, inconsistencies in findings still occur. For example, in Hawaii, Brandon, Newton, and Hammond (1987), analyzing mathematical data from 1982 to 1983 produced by the administration of the *Stanford Achievement Test* (SAT), found that girls obtained systematically better results than boys. Brandon and Jordan (1994), analyzing the results of the grade 8 mathematics section of the *National Assessment of Educational Progress* (NAEP), found as well that girls obtained significantly higher scores than boys. The Education Department of the U.S. reported the same results for the year 2000. Nevertheless, no explanation concerning possible reasons for these outcomes was given. In a recent study Bjorg and Sriraman (2008) argued that the economic and socio-cultural environment could explain why in Iceland girls performed better than boys in mathematics when the PISA 2006 test was administered.

Several studies analyzing gender differences in mathematics achievement have focused on mathematics in general. But, in addition, there have been studies focusing on gender differences in relation to specific mathematics topics. For example, in relation to proportional reasoning skills, Linn and Pulos (1983) observed that females tend to be less successful than males. Rubinstein (1985), working with grade 8 students, examined difficulties they had with order and estimation, and found that male students obtained better results than their female peers in all tasks relating to estimation and order of quantities, but that female students obtained better results in mental calculations. In a test of geometry, measurement, and probability, Meece et al. (1982), and Hunt (1985) found that males obtained better results than females. Nonetheless, some researchers reported that women obtained better scores in arithmetical reasoning tests because they require the use of verbal strategies. Belenky, Clinchy, Golberger and Tarule (1986) recognized that cognitive processes when organizing and processing information were more favorable to women.

Across the review of research literature in the last twenty years, we found only a few studies focusing on gender differences in relation to the understanding of algebra, and the results seem to be inconsistent. Swafford (1980), for example, analyzing mathematical achievements and skills in solving algebraic problems, reported that female students obtained the same and, in some cases, better results than their male classmates on a standardized algebra test. In contrast, Hanna (1989) and Ethington (1990), when analyzing data from the *Second International Mathematics Study*, did not find significant differences in algebra, but observed that more female than male students left questions unanswered. Kirshner (1989) investigated students' syntactic knowledge representations in algebra. He argued that visual characteristics of algebraic notations may be highly correlated with propositional rules of algebraic syntax. He did not find significant differences in the average performance of males and females on an algebra test; however, differences favoring males were found on the unspaced complex algebra items. The conclusion was that proportionately more females than males depend on visual cues for syntactic decision making.

In Mexico, although the past 30 years have seen an increasing interest in gender issues focusing on equity, violence, discrimination, health, and education in general, only in recent years have studies addressed the topic of gender and mathematics. However, no study has focused on gender differences in relation to algebra. Therefore, we consider it interesting and necessary to develop research in this direction and contribute to filling this gap. In this chapter we report the findings of a study investigating possible differences between males and females in relation to the understanding of the concept of variable. We analyzed ninth grade students' capability to interpret, symbolize, and manipulate variables in different algebraic contexts.

This chapter is organized in four sections. In the first one we briefly present the research on gender and mathematics developed in Mexico. After this, we present the *3UV Model* (Ursini & Trigueros, 1998; Ursini & Trigueros, 2001; Trigueros & Ursini, 2001), the theoretical framework we used to analyze our data. In the third section we describe the study and discuss the results obtained. Finally, there is the conclusion section.

GENDER AND MATHEMATICS IN MEXICAN STUDIES

In the past 30 years a considerable number of studies on gender issues have been developed in Mexico. They have analyzed problems related to health, violence, discrimination, and equity, as well as education and science. Only a few studies until now have addressed the topic of gender and mathematics. Some of them have analyzed gender differences related to results attained on standardized tests designed to evaluate performance in mathematics (Bosch & Trigueros, 1996; González, 2003), while other have focused on more specific topics: gender differences in spatial visualization (Rivera, 2003); the differential relations that mathematics teachers establish with female and male students (Ramírez, 2006); the distinct attitudes of girls and boys towards mathematics in general (Campos, 2006) and, more specifically, towards the use of technology as an aid in teaching and learning mathematics (Ursini, Sanchez, & Orendain, 2004; Ursini, Ramírez, & Sánchez, 2007); and how these aspects change during schooling (Ursini & Sanchez, 2008).

A first mention of gender and mathematics in Mexico can be found more than 10 years ago, when Bosch and Trigueros (1996) commented on the lack of research in this field in our country and provided some interesting data from the Mathematical Olympiads in which high school students participated. They commented that during the Mathematical Olympiads carried out between 1987 and 1992, out of a total of 36 winners, only two were women. Bosch and Trigueros (1996) also analyzed the results attained by students on high school and university admittance examinations. They identified gender inequality in access to higher education that was related to the degree of economic development and suggested that studying mathematics and being successful in this field were still related strongly to the male sphere. González (2003), analyzing the performance of grade 7–9 students in mathematics, found that boys' performance was significantly better than girls' for items involving spatial imagination and mathematical reasoning. Rivera (2003), in contrast, found no significant gender differences for items involving spatial imagination when studying 231 grade 8 low socioeconomic level students' performance. However, he determined that girls' and boys' errors and misconceptions were different. Studying classroom interactions in primary school, Ramírez (2006) found that female teachers tended to consider mathematics as a male domain, assuming that boys were "naturally" better than girls at mathematics and that girls had to work harder and make a greater effort in order to be successful in this subject.

Attitudes towards mathematics have been addressed in several Mexican studies (Ursini, Sanchez & Orendain, 2004; Campos, 2006; Ursini, Ramírez & Sánchez, 2007; Ursini & Sanchez, 2008). They all consistently found through statistical analyses of data that only few gender differences appear in relation to attitudes to mathematics. Campos (2006), studying how attitudes to mathematics change from grades 6 to 9, reports that while in grade 6 more girls than boys had positive attitudes towards mathematics, more boys than girls in grade 9 showed such positive attitudes. In a study with 439 secondary school students (aged 12–15), Ursini, Sanchez and Orendain (2004) found no significant gender difference in attitudes to mathematics or to computer-based mathematics; the majority of girls and boys were neutral to mathematics, but positive to computer-based mathematics. In a comparison of two groups of students using technology (458 students, who used spreadsheets or *Cabri- Géomètre*, and 221 who did not), Ursini, Ramírez, and Sanchez (2007) found that attitudes towards mathematics were more positive among those who used technology. They also found that the vast majority of students, regardless of technology use and gender, felt positive towards computer-based mathematics. A comparative longitudinal study investigating changes in girls' and boys' attitudes towards mathematics, computer-based mathematics, and self-confidence in mathematics was developed by Ursini and Sanchez (2008). A total of 430 students using technology for mathematics and 109 students not using it were monitored for three years. The statistical analyses of data showed few gender differences in the way students' attitudes and self-confidence changed over the three years. They found that use of technology did not have a positive impact on students' self-confidence. These results were enriched by interviewing 25 students (12 girls and 13 boys) at the end of the study. The analysis of the arguments they presented to explain and justify their attitudes towards mathematics, computer-based mathematics, and their self-confidence in working in mathematics provided evidence of important gender differences in the ways in which boys and girls construct their attitudes, indicating how their constructions reflect the gender stereotypes within Mexican society.

As already mentioned above, there have been no studies in Mexico focusing on gender differences in relation to the understanding of algebra. The only reference to gender and algebra can be found in Juárez (2002), who studied mathematics teachers' understanding of the concept of variable. He claimed that when analyzing the percentages of correct answers given to a questionnaire testing the capability to work with specific unknowns, general numbers, and related variables, male teachers obtained better results than female teachers.

In Mexico, the perspective of gender has evolved over time. An important number of researchers consider that gender determines human behavior, including the way we dress, talk, and behave (Barberá, 1998). It affects the development of psychological activity through expectations, desires, norms, and values, which play a role in the way we reason, judge, and construct our self-concept. The process of gender construction is constantly interacting with demographic variables, social factors, and specific situational components. Likewise, there is a relationship with age, environmental opportunities, socio-economic levels, and cultural issues. Gender is conceived as a socio-cultural construct that organizes social relationships between men and women, their social practices, and the way in which people interact with psychological objects and cognitive constructs (Flores, 2000). This conception of gender emphasizes its socio-cultural character and implicitly advocates for a qualitative approach in order to explain gender differences. Although we totally agree with this conception, the study we present in this chapter aimed not to provide an explanation of gender differences in relation to the understanding of algebraic variable, but only to obtain a first, although reduced, panorama of these possible differences. Therefore, in the first instance, data were analyzed only using descriptive statistics, and no interviews were done. In the near future, based on the knowledge obtained by the outcomes of this study, we would like to conduct another one in which a qualitative approach could help us identify some of the causes underlying the gender differences detected.

THE CONCEPT OF VARIABLE AND THE 3UV MODEL

The concept of variable is fundamental for the development of algebraic knowledge and of mathematical literacy. The understanding of this concept is central to elementary mathematics but it is central as well to the understanding of more advanced mathematics topics. Nevertheless, the versatility of this concept makes it very difficult to be mastered by students. These difficulties are at the basis of many of the problems they face when learning elementary algebra and other branches of mathematics.

During the last thirty years, many research studies have focused on the difficulties students have when working with variables in elementary algebra (Küchemann, 1981; MacGregor & Stacey, 1996; Matz, 1980; Phillip, 1992; Schoenfeld & Arcavi, 1988; Usiskin, 1988; Ursini & Trigueros, 1998; Wagner, 1983; Warren, 1999). Research has shown that each use of variable is linked to specific epistemological and didactical obstacles (Bednarz & Dufour-Janvier, 1991; Bolea, Bosch, & Gascon, 1998; Chevallard, 1983; English & Sharry, 1996; Filloy & Rojano, 1989; Herscovics & Linchevski, 1991; Kieran, 1984). It has been stressed as well that many of the difficulties are linked to the multifaceted character of variable (e.g., Matz, 1980; Trigueros & Ursini, 2003; Usiskin, 1988; Ursini & Trigueros, 2001; Wagner, 1983) and that after various algebra courses the majority of students are still unable to properly interpret variables depending on the context of a problem (Ursini & Trigueros, 2006).

Trigueros and Ursini (2001) consider that in almost any algebra problem, variable is present with its multifaceted character and it is fundamental for students to develop the capability to use it in a flexible way. These researchers state that for a richer comprehension of algebra and mathematics in general a deep understanding of the concept of variable is fundamental. They describe this understanding in terms of the following basic capabilities:

- to perform simple calculations and operations with literal symbols
- to develop a comprehension of why these operations work
- to foresee the consequences of using variables
- to distinguish between the different uses of variable
- to shift between the different uses of variable in a flexible way
- to integrate the different uses of variable as facets of the same mathematical object

An analysis of students' work with problems involving different uses of variable led these researchers to the design of a theoretical framework called the *3UV Model* (Three Uses of Variable Model) (Ursini & Trigueros, 1997, 1998; Trigueros & Ursini, 2001, 2003). The *3UV Model* considers three uses of variable, the most frequently used in elementary algebra: specific unknown, general number and variables in functional relationship.

The idea of specific unknown refers to a specific value that may be calculated using given data. General numbers are linked to generalization, patterns, regularity and general method. The idea of variable in a functional relationship appears when two related quantities are involved. In this last use of variable two aspects are fundamental, the static and the dynamic one (Ursini, 1994).

For each of these uses of variable, there are different aspects that require interpretation, symbolization, and manipulation. In the *3UV Model* it is considered that in order to work successfully with problems and exercises which involve the variable as unknown, it is necessary to:

U1. Recognize and identify in a problem situation the presence of something unknown that can be determined by considering the restrictions of the problem.

U2. Interpret the symbols that appear in an equation as representing specific values that can be determined by considering the given restrictions.

U3. Substitute for the variable the value or values that make the equation a true statement.

U4. Determine an unknown quantity that appears in equations or problems by performing algebraic and/or arithmetic operations.

U5. Symbolize the unknown quantities identified in a specific situation and use them to represent the situation by an equation

To work successfully with variable as a general number it is necessary to:

G1. Recognize patterns; perceive rules and methods in sequences and in families of problems.

G2. Interpret a symbol as representing a general, indeterminate entity that can assume any value.

G3. Deduce general rules and general methods by distinguishing the invariant aspects from the variable ones in sequences and families of problems.

G4. Manipulate (simplify, expand) the symbols.

G5. Symbolize general statements, rules or methods.

Working with the variables in a functional relationship (related variables) implies the capability to:

F1. Recognize the correspondence between related variables independently of the representation used: tables, graphs, verbal expressions, analytic representations.

F2. Determine the values of the dependent variable given the value of the independent one.

F3. Determine the values of the independent variable given the value of the dependent one.

F4. Recognize the joint variation of the variables involved in a functional relationship independently of the representation used (tables, graphs, analytic expressions).

F5. Determine the interval of variation of one variable given the interval of variation of the other one.

F6. Symbolize a functional relationship based on the analysis of the data of a problem.

A good understanding of variable implies the comprehension of all these aspects and the possibility to relate both the aspects and the uses of variable in different problems. The ability to interpret the different uses of variable as required by specific problematic situations permits students to acquire flexibility in shifting between them and integrating them as facets of the same mathematical object. The ability to manipulate symbolic variables helps them to perform simple calculations. Both these abilities help students develop a comprehension of why the operations work. The ability to symbolize rules and relationships in different problem situations leads them to foresee the consequences of using variables. Beginning algebra students should develop these basic capabilities that are related to a deep understanding of elementary algebra in a gradual fashion. It is considered that an adequate solution of algebraic problems requires a flexible use of these three uses of variable and the aspects characterizing each one of them, as well as their integration in one multifaceted concept and the capability to differentiate them (Ursini, Escareño, Montes & Trigueros, 2005).

THE STUDY

The purpose of this study was to have a first glimpse at gender differences in relation to the understanding of the concept of variable of Mexican secondary school students. Through the responses given to a questionnaire, we analyzed ninth graders' capability to interpret, symbolize, and manipulate variables in different algebraic contexts.

The Questionnaire

A 33-item questionnaire was designed to test students' capability to work with the three uses of variable and their aspects. The great majority of the items were borrowed from the diagnostic questionnaire developed by Trigueros, Reyes, Ursini and Quintero (1996). Three items were borrowed from the algebra test of the research titled *Children's Understanding of Mathematics: 11–16* (Hart, Brown, Küchemann, Kerslake, Ruddock, & McCartney, 1976). All the items required interpretation, symbolization, and/or manipulation of each one of the three uses of variable (specific unknowns, 7 items; general numbers, 12 items; related variables, 13 items; see Table 8.1). Only one item required pattern recognition and its algebraic symbolization, and one item required the flexible use of the three uses of variable. All the items involving specific unknown required its manipulation in order to realize that the variable involved was representing an unknown and not a general number.

As can be seen in Table 8.1, the questionnaire does not use the same number of items for each one of the different uses of variable. Fewer questions explore the understanding of variable as unknown. The reason for this is due to the results obtained in previous studies. It was observed that in grades 7 and 8 in Mexico this use of variable was strongly emphasized (Benitez, 2004), and students tend to work better with this characterization

TABLE 8.1 Number of Items per Use of Variable

	Variable as unknown	Variable as general number	Variable in functional relationship	Three uses of variable
Interpretation	2	4	3	
Manipulation	3	4	7	1
Symbolization	2	4	3	
Total items	7	12	13	1

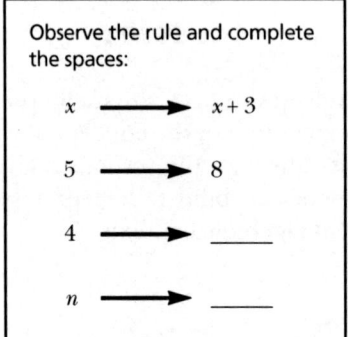

Figure 8.1 Question 19a first version.

x	$x + 3$
5	8
4	
n	

Figure 8.2 Question 19a modified version.

of variable. The questionnaire was piloted with 30 students. The results led us to make small modifications. For instance, we observed that students had difficulties making sense of the arrow notation used to represent correspondence (Figure 8.1) and were more used to a table representation (Figure 8.2), so we eliminated the arrow notation.

Nevertheless, in one of the questions (item 13), we maintained the arrow notation with the purpose of comparing students' answers. All the items were analyzed from *3UV Model* (Table 8.2).

TABLE 8.2 Uses and Aspects of Variable for Each Item

Number of Item	Use of Variable	Aspects of variable involved
1	Variable as Unknown	U1 and U5
2	Variable as General Number	G5
3	Variable as General Number	G2, G4 and G5
4	Variable in a Relationship	F1, F4, F3 and F2
4a	Variable in a Relationship	G2 y F6
5a	Variable as General Number	G2
5b	Variable as Unknown	G2, G4, and U2
5c	Variable as General Number	G2
5d	Variable as General Number	G2
5e	Variable as Unknown	G2, G4 and U2
5f	Variable as General Number	G2
6	Variable in a Relationship	F1 and F5
6a	Variable in a Relationship	F1 and F5
7	Variable in a Relationship	F1 and F4
8	Variable as General Number	G2 and G4
8a	Variable as General Number	G2 and G4
9	Variable as Unknown	G2, G4, U2, U4 and U3
9a	Variable as Unknown	G2, G4, U2, U4 and U3
10	Variable as Unknown	U1 and U5
11	Variable in a Relationship	F1 and F6
11a	Variable in a Relationship	F1 and F2
12	Variable as General Number	G1, G2 and G3
12a	Variable as General Number	G4 and G5
13	Variable in a Relationship	G2, F2, G4 and G5
14	Variable as Unknown	U2, U4 and U3
15	Variable in a Relationship	F1, U4 and F2
15a	Variable in a Relationship	F1, U4 and F2
16	Variable as General Number	G2 and G5
17	Variable as Unknown, General Number and in a Relationship	G2, F1, G4, U2 and U4
18	Variable in a Relationship	F1, G2, F4 and F6
18a	Variable in a Relationship	F2
19	Variable in a Relationship	G2, F1 and F2
19a	Variable in a Relationship	F1, G2, F2 and F6

A more complete analysis of the items borrowed from the diagnostic questionnaire developed by Trigueros, Reyes, Ursini and Quintero (1996) can be found in Trigueros and Ursini (2003). This analysis also favored the qualitative analysis stage of the results. An example of a detailed analysis of some items from the CSMS study is shown in Table 8.3.

TABLE 8.3 Analysis of Three Questions Borrowed from CSMS Study

Use of variable	Item	Analysis
Variable as Unknown	If $y + 5 = 8$, how many values can you give to "y"?	The interpretation of variable as a specific unknown (U2) is required. In order to determinate its value it is necessary to perform arithmetic operations (U4) and substitute to variable the value that makes the equation a true statement (U3).
Variable as General Number	Joan has j marbles and Peter has p marbles. Write an expression that expresses the quantity of marbles that they have in total.	First j and p need to be interpreted as general numbers (G2), after this it is required to operate with these general numbers adding them (G5).
Variable in a relationship	Solve the following system of equations: $u = v + 3 \quad m = 3n + 1$ $v = 1 \quad n = 4$	It is necessary to consider that the values of u and m depend of the values of v and n (F1), to substitute the values for v and n and determine the values of u and m (U4, F2).
Variable as unknown, as general number in a relationship.	If we know that $r = s + t$, and that $r + s + t = 30$, what value can "r" be?	The variables have to be interpreted as general numbers (G2) and one has to recognize the correspondence between related variables (F1). "r" is the addition of "s" and "t." Manipulate the variable substituting "r" to $r + s$ (G4), subsequently recognize "r" as an unknown quantity and determine its value (U2 and U4).

Participants

The questionnaire was administered to 133 (67 female, 66 male) ninth grade students (aged 14–15 years old) attending a public school in Mexico City. The school was located in a lower-middle class socio-economic zone. According to the Mexican curriculum (Plan y Programas de Estudio. Educación Básica. Secundaria, 1993), in grade 9 students are expected to have enough experience in order to be able to face problems and exercises requiring symbolization, interpretation and manipulation of variables. This is expected because following the curriculum, in grade 7 they should have covered pre-algebra subjects, using a teaching methodology based on making the most of the opportunities offered by arithmetic and geometry so that students gradually start to use literal symbols. In grade 8 the approach to algebra begins with a brief review of syntax rules and the handling of linear equations, followed by monomial and polynomial operations and the solution of linear equations.

Data Analysis

The responses to the questionnaire were analyzed using descriptive and inferential statistics. The percentages of unanswered, correct, and incorrect responses were calculated per student and per question. After this, students' responses were organized according to gender. In order to identify significant differences between males and females the χ^2 ($p < 0.01$ and $p < 0.05$) was calculated. In order to obtain more information about gender differences, we investigated the solution processes followed by boys and girls to items for which a significant gender difference was found. We looked for possible response patterns and procedural errors common to each gender.

Subsequently, the percentages were calculated for the different categories (unanswered, correct, and incorrect responses) and a general comparison was made of results, as well as another analysis in which the different uses of variable were taken as reference. A comparison of results from the statistical analysis was also carried out, taking into consideration the different characterizations of variable.

Results

The analysis of data aimed to highlight possible gender differences in students' capability to symbolize, interpret, or manipulate variables. In the first approach, the percentages of correct and incorrect answers and unanswered questions were calculated for each item. Figure 8.3 shows the percentages of correct answers for male and female students. We can see that no question was correctly answered by everyone and that male and female students obtained higher percentages in the same questions. We can also see that a higher percentage of men than women correctly answered 32 out of the 33 questions. Only in questions 16 and 6a did males and females obtain the same percentage of correct answers (24.24% and 75.76%, respectively).

As shown in the graph, five exercises (items 1, 12, 14, 16 and 18a) were correctly answered by more than 65% of male and female students. And three of these items (1, 12, and 16) were answered correctly by more than 70%. Questions 1 and 16 involve the symbolization of variable: in the first one as unknown quantity (I5), and in the second as general number (G5). Question 12 requires the recognition of a pattern, clearly stating the first

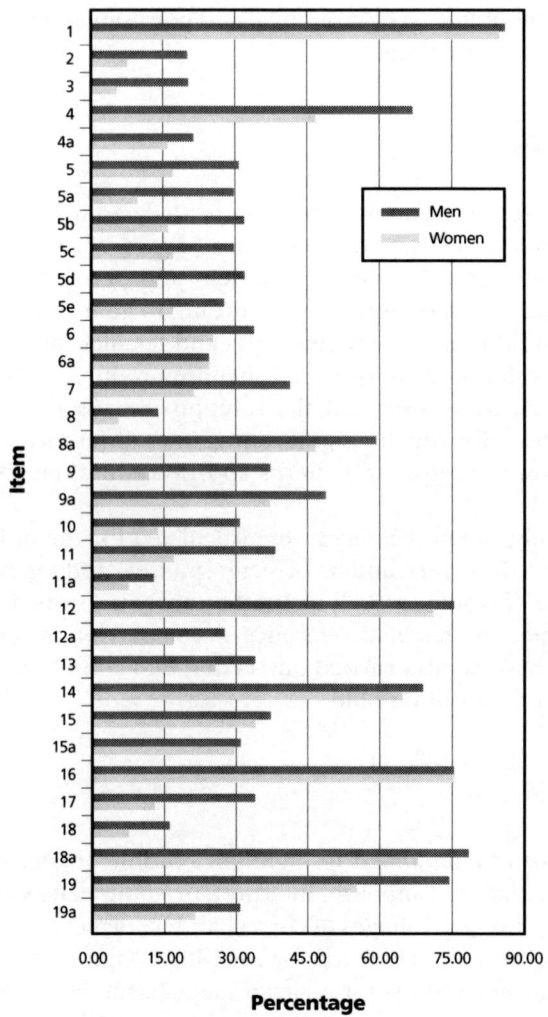

Figure 8.3 Percentage of correct answers.

steps to be taken in order to create a general rule. In exercise 14, the interpretation of variable as a specific unknown (U2) is required, in order to determinate its value performing arithmetic operations (U4) and substituting the value obtained which makes the equation a true statement (U3). Male and female students obtained high percentages in exercises 1, 12, and 16, which indicates that the majority of students can identify the presence of something unknown (I1) and symbolize an equation $ax = b$ (I5). They recognized the pattern that rules the squared numbers (G1) and could distinguish between that which varies and that which remains constant (G3).

They were also able to interpret the variable as general number (G2) and symbolize a sum of two expressions with those characteristics (G5).

Female students had more difficulties than male students in manipulating variables (G4) and in functional relationships (items 18a and 19), although the differences were not significant. While 79% and 74% of male students correctly answered questions 18a and 19, respectively, only 68% and 54%, respectively, of female students were able to do so. Difficulties were observed for male and female students in relation to 11 items (2, 3, 4a, 5b, 5d, 5f, 6a, 8, 11a, 12a and 18), which obtained less than 30% of correct answers. Table 8.4 shows the uses and aspects of variable involved in these questions.

In general, girls and boys had more difficulties with general numbers and related variables than with specific unknowns confirming our expectancies in the design of the questionnaire. In particular, they both had difficulties with the symbolization and interpretation of those two uses of variable.

Girls obtained less than 30% correct in 23 (2, 3, 4a, 5a, 5b, 5c, 5d, 5e, 5f, 6, 6a, 7, 8, 9, 10, 11, 11a, 12a, 13, 15a, 17, 18, and 19a) of the 33 items. Within those questions, nine required interpretation, six required symbolization, and seven required manipulation of variable in its three different uses. In contrast, males obtained less than 30% correct for only 11 items (2, 3, 4a, 5b, 5d, 5f, 6a, 8, 11a, 12a, and 18). Two of these items required interpretation of variable as unknown, three involved symbolization of variable as general number and one interpretation of general number, one required interpretation of variable in a functional relationship, while the others required the manipulation of related variables and the symbolization of variable.

One of the gender similarities was that both boys and girls obtained less than 30% correct in 9 items (2, 3, 4a, 5b, 5d, 5f, 6a, 8, and 11a). Five of these questions implied interpretation of variable as general number (G2), and four required manipulation of variable as general number (G4) in the solution process. Both females and males had difficulties with these aspects of variable.

With respect to the percentage of unanswered questions, Figure 8.4 shows that only exercise 4 was answered by all the students. This question

---

TABLE 8.4 Use and Aspect of Variable to Questions with Less Than 30% Correct Answers

Number of question	Use and aspect of variable
2, 3 and 12a	Symbolization of variable as general number (G5).
4a, 11a and 18	Symbolization of variable in a functional relationship (F6).
5b	Interpretation of variable as unknown quantity (I2).
5d y 5f	Interpretation of variable as general number (G2).
6a	Interpretation of variable in a functional relationship (F1).
8	Manipulation of variable as general number (G4).

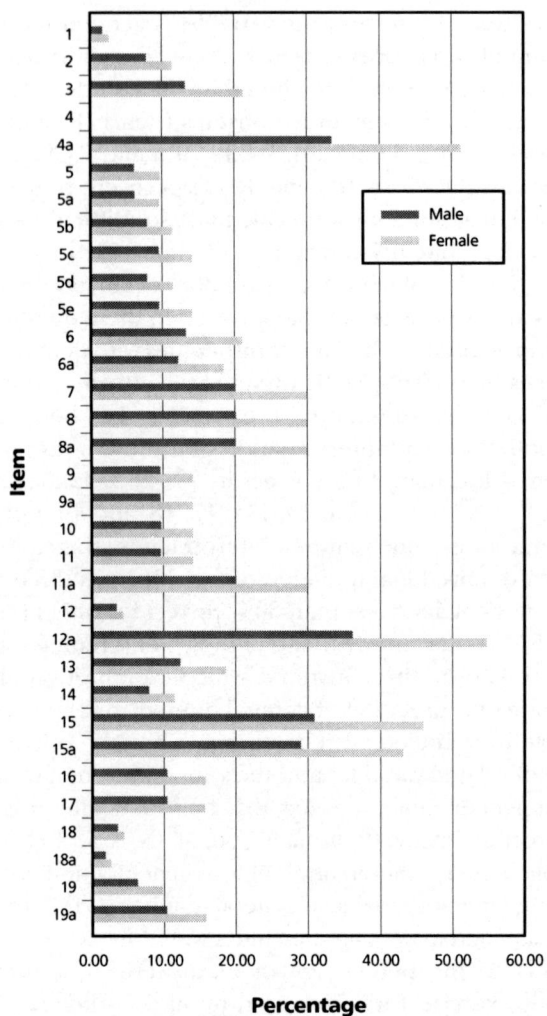

Figure 8.4   Percentage of non-answered questions.

required completion of a table in which values of the dependent variable and values of the independent variable needed to be determined. It can be observed that for all the other items, more female students decided not to answer. This result agrees with the one reported by Hanna (1989). An explanation might be that ability related to mathematical thinking that has been socially attributed to males has been internalized in such a way that boys feel the necessity to answer, or they feel obligated to do it. On the other hand, in the case of girls it is often considered that they do not need to be efficient in mathematics just because they are girls. As a result, low self-confidence in working with mathematics seems to be involved in their perception of their capability, skill, or competence in mathematics.

Although for the majority of questions, the difference in percentages is not remarkable, significant differences were found for items 4a, 12a, 15, and 15a ($p < 0.01$). Questions 4a and 12a require the symbolization of a functional relationship and of a general rule, respectively, while items 15 and 15a require the solution of systems of equations. A gender similarity was that these questions had the highest percentages of no answers. This indicates that both males and females had difficulties working with variables in a functional relationship, because in three of these items the work with this use of variable is involved.

In items 5a, 5b, 5c, 5d, 5e, and 5f we observe that the difference between percentages is greater (6.06% vs. 19.70%, 6.06% vs. 22.73%, 7.58%, vs. 24.24%, 9.09% vs. 25.76%, 7.58% vs. 27.27%, and 9.09% vs. 27.27, respectively), these questions implied the interpretation of variable in three different uses and it remarked significant differences.

The calculation of percentages of correct and incorrect answers and unanswered questions provided the opportunity to establish which were the easiest questions and which the most difficult for students. In order to obtain more information regarding the difficulties they experienced with the different uses of variable and its aspects, the results were analyzed per use of variable.

About the work with variable as specific unknown, if we consider the differences in the percentage of correct answers of the seven items that utilize this characterization of variable, we observe differences favoring males in five items (5b, 5e, 9, 9a, and 10). Three of them require as well the interpretation and manipulation of variable as general number (G2 and G4); these results show that girls had more difficulties than boys with these aspects. Only for questions 1 and 14 did females and males obtain more than 60% correct, showing fewer difficulties in symbolizing an equation such as $ax = b$ and determining the unknown quantity in an equation of the type $x + a = b$.

When they worked with variable as general number, in 8 of the 12 questions (2, 3, 5a, 5c, 5d, 5f, 8 and 8a), notable differences were observed in the percentages correct in favor of boys. Female students experienced difficulties in the interpretation and the manipulation of variable as a general number in these items.

When males and females worked with problems involving variable in a functional relationship, they had more difficulties than when working with a specific unknown or a general number (gender similarity). The percentages obtained show that girls had more difficulties than boys with this use of variable. In 9 of 13 questions, girls obtained less than 30% correct answers. Three problems (4, 18a, and 19) were answered correctly by more than 60% of male students. These problems required determining the value of the independent variable or dependent variable given the value of the other one (F2, F3).

For questions 4a, 6, 6a, 7, 11, 11a, 15, 15a, 18, and 19a, a higher percentage of correct answers favored males, with the exception of item 4a. These questions required the recognition of correspondence between all variables. Girls had difficulties in interpreting variables in a functional relationship. Questions 4a, 18, 19, and 19a required interpreting a variable as a general number together with correspondence between variables. This is probably another reason for girls' difficulties.

After calculating the percentage of correct answers, incorrect answers, and unanswered questions, the percentage of correct answers was calculated for each student (Table 8.5). This allowed us to observe the results from a different perspective. It can be observed (Table 8.5) that 16 males but only 5 females had more than 60% of their answers correct. The highest percentage for boys was 97%, while it was 94% for girls. This analysis showed that the majority of students obtained a percentage between 0 and 29%, followed by those who had between 30% and 59% of correct answers, while a small number of students had between 60% and 100%.

Although the majority of students (both boys and girls) obtained less than 30% of answers correct, 47% of males answered correctly less than 30% of the questionnaire, while more than 60% of females answered incorrectly 24 to 25 items.

Significant differences ($p < 0.01$) in favor of males were found in questions 5a, 5b, 5d, and 9. The items 5a and 5b require the interpretation of variable as an unknown, 5b requires the interpretation of variable as a general number, and question 9 implies obtaining the value of an unknown quantity in a linear equation in which the syntactical representation of the unknown value appears in both sides of the equation.

It was very interesting to observe that there were significant differences in the equation $13x + 27 - 2x = 30 + 5x$ (question 9), while no differences were found when solving the equation $4 + x = 2$ (item 9a), in which the variable does not need to be interpreted and manipulated as a general number, although the same aspects of variable as an unknown quantity are implicit (G2, G4, I2, I4, I3). Although significant differences were not found in the solution of item 9a, less than 50% of males and less than 40% of females solved the equation correctly.

TABLE 8.5 General Results on the Percentage of Correct Answers

	0–29%	30–59%	60–100%
Male students	31	19	16
Female students	42	20	5

The difficulties experienced by girls could be related to the integration of the different uses of variable in the same solution process. Three of these four questions (5a, 5b, 5d, and 9) require shifting between an unknown and a general number. The multiple appearances of the unknown in an equation could also have been a cause of difficulties because of the necessity to operate with it (Filloy & Rojano, 1989; Herscovics & Linchevsky, 1991).

To summarize, we want to stress that significant differences were found in twelve questions, of which half involved the interpretation of variable. Four of these items require the interpretation of variable as a general number and two imply the interpretation of variable as a specific unknown. One of these questions combines the three aspects (interpretation, manipulation and symbolization) of variable as a general number. Another one requires the interpretation of the variable in a functional relationship, focusing on the variation of variables and on the range of variation. The others combine two or more uses or aspects of variable where shifting between them and integrating them is required. This allows us to claim that we found significant gender differences when students worked with the three uses of variable, and in particular, the differences were significant when interpretation of variable in its three different uses was involved and when students had to interpret, symbolize, and manipulate variables in a functional relationship. Gender differences were remarkable in those exercises in which it was necessary to shift between one use of variable to another, for instance, in those exercises in which the variable needs to be interpreted as a general number and manipulated and subsequently reinterpreted as an unknown.

Analysis of Some Students' Responses

The purpose of this part of the analysis is to explore gender similarities and gender differences in relation to incorrect strategies of solution processes. Students' answers were analyzed using the theoretical framework.

We decided to analyze those questions for which we found significant differences in the statistical analysis; the first approach examined the percentages of correct answers. In Table 8.6 we show those data.

These exercises require interpreting the variable as an unknown, but initially the letter needs to be interpreted as a general number in order to manipulate it. One of the most frequent errors was to interpret the letter only as a general number. Both males and females answered in a similar way; 6% of them interpreted the variable as a general number and answered "∞," "any number," "infinity," or they gave a numerical sequence with ellipsis (see Figures 8.5, 8.6, 8.7, and 8.8).

TABLE 8.6 Percentages of Correct Answers by Gender

Aspect of variable and item	Analysis	Percentages of correct answers		Significant proof
		Males	Females	
(Interpretation of variable as unknown) For the following expression, how many values can the letter have? $3 + a + a = a + 10$	The variable given as general number (G2) needs to be interpreted, manipulated (G4) and finally considered as the representation of a specific value (I2).	29%	9%	Significant difference was found ($p < 0.01$).
(Manipulation of variable as unknown) Find the value of "x" in the following equation: $13x + 27 - 2x = 30 + 5x$	To find the unknown value, one has to interpret the variable as a general number (G2), manipulate it (G4) in order to reduce the expression and subsequently interpret it as representation of a specific value (I2); establish the unknown quantity by carrying out the algebraic and/or arithmetic operations (I4).	36%	11%	Significant difference was found ($p < 0.01$).

$3 + a + a = a + 10$ _____ ⌐

Figure 8.5 Interpreting letter as a general number.

$3 + a + a = a + 10$ _infinitos_

Figure 8.6 Infinity response.

$3 + a + a = a + 10$ _6, 5, 4, 3, 2, 1, ..._

Figure 8.7 Numerical sequence with ellipsis.

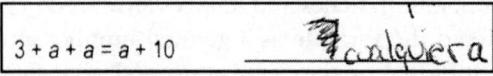
$3 + a + a = a + 10$ _____ cualquera

Figure 8.8 Response of any number.

The difficulty might be related to the multiple appearance of the variable as well as to the necessity to shift between viewing it as a general number that should be manipulated without knowing its value and viewing the variable as an unknown whose value has to be determined.

Another of the most common answers to this question was 3 or 3 values (see Figures 8.9 and 8.10); this answer may be related to assignation of the value (Küchemann, 1981) of the unknown quantity depending on the number of times it appears in the expression. We do not have evidence concerning their interpretation of the three quantities as the same quantity or as different quantities.

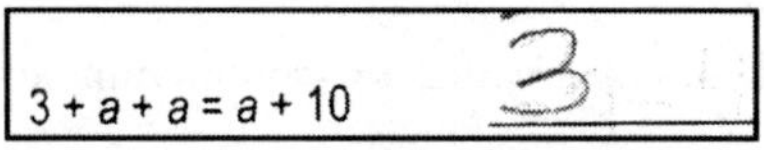

Figure 8.9 Response of 3.

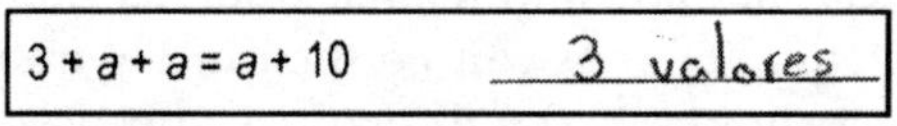

Figure 8.10 Response of 3.

In the same question two students (one male and one female) answered $3a = 10a$ showing a concatenation of the elements of the expression (see Figure 8.11). This shows students' difficulty in making sense of and operating with a variable. One boy who interpreted the variable as a general number contemplated negative numbers to indicate the symbolic representation of any quantity (see Figure 8.12). In contrast, one girl tried to manipulate the variable, considering it as a general number, although not appropriately (see Figure 8.13).

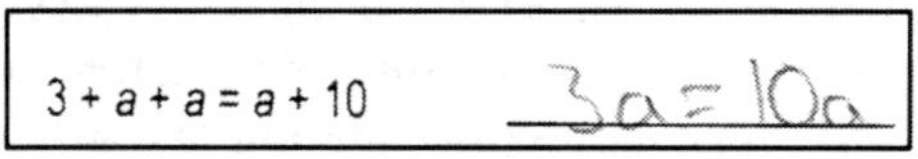

Figure 8.11 Concatenation.

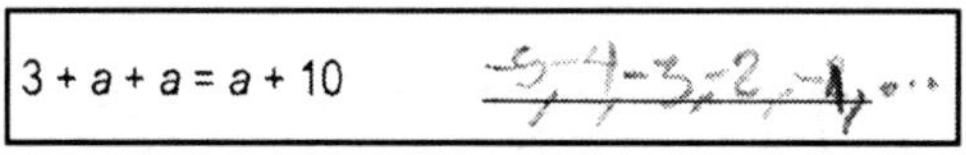

Figure 8.12 Sample of negative number use.

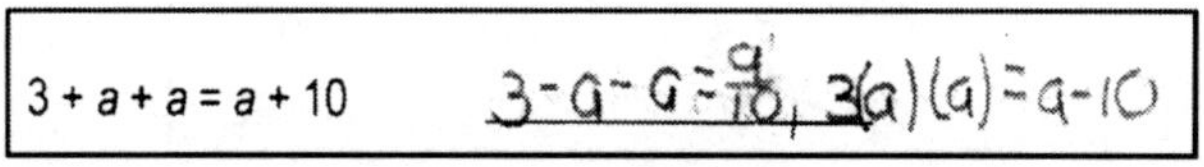

Figure 8.13 Use of variable as a general number.

In exercise 5d ($x + 5 = x + x$) about the interpretation of variable, the most frequent answer was to interpret the variable only as a general number without interpreting it as an unknown quantity. (Also we observed the concatenation of the elements of the equation obtained $5x = 2x$ females and males and the sameness was ignored.) It is possible that they considered different quantities for each letter in the same expression or 3 values because the variable appears 3 times. Once again, one female manipulated the variable as a general number but not properly (see Figure 8.14).

$$\frac{x}{5} = (x)(x), \quad x - 5 = \frac{x}{x}, \quad (x)5 = x - x$$

Figure 8.14 Interpretation of variable as a general number.

In question 9, where one needed to determine an unknown quantity of an equation in which the variable appears on both sides of an equation, the same percentage of male and female students showed knowledge in relation to linear equations. As can be seen in Figure 8.15, the solution process was correct; however, the difficulties appeared when they finalized the procedure.

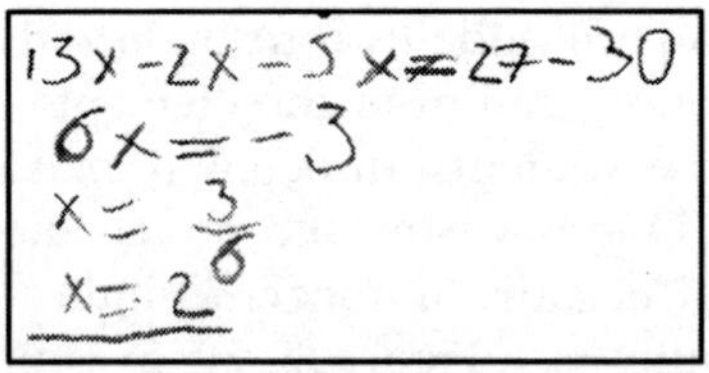

Figure 8.15 Final answer incorrect although correct procedure.

In this exercise, 4 males and 8 females changed the place of 3 and 6 to obtain a whole number (see Figure 8.16). In both cases the students were interpreting the variable as a general number (G2), manipulating it (G4), subsequently interpreting it as a representation of a specific unknown (I2), and determining its value with algebraic operations. Difficulties were observed when they carried out an arithmetic operation in the last step of the solution process.

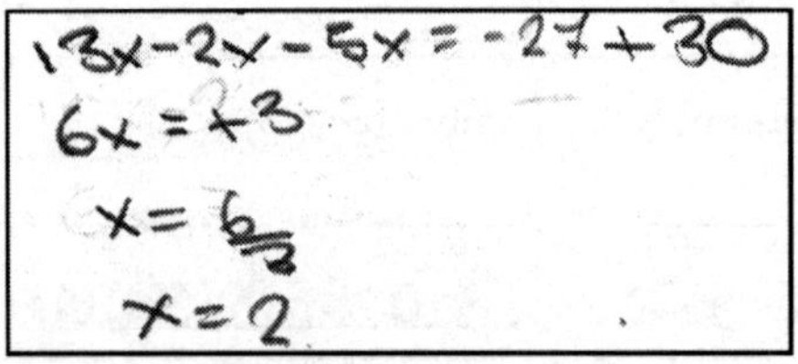

Figure 8.16 Interchanging of places of 6 and 3.

This suggested to us that they had difficulties accepting a fraction as the solution to an equation (probably because they usually worked with equations in which the results were wholes numbers or decimal numbers).

Question 3 required a symbolic representation of the perimeter of a geometric figure. The base is indicated by a number and one letter; students had to interpret the variable as a general number, manipulate it, and use it as an algebraic representation of perimeter. For this problem, males and females obtained different answers; this indicates to us that they had difficulties representing the perimeter of the figure algebraically; possibly they expected whole numbers and operations with them. Two males and four females answered $2h + 2b$, an indication that they used a formula to obtain the perimeter (double the height plus double the base) (see Figure 8.17). Another interesting answer given by females (not males) was $l + l + l + l$, using the expression for the perimeter of a square (see Figure 8.18). If females considered "l" as a general number, this answer showed a wrong conception because the letter can assume different quantities. Therefore we consider that they neglected the problem data.

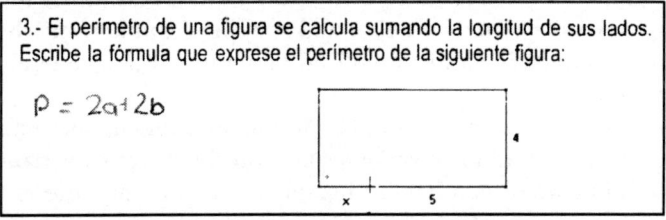

3.- El perímetro de una figura se calcula sumando la longitud de sus lados. Escribe la fórmula que exprese el perímetro de la siguiente figura:

$P = 2a + 2b$

Figure 8.17 Use of formula.

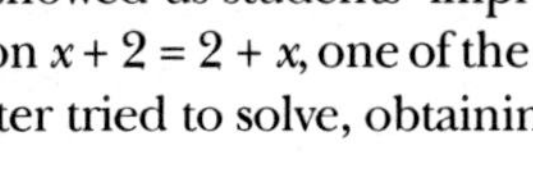

$P = L + L + L + L$

Figure 8.18 Use of formula for a square.

In those questions where the interpretation of variable as a general number was involved, the answers showed us students' improper sense of variable. For instance, with the expression $x + 2 = 2 + x$, one of the strategies was to consider it as an equation that they later tried to solve, obtaining $0 = 0$ (see Figure 8.19).

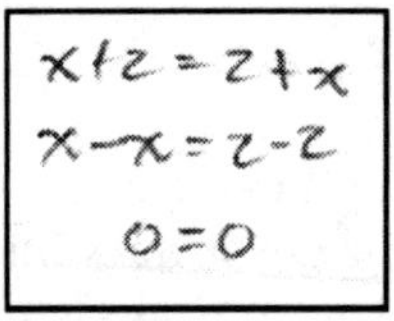

$x + 2 = 2 + x$
$x - x = 2 - 2$
$0 = 0$

Figure 8.19 Solution of equation with general number.

Finally, with respect to use of a variable in a relationship, more females than males did not recognize the relation between different values.

The purpose of this study was to explore gender differences working with three uses of variable. One of the most important findings of the last analysis was the difference between males and females in strategy solution; in addition, some algebraic difficulties were apparent.

CONCLUSIONS

All the students had more difficulties with variable as a general number and with related variables than with the variable as specific unknowns. In particular, they had difficulties with the symbolization and interpretation of these two uses of variable. We found gender differences between male and female students in 31 of the 33 questions. Males obtained higher scores than females, but these differences were significant only for 12 items.

The percentage of unanswered problems was higher for females than males. Significant differences favoring boys were found in the interpretation of variable as a general number and as a specific unknown, when interpretation, manipulation and symbolization of variable as a general number were required.

Considering the interpretation of variable in a functional relationship, focusing on the variation of variables and on the range of variation, girls had more difficulties than boys to flexibly move between different uses of variable. This finding allows us to establish that gender differences exist when students work with two or three uses of variable in the solution procedure. The differences were related to interpretation of variable in its three different uses and aspects. Clearly the gender differences were more significant in those exercises in which students need to shift between different uses of variable as facets of the same mathematical object and when they are required to integrate these different uses.

Male and female students tended to interpret the variable incorrectly. Frequently they interpreted the variable as an unknown when the interpretation of variable as a general number was required. For tautological expressions, female students interpreted the variable as unknown. The incorrect interpretation of variable in a specific mathematical context may lead to errors such as the concatenation of algebraic terms. Students in general had difficulties accepting a negative number or a fraction as a valid solution to an algebraic equation.

Students' difficulties can be linked to many factors. The complexity of the concept of variable, the type of questions used, and students' socio-economic and academic background surely influenced these results. However, the deep cultural gender differences that permeate Mexican society have, from one perspective, an important role in the differences we found between girls and boys. The dominant culture tends to put males into adaptative-instrumental roles and females into integrative-expressive roles for many years. Males and females are thought to have personality traits to go with these roles. Thus men who have to manipulate the society are stereotypically seen as assertive, rational, logical and competent, good at problem-solving, and interested in the world of objects and phenomena. On the other hand, women have to ensure good social relationships and are stereotypically seen as sympathetic, emotional, aware of others' feelings, tactful, good at expressing feelings, and, above all, interested in people and their concerns (Bem, 1974; Broverman, Broverman, Clarkson, Rosenkranz, & Vogal, 1972; Whitehead, 1994).

That conception of gender has not changed in the majority of Mexican communities and has surely influenced the results of this study. It is possible that ability relating to mathematical thinking, which has been socially attributed to males, has been internalized in such a way that men feel the necessity to answer all the items of the questionnaire while women, usually considered not efficient in mathematics, do not. Low self-confidence in mathematics seems to be involved in women's perception of their capability, skill or competence in mathematics.

Our findings have shown that there are gender differences when working with variables. More research is needed in this direction nationally and internationally.

REFERENCES

Atweh, B., & Cooper, T. (1995). The construction of gender, social class and mathematics in the classroom. *Educational Studies in Mathematics, 23*(3), 293–310.

Barberá, E. (1998). *Psicología del género.* Barcelona, España: Ariel Psicología.

Bednarz, N. & Dufour-Janvier, B. (1991). A study of external representations of change developed by young children. *Proceedings of the Thirteenth PME-NA Conference* (pp. 140–146). Blacksburg, VA: Virginia Tech.

Belenky, M.F., Clinchy, B.M., Goldberger, N.R., & Tarule, J.M. (1986). *Women's ways of knowing: The development of self, voice, and mind.* New York: Basic Books.

Bem, S. (1974). The measurement of psychological androgyny. *Journal of Counselling and Clinical Psychology, 42,* 155–169.

Benitez, E. (2004). *Los usos de la variable en algunos libros de texto de matematicas para la escuela secundaria.* Tesis de Maestria en Ciencias, DME-Cinvestav, Mexico.

Bjorg, O., & Sriraman, B. (2008). Exploring gender factors related to PISA 2003 results in Iceland: A youth interview study. *Zentralblatt fuer Didaktik der Mathematik, 40,* 591–600.

Bolea P., Bosch M., Gascon P. (1998) The role of algebrization in the study of a mathematical organization. *Proceedings of CERME 1, 2* (pp. 135–145). Osnabrueck, Germany.

Bosch, C., & Trigueros, M. (1996). Gender and mathematics in Mexico. In G. Hanna (Ed.) *Towards gender equity in mathematics education. An ICMI study* (pp. 277–284). Kluwer Academic publishers, Dordrecht, Boston, London.

Brandon, P. R., & Jordan, C. (1994). Gender differences favoring Hawai'i girls in mathematics achievement: Recent findings and hypotheses. *Zentralblatt fuer Didaktik der Mathematik, 94*(1), 18–21.

Brandon, P. R., Newton, B. J., & Hammond, O. W. (1987). Children's mathematics achievement in Hawaii: Sex differences favoring girls. *American Educational Research Journal, 24*(3), 437–461.

Burton, L. (Ed). (1990). *Gender and mathematics.* London: Cassell.

Broverman, I. K., Broverman, D. M., Clarkson, F. E., Rosenkranz, P. S., & Vogal, S. R. (1972). Sex-role stereotypes: A current appraisal. *Journal of Social Issues, 28,* 59–78.

Campbell, P. B. (1995). Redefining the "the girl problem in mathematics." In W.G. Secada, E. Fennema, & L. B. Adajian (Eds.), *New directions for equity in mathematics education* (pp. 225–241). New York: Cambridge University Press.

Campos, C. (2006). *Actitud hacia las matemáticas: Diferencias de género entre estudiantes de sexto de primaria y tercer grado de secundaria* (unplublished masters thesis). Cinvestav-IPN, Mexico: Department of Mathematics Education.

Chevallard, Y. (1983). Le passage de l'arithmétique á l'algébrique dans l'enseignement des mathématiques au Collège. *Petit X, 5,* 51–94.

English, L., & Sharry, P. (1996). Analogical reasoning and the development of algebraic abstraction. *Educational Studies in Mathematics, 30,* 135–157.

Ercikan, K., McCreith, T., & Lapointe, V. (2005). Factors associated with mathematics achievement and participation in advanced mathematics courses: An examination of gender differences from an international perspective. *School Science and Mathematics, 105*(1), 5.

Ethington, C. A, (1990). Gender differences in mathematics. An international perspective. *Journal for Research in Mathematics Education, 21,* 74–80.

Eudave, D. (1994). Las actitudes hacia las matemáticas de los maestros y alumnos de bachillerato. *Educación Matemática, 6*(1), 46–58.

Fennema, E. (1974). Mathematics learning and the sexes: A review. *Journal for Research in Mathematics Education, 5*(3), 126–139.

Fennema, E. (1990). Teachers' beliefs and gender differences in mathematics. In E. Fennema & G. C. Leder (Eds.), *Mathematics and gender* (pp. 169–187). New York: Teachers College Press.

Fennema, E., & Sherman, J. A. (1976). Fennema-Sherman mathematics attitude scales. Instruments designed to measure attitudes towards the learning of mathematics by females and males. *Catalog of Selected Documents in Psychology, 6*(1), 31.

Fierros, E. G. (1999, April). *Examining gender differences in mathematics achievement on the Third International Mathematics and Science Study (TIMSS).* Paper presented at the Annual Meeting of the American Educational Research Association, Montreal, Quebec, Canada.

Filloy, E., & Rojano, T. (1989). Solving equations: The transition from arithmetic to algebra. *For the Learning of Mathematics, 9*(2), 19–25.

Flores, F. (2000). *Psicología social y género. El sexo como objeto de representación social.* México City: McGraw-Hill.

Forgasz, H. J. (2002). Computers for learning mathematics: Equity factors. In B. Barton, K. C. Irwin, M. Pfannkuch, & M. O. J. Thomas (Eds.), *Mathematics education in the South Pacific. Proceedings of the 25th annual conference of the Mathematics Education Research Group of Australasia* (pp. 260–267). Auckland: MERGA Inc.

Forgasz, H. J., & Leder, G. C. (2000). The mathematics as a gendered domain scale. In T. Nakahara & M. Koyama (Eds.), *Proceedings of the 24th Conference of the International Group for the Psychology of Mathematics Education 2* (pp. 273–279). Hiroshima: Hiroshima University.

Francis, B., & Skelton, C. (2005). *Reassessing gender and achievement. Questioning contemporary key debates.* London and New York: Routledge.

González, R. M. (2003). Diferencias de género en el desempeño matemático de estudiantes de secundaria. *Educación Matemática, 15*(2), 129–161.

Hanna, G. (1989). Mathematics achievement of girls and boys in grade eight. Results from twenty countries. *Educational Studies in Mathematics, 20* (3), 225–232.

Harris, M. (1987). An example of traditional women's work as a mathematics resource. *For the Learning of Mathematics, 7*(3), 26–28.

Hart, K., Kerslake, D., Brown, M. L., Ruddock, G., Kůchemann, D. E., & McCartney, M. (1976). *Children's Understanding of Mathematics: 11–16.* London: John Murray.

Herscovics, N. & Linchevski, L. (1991). Crossing the didactic cut in algebra: grouping like terms in an equation. In R. G. Underhill (Ed.), *Proceedings of the Thirteenth PME-NA Conference* (pp. 196–202). Blacksburg, VA: Virginia Polytechnic Institute.

Hunt, F. (Ed.)(1987). *Lessons for life. The schooling of girls and women 1850–1950.* Oxford: Blackwell.

Johnson, R. M. (2000). *Gender differences in mathematics performance: Walberg's educational productivity model and the NELS:88 Database.* Paper presented at the Annual Meeting of the American Educational Research Association, New Orleans, LA, USA.

Juárez, J. (2002). *La comprensión del concepto de variable en profesores de secundaria.* Tesis Maestría en Ciencias. DME-CINVESTAV, México.

Jungwirth, H. (1991). Interaction and gender-findings of a microethnographical approach to classroom discourse. *Educational Studies in Mathematics, 22,* 263–284.

Kieran, C. (1984). A comparison between novice and more-expert algebra students on tasks dealing with the equivalence of equations. In J. Moser (Ed.). *Proceedings of the sixth annual meeting of PME-NA* (pp. 83–91). Madison: University of Wisconsin.

Kirshner, D. (1989). The visual syntax of algebra. *Journal for Research in Mathematics Education, 20,* 274–287.

Kůchemann, D. (1981). Algebra. In K. Hart (Ed.), *Children's Understanding of Mathematics: 11–16,* (pp. 102–119). London: John Murray.

Leahey, E. & Guo, G. (2001). Differences in mathematical trajectories. *Social Forces, 80*(2), 713–732.

Leder, G. (1974). Sex differences in mathematics problem appeal as a function of problem context. *Journal of Educational Research, 67*(8), 351–353.

Leder, G. (1992). Mathematics and Gender: Changing perspectives. In D.A. Grows (Ed.), *Handbook of Research on Mathematics Teaching and Learning* (pp. 597–633). Reston, VA: National Council of Teachers of Mathematics.

Leder, G. (1996). Equity in the mathematics classroom: beyond the rhetoric, In L.H. Parker (Ed.), *Gender science and mathematics* (pp. 95–104). Berlin: Kluwer Academic Publishers.

Leder, G. C. (2001). Pathways in mathematics towards equity: a 25 year journey. In M. van den Heuvel-Panhuizen (Ed.), *Proceedings of the 25th Conference of the International Group for the Psychology of Mathematics Education,* 1 (pp. 41–54). Utrecht, Netherlands: University of Utrecht.

Leder, G. C., & Fennema, E. (1990). Gender differences in mathematics. A synthesis. In E. Fennema & G. C. Leder. (Eds.), *Mathematics and gender* (pp. 188–200). New York: Teachers' College Press.

Leder, G., & Forgasz, H. (1992). Gender: A critical variable in mathematics education. In B. Atweh & J.M. Watson (Eds.), *Research in mathematics education in Australasia: 1988–1991* (pp. 67–95). Brisbane: Mathematics Education Research Group of Australasia.

Linn, M. C., & Pulos, S. (1983). Aptitude and experience influences on proportional reasoning during adolescence: Focus on male-female differences. *Journal for Research in Mathematics Education, 14*, 30–46.

Maccoby, E. E., & Jacklin, C. N. (1974). *The psychology of sex differences.* Stanford, CA: Stanford University Press.

Macgregor, M. & Stacey, K. (1996). Origins of students' interpretations of algebraic notation. In L. Puig & A. Gutiérrez (Eds.), *Proceedings of the 20th annual conference of the International Group for the Psychology of Mathematics Education 3* (pp. 297–304). Valencia, Spain: PME.

Macgregor, M., & Stacey, K. (1997). Students' understanding of algebraic notation: 11–15. *Educational Studies in Mathematics, 33*, 1–19.

Matz, M. (1980). Towards a computational theory of algebraic competence. *Journal of Mathematical Behavior, 3*(1), 93–166.

Meece, J. L., Parsons, J. E., Kazcala, C. M., Goff, S. B., & Futterman, R. (1982). Sex differences in math achievement: Toward a model of academic choice. *Psychological Bulletin, 92*(2), 324–348.

Mullis, I.V.S., & Stemler, S.E. (2002). Analyzing gender differences for high-achieving students on TIMSS. In D.F. Robitaille & A.E. Beaton (Eds.), *Secondary Analysis of the TIMSS Results* (pp. 277–290). Berlin: Kluwer Academic Publishers.

Phillip, R. (1992). The many uses of algebraic variables. *Mathematics Teacher, 85*(7), 55–561.

Ramírez. M. P. (2006). *Influencia de la visión de género de las docentes en las interacciones que establecen con el alumnado de en las clases de matemáticas.* Unpublished masters thesis, Cinvestav-IPN, Mexico, Department of Mathematics Education.

Rivera, M. (2003). *Diferencias de género en la visualización espacial; un estudio exploratorio con estudiantes de 2° de secundaria.* Unpublished masters thesis, Cinvestav-IPN, Mexico, Department of Mathematics Education.

Rubinstein, R. N. (1985). Computational estimation and related mathematical skills. *Journal for Research in Mathematics Education, 16,* 106–119.

Schoenfeld, A. H., & Arcavi, A. (1988). On the meaning of variable. *Mathematics Teacher, 74,* 418–427.

Swafford, J. (1980). Sex differences in first-year algebra. *Journal for Research in Mathematics Education, 11,* 335–659.

Trigueros, M., Reyes, A., Ursini, S., & Quintero, R. (1996). Diseño de un cuestionario de diagnóstico acerca del manejo del concepto de variable en el álgebra. *Enseñanza de las Ciencias, 14*(3), 351–363.

Trigueros, M., & Ursini, S. (1999). Does the understanding of variable evolve through schooling? In O. Zaslavsky (Ed.), *Proceedings of the 23rd annual conference of the International Group for the Psychology of Mathematics Education* (pp. 4-273–4-280). Haifa, Israel: Israel Institute of Technology.

Trigueros, M. & Ursini, S. (2001). Approaching the study of algebra through the concept of variable. In H. Chick, K. Stacey, J. Vincent & J. Vincent (Eds.) *Proceedings of the 12th ICMI Study Conference. The future of the teaching and learning of algebra 2* (pp. 598–605). Melbourne: The University of Melbourne.

Trigueros, M. & Ursini, S. (2003) First-year undergraduates' difficulties in working with different uses of variable. In A. Selden, E. Dubinsky, G. Harel & F. Hitt (Eds), *CBMS issues in mathematics education, Vol. 12* (pp. 1–29). Washington, DC: American Mathematical Society in cooperation with Mathematical Association of America.

Ursini, S. (1994). *Pupils' approaches to different characterizations of variable in Logo.* Unpublished doctoral dissertation, University of London, Institute of Education.

Ursini, S. (1996). Una perspectiva social para la educación matemática. La influencia de la teoría de L.S. Vygostky. *Educación Matemática, 8*(3), 42–49.

Ursini, S., Escareño F., Montes, D. & Trigueros, M. (2005). *Enseñanza del álgebra elemental: Una propuesta alternativa.* México City: Trillas.

Ursini, S., Ramírez, M. P., & Sánchez, G. (2007). Using technology in the mathematics class: How this affects students' achievement and attitudes. *Proceedings of the 8th ICTMT,* (Integration of ICT into Learning Processes) (CD-ROM). Czech Republic: University of Hradec Králové.

Ursini, S., & Sánchez, G. (2008). Gender, technology and attitude towards mathematics: a comparative longitudinal study with Mexican students. *ZDM Mathematics Education, 40,* 559–577.

Ursini, S., Sánchez, G., & Orendain, M. (2004). Validación y confiabilidad de una escala de actitudes hacia las matemáticas y hacia las matemáticas enseñada con computadora. *Educación Matemática, 16*(3), 59–78.

Ursini, S., & Trigueros, M. (1997). Understanding of different uses of variable: A study with starting college students. In E. Pehkonen (Ed.), *Proceedings of 21st Conference of the International Group for the Psychology of Mathematics Education 2,* (pp. 161–168). Finland: University of Helsinki.

Ursini, S., & Trigueros, M. (1998). Dificultades de los estudiantes universitarios frente al concepto de variable. In F. Hitt (Ed.), *Investigaciones en Matemática Educativa* (pp. 445–459). México City: Grupo Editorial Iberoamérica.

Ursini, S., & Trigueros, M. (2001). A model for the uses of variable in elementary algebra. In M. Van den Heuvel-Panhuizen (Ed.), *Proceedings of the 25th*

Conference of the International Group for the Psychology of Mathematics Education (pp. 4-327–4-334). Utrecht: Freudenthal Institute.

Ursini, S., & Trigueros, M. (2006). ¿Mejora la comprensión del concepto de variable cuando los estudiantes cursan matemáticas avanzadas? *Educación Matemática, 18*(3), 5–38.

Usiskin, Z. (1988). Conceptions of school algebra and uses of variables . In A. F. Coxford and A. P. Shulte (Eds.), *The Ideas of Algebra, K–12* (pp. 8–20). Reston, VA: National Council of Teachers of Mathematics.

Wagner, S. (1983). What are these things called variables? *Mathematics Teacher, 76,* 474–479.

Warren, E. (1999). The concept of a variable; gauging students´ understanding. In O. Zaslovsky. (Ed.), *Proceedings of the 23rd annual conference of the International Group for the Psychology in Mathematics Education* (pp. 4-313–4-320). Haifa, Israel: Israel Institute of Technology.

Whitehead, J. (1994). Academically successful schoolgirls: A case of sex-role transcendence. *Research Papers in Education, 9,* 53–80.

Zhang, L. & Manon, J. (2000, July). *Gender and achievement—Understanding gender differences and similarities in mathematics assessment.* Paper presented at the annual meeting of the American Educational Research Association, New Orleans, LA.

CHAPTER 9

FACTORS CONTRIBUTING TO GENDER DIFFERENCES IN MATHEMATICS PERFORMANCE OF UNITED STATES HIGH SCHOOL STUDENTS

Pamela L. Paek
*National Center for the Improvement
of Educational Assessment*

It is critically important that the mathematical capacity of students in the United States be improved. As the latest performance of American 15-year-olds on the 2003 Programme for International Student Assessment (PISA) demonstrates, the U.S. ranked 24th out of 29 developed nations in mathematics literacy and problem solving (National Academy of Sciences, National Academy of Engineering, Institute of Medicine, 2007). One reason for the poor performance of U.S. students may have been that they were not provided with appropriate instruction that successfully integrated mathematical content learning with experiences fostering understanding

International Perspectives on Gender and Mathematics Education, pages 203–224
Copyright © 2010 by Information Age Publishing
203

about self-regulation of their learning processes (National Research Council, 2006). Interestingly, only one of these 29 countries (Iceland) had results where girls outperformed boys (Lemke et al., 2004).

In addition, a major challenge for mathematics teachers is to understand what students know and what misconceptions deter them from solving problems correctly. We can infer that students who are higher achieving (as measured by grades and test scores) understand more than do students who are lower achieving, but that inference merely allows us to stratify them, not to deeply understand how they are engaging with the mathematics.

To improve teaching and learning, a first step is for teachers and students to better understand specifically what students do and think when they solve problems such as those that appear on large-scale assessments like the PISA, especially when there are consistent gender differences in performance.

Strengthening students' self-regulation of learning requires that teachers actively engage them in complex mathematical tasks that require them to construct their mathematics understanding in meaningful ways, as well as addressing their individual needs and learning styles (National Research Council, 2006). Otherwise, performing a mathematics task does not necessarily indicate that a student thoroughly understands it (Stevens & Casillas, 2006; Stevens & Soller, 2005; Stevens, Soller, Cooper, & Sprang, 2004).

Students can perform mathematical steps without making the connections necessary to transfer or apply those steps to a different context. To understand and solve a problem, students not only need to perform the correct steps but also to know a problem's content and a strategy for solving it (Polya, 1945). They must be motivated to tackle the problem and monitor their progress through the entire problem-solving process (Marshall, 1995). Instructional approaches that emphasize students taking responsibility for their own learning—that is, approaches that provide students with opportunities to choose and guide their own strategies and assess their progress—are key to student success in problem solving and in making the connections necessary to regulate their learning in new contexts (Inquiry Synthesis Project, 2006; National Research Council, 2006).

In the study reported here, I used an online computer system to examine males' and females' problem solving processes on SAT[1] mathematics items in order to capture what happens between presentation of an item and a student's final answer. I focused on SAT mathematics items because of the SAT's centrality and importance to college admission in the United States, and because research continues to confirm that males perform better than females on this assessment. This study investigated potential gender differences in the SAT test-taking process, to explain what differences in process and problem-solving strategies may be involved in these differences.

To understand the specific steps that students take when tackling different kinds of mathematical problems, Integrated Multi-Media Exercises

(IMMEX; www.immex.ucla.edu), an online library of multimedia problem-solving exercises, was used in this study. In addition to providing students with opportunities to engage in problem solving, IMMEX has a refined set of tools for monitoring students' performance and progress (Soller & Stevens, 2007; Stevens & Soller, 2005; Stevens et al., 2004). Students were presented with 31 SAT mathematics items on the IMMEX platform. For each problem, IMMEX provided students with various menus to choose from in a problem space—that is, a finite but representative set of concepts, numbers, and equations that students could combine to create a solution path. With these menus, students identified, defined, and represented the steps they took to solve each problem. This case-based paradigm provided sufficient information, with the same interface and resources, to ensure that students with diverse experiences could successfully synthesize information and thus solve the problems (Stevens & Palacio-Cayetano, 2003). By documenting the sequence of steps students took, and the time they spent on each step, IMMEX enabled a detailed investigation of procedures that students used to complete a task. How students were planning and regulating their problem solving could then be determined (Stevens & Casillas, 2006; Stevens & Soller, 2005; Stevens et al, 2004).

The purposes of this study were:

1. to identify the strategies that high school students use when solving mathematics problems
2. to analyze gender differences in strategy use on the SAT.

This study enabled a more thorough understanding of student test-taking behavior by providing a more detailed look at what students did before they chose a final answer. Ultimately, determining what enables higher-performing students to respond correctly and why students perform differently by gender can inform new ways of conceptualizing both assessment and instruction.

CONCEPTUAL FRAMEWORK

Gender Differences

Historically, studies document that high school males tend to take more—and more advanced level—mathematics classes than do females in the United States (American Association of University Women, 1992, 1998; U.S. Department of Education, 1997). However, more recent studies (Bozick & Ingels, 2008; Dalton, Ingels, Downing, & Bozick, 2007; Perkins, Kleiner, Roey, & Brown, 2004) have shown that this gender gap is indeed closing, with comparable course-taking by gender and actually higher performance (as measured by grade point averages) for females in these higher mathematics courses.

Regardless of comparable mathematics course-taking, males continued to outperform females on standardized assessments as shown by the 1997 Third International Mathematics and Science Study (TIMSS) (Mullis et al., 1998) and all years of the U.S. National Assessment of Educational Progress (NAEP) (Braswell et al., 2001; Mullis, Martin, Fierros, Goldberg, & Stemler, 2000; Grigg, Donahue, & Dion, 2007; U.S. Department of Education, 2007; U.S. Department of Education, Institute of Education Sciences, National Center for Education Statistics, National Assessment of Educational Progress [NAEP], n.d.). On both TIMSS and NAEP, these gender differences increase with age, and more males than females receive top scores in mathematics—trends most prominently seen in grade twelve (Braswell et al., 2001; Grigg et al., 2007; Mullis, et al., 2000; U.S. Department of Education, 2007; U.S. Department of Education, Institute of Education Sciences, National Center for Education Statistics, National Assessment of Educational Progress [NAEP], n.d.). More generally, research on gender differences in mathematics achievement has established that males generally perform better than females on standardized mathematics tests in the U.S. (American Association of University Women, 1992, 1998; Braswell et al., 2001; Coley, 2001; Gallagher & Kaufman, 2005; Mullis et al., 2000; Perie, Grigg, & Dion, 2005; Santapau, 2001). Geary, Saults, Liu, and Hoard (2000) also found that males outperformed females on more mathematics-fact retrieval tests like the SAT. However, more recent studies have indicated that these differences are negligible and that the mathematics capabilities of females and males are comparable (Hyde & Linn, 2006).

In an effort to explain gender differences in mathematics performance, researchers have investigated the mathematics problem-solving processes used by males and females. For instance, in some studies it was found that females' approaches to solving SAT mathematics problems were more algorithmic than the approaches taken by males (Gallagher, 1990, 1992; Gallagher & DeLisi, 1994; Gallagher et al., 2000; Gallagher & Kaufman, 2005). Gallagher's research showed that males tended to use innovative strategies more often than algorithmic processes taught in classrooms. These strategies accounted for males' higher performance on a wider variety of such mathematics items.

Standardized Testing

Standardized assessments are widely used in the U.S. to measure student achievement. Many of these assessments are considered "high stakes" because they publicly rank the performance of schools and various student groups, and because individual students' future education and career options can be affected by these test results. Because many programs and universities require students to achieve a minimum score to be considered for admission, high-stakes assessments are considered by some as gatekeepers for student admission to college, even more so for the most elite universities, or to specialized academic programs. As a result, standardized assessments are not reviewed favorably in some quarters, as reducing student knowledge into a single set of quantitative scores seems to deny the complexity of what students know. Nonetheless, standardized assessments are the most common tools for evaluating student performance in the United States.

The SAT is a primary example of a high-stakes assessment, as it plays a central role in many U.S. college admissions processes and decisions for granting scholarships. The SAT® I[2] consisted of a verbal and a quantitative component, each with three sections. Scores were reported in three numbers: a verbal score, a quantitative score, and a total score, which was the sum of the verbal and quantitative scores. The verbal and quantitative scores each had a minimum score of 200 and a maximum of 800, which meant the total score ranged from 400 to 1600.

Because the SAT has been used to rank students applying to college, much research has been conducted to ensure the SAT items discriminate well, meaning that the items correctly distinguish between those students who show they have the knowledge and skills to answer the item correctly and those who do not (e.g., Angoff, 1974; Katz, Friedman, Bennett, & Berger, 1996; Wainer, 1983). Thus, it is assumed that a higher score on the SAT is a valid indication of the more able students (Dorans & Lawrence, 1987). There has been much debate, however, regarding the validity of performance on the SAT items as indicators of how students will perform in *college* (American Association of University Women, 1992, 1998; Atkinson, 2001; Geiser & Studley, 2001), mainly due to differential performance on the assessment by various subgroups.

Problem Solving

When approaching mathematics problems, students rely on various resources and types of information (Chi & Glaser, 1985; Ericsson & Simon, 1993; Schoenfeld, 1988). These personal tools form the frameworks that students use to interpret and solve different problems. Ideally, students are able to identify and interpret a problem sufficiently well to choose an appropriate framework for solving it, resulting in a correct response. In reality, however, many students do not always know which framework to choose, or choose inappropriate frameworks, which results in inconsistent patterns of correct and incorrect responses (Marshall, 1995). Further, choosing an appropriate framework does not necessarily lead to a correct answer, because computational errors can also yield an incorrect response (Tatsuoka, 1990).

Every mathematics problem contains a host of concepts that can be linked to a variety of steps and processes to successfully solve the problem. When students choose to follow certain steps in a particular order, they demonstrate a pathway for organizing information to solve that problem. By assembling detailed information on students' responses to multiple problems, researchers can trace the steps that students take to solve each problem and evaluate how well their approaches worked on different types of items. This can be done at the individual student level and across various groups of students (e.g., by classrooms). Most students refine their problem-solving strategies over time, consistent with models of skill acquisition (Ericsson, 2004), gradually using fewer steps and eventually settling on a preferred approach (Stevens & Soller, 2005; Stevens et al., 2004). Researchers can analyze student self-regulation and self-monitoring of strategies by investigating these steps to provide a better understanding of the complexity of students' approaches to different mathematics items (Hartley & Bendixen, 2001; Song & Hill, 2007). This is especially important on high-stakes assessments, which tend to have huge accountability and classification outcomes, but tend to look only at students' correct and incorrect responses, rather than a detailed, empirical investigation of how students organize information.

How Students Self-Regulate Their Learning

Students who plan and self-regulate their learning know which strategies to employ more often than those who do not (Berardi-Coletta, Buyer, Dominowski, & Relinger, 1995; Gentner & Stevens, 1983; Halford, 1993; Hartley & Bendixen, 2001; Petrides, 2002; Shapley, 2000; Song & Hill, 2007). Students regulating their learning tend to use fewer unproductive steps and look ahead by assembling their procedures before acting on them (Barba, 1990; Hannafin, Hill, Oliver, Glazer, & Sharma, 2003; Petrides, 2002). These students tend to take a larger problem and break it down into smaller, more familiar tasks, thus simplifying its complexity. This strategy allows students to group or chunk information into concepts and hierarchies that facilitate good problem solving (Brown, 1978; Glaser, 1996; Siegler, 1988); thus, these students perform at a higher level and tend to receive higher test scores and grades.

Students with lower test scores and grades, on the other hand, exhibit more fragmentary knowledge and perform steps that are not conducive to, or consistent with, correct solutions. These students are not adept at regulating their learning mainly because they do not know which knowledge structures to combine (Brown, 1978; Glaser, 1996; Siegler, 1988). Students who do not regulate their learning usually do not reduce a problem into smaller, more manageable tasks (Barba, 1990). They tend to jump into the problem quickly, with little forethought given to how they will solve it. Their steps usually involve inconsistent applications or use of information irrelevant to solving the problem. Having few regulatory skills to monitor their performance, such students go through the steps of solving a problem without checking on the effectiveness of their strategy (Brown, 1978; Glaser, 1996). Students who are not strong regulators of their learning do not spend as much time trying to organize their thoughts before taking steps, resulting in many unproductive steps (Barba, 1990). Rather than approaching each problem individually and cognizant of potentially new and different constraints, these students make decisions on the basis of past experience. This leads to strategies being applied haphazardly and only occasionally resulting in a correct solution (Tatsuoka, 1990, 1993).

Creating a Context for Analyzing Problem Solving

Employing problem-solving contexts that allow for a detailed analysis of the strategies and steps students use provides more and stronger evidence of students' understanding of mathematics and self-regulation of their thinking (Lave, Smith, & Butler, 1988; Noddings, 1988; Schoenfeld, 1988). Analyzing students' problem-solving strategies and processes provides much greater insight into student understanding than does simply analyzing a right or wrong answer in isolation.

The process of a detailed task and error analysis involves gathering evidence of various solution paths, distractors, and misconceptions (Glaser, 1990; Marshall, 1995; Mislevy, Yamamoto, & Anacker, 1991). Given the myriad ways that students can organize information and regulate the steps they take to solve problems, an effort to truly understand what students are thinking and doing as they work through problems can seem insurmountable. Advances in technology, however, have provided one way to track in detail the actual steps students take to solve problems and the time they spend on each step (Hartley & Bendixen, 2001).

One pioneering technology that enables better understanding of student thinking is the Integrated Multi-Media Exercises (IMMEX) program, which draws on case-based (Schank, 1990) and production system (Newell & Simon, 1972) models of problem solving. Such a tool provides a more qualitative look at how students solve problems, since it captures the variety of approaches they take in intricate detail. This captured information opens the door to a deeper understanding of the comprehensive nature of student thinking. Rather than only analyzing a student's final answer, this tool allows researchers to look at each step that led to that answer.

IMMEX presents a problem to be solved in a *problem space*, that is, a space with a finite set of concepts, numbers, and equations that students must combine in order to create a solution path. Within the IMMEX problem space, various drop-down menus provide pathways from which the students can choose. A problem space typically will not encompass all the combinations that students could use to solve problems, but it does provide an essential preliminary view for better estimating what they are able to do (Tatsuoka, 1990, 1995).

Working in IMMEX, students can assess the *problem structure*—the key problem elements and their relationships needed to solve the problem— and then organize a mathematical *representation*—an arrangement of that information into a series of steps that can be used to solve the problem (Bennett, Morely, & Quardt, 1998). Most students want to arrive at an answer and will follow some process to produce one. If their chosen process leads to a wrong answer, they will probably try a different process if given a chance to try again. The standard default on IMMEX is to provide students with unlimited time and up to two chances to solve the problem before moving the student onto another problem. IMMEX software allows researchers to track all the steps—forward, backward, and sideways—that students take as they attempt to solve problems. IMMEX also records and displays the sequence of steps and the time spent on each one (Ericsson & Simon, 1993; Stevens, 1991).

METHOD

Although this study is primarily about online problem-solving processes, the SAT has traditionally been administered on paper. Moreover, this study analyzed—and compared by gender—the problem-solving strategies of 122 high school students on a set of SAT mathematics problems administered both on paper and in IMMEX. The researcher also interviewed a subset of students (4 females and 4 males) for their feedback and impressions about solving problems online, a mode that forced them to show their steps.

Participants

A total of 122 high school students (72 females and 50 males) from a single high school in Orange County, California, participated in this study. Most of the students were enrolled in Advanced Placement (AP)[3] Chemistry and were either in grade 11 or 12. Mathematics course-taking varied, including Algebra II, Pre-Calculus, or AP Calculus. It is important to note that students who take four years of mathematics in high schools using traditional curricula usually start with either Algebra I or Geometry in their freshman year, and tend to progress through the following sequence of courses: Algebra I, Geometry, Algebra II, Pre-Calculus, and Calculus.

Mathematics Grades in the Studied Population

Previous studies have shown that high school females receive better grades than males in the United States' mathematics classroom (Dwyer & Johnson, 1997; Kenney-Benson, Pomerantz, Ryan, & Patrick, 2006; Pomerantz, Altermatt, & Saxon, 2002). In this study, I analyzed grades for participating students in Algebra I and Geometry in both the first and second semesters to see if there were gender differences. We selected these two courses because most SAT mathematics items are based on students' knowledge of Algebra I and Geometry.

Two-tailed t-tests were used to compare differences by gender. As seen in Table 9.1, the females tended to have higher grades than the males in the two mathematics courses. In all four semesters, females outperformed males, although the differences were not statistically significant ($p > 0.05$). In addition, males exhibited more variability in grades than females, as shown in the higher standard deviation (SD) for all four semesters.

TABLE 9.1   Algebra I and Geometry Grades in the Study's Student Population

Mathematics Course	Gender	N	Mean	SD
Algebra I, first semester	Female	71	3.69	0.49
	Male	49	3.51	0.74
Algebra I, second semester	Female	71	3.64	0.58
	Male	49	3.50	0.67
Geometry, first semester	Female	71	3.64	0.57
	Male	49	3.61	0.62
Geometry, second semester	Female	71	3.62	0.58
	Male	49	3.53	0.66

PROCEDURE

Participants first completed a released SAT from the 10 Real SATs (College Board, 1997) under standard SAT time constraints using paper and pencil. Next, within a month of the paper SAT, students used IMMEX to solve 31 SAT mathematics problems that came from the same SAT the students had completed for the first part of the study. These items were chosen as representative of the content in the 60 items students had to solve on the paper version. Students were told they were not under time constraints to complete the IMMEX problems. The reason we did not use all 60 SAT items was because of teachers' input indicating that students would not complete all 60 items. Thus, we gave a number we thought most students would be willing to complete.

Developing the IMMEX Problem Space

I created a specific problem space in IMMEX for solving each of the 31 mathematics items along with a method for using and coding the steps that students used to problem solve. The problem space included formulas, definitions of concepts, and a breakdown of the process for arriving at a solution. The problem space for this study was developed based on talk-aloud protocols that the researcher had already adopted with 20 students who discussed and/or wrote down the steps they used to solve certain mathematics items. Additionally, I incorporated into the problem space common errors that students make in arithmetic, basic algebra, and geometry (Tatsuoka, 1990, 1995). These two elements (the formal task analysis and the incorporation of common errors) helped to determine the menus and submenus needed for the IMMEX platform in this study so that the majority of students could solve the problems using the information given (Mislevy, Yamamoto, & Anacker, 1991).

The problem space was the same for each of the 31 items in IMMEX, except for some of the submenus, which were changed to correspond with associated numbers and equations related to each problem. The problem space included the mathematics concepts necessary for a correct solution, as well as bugs and distractors that were included to track where students made arithmetic errors or had misconceptions about the problem. Students could easily navigate through all the menus and submenus and still not be able to correctly solve a problem. To reach an accurate solution, they needed to know what kinds of information were pertinent and be able to order that information correctly. Within the IMMEX problem space, each problem was presented with five main menus. The problem and the five main menus were always at the top of the screen, even when students navigated through the submenus. Each main menu represented one of five mathematics concepts: arithmetic, angles, area, perimeter, and solving equations. Clicking on one of these menus revealed a host of submenus also representing math concepts. Clicking on a submenu led to a series of equations and/or numbers. These equations/numbers were represented in expanded form, so that the student had to decide where to combine or collapse terms. The menu structure included shortcuts so that students could collapse several steps into one. For example, consider the equation $2x + 10 = 5 - 3x$. The traditional way of solving it would be to either move the numbers to one side or the variables to one side as a first step. To illustrate the former, a student could first move the numbers to one side by subtracting 10: $2x + 10 - 10 = 5 - 10 - 3x$. To get the variables on one side of the equal sign, the next move would be to add $3x$ to both sides: $2x + 3x = -5 - 3x + 3x$. Then both sides of the equation would be divided by 5: $5x/(-5) = -5/5$ to arrive at the final response of $x = -1$. In IMMEX, the menu shortcuts enabled students to collapse these three steps, computing the information in their heads so they could move to the answer in one step.

The IMMEX problem space also included common arithmetic errors students could make that were associated with the distractors offered on the paper-and-pencil test. In the problem above, for example, students could incorrectly subtract $3x$ so that the response would be $2x - 3x = -5$, resulting in a final answer of $x = 1$. These types of mistakes were included as options in the submenus and the equation structures. The purpose of these incorrect paths was to document where students might go wrong in determining their final answer choice, to find out where errors may occur, not just that the final answer was incorrect or correct.

After completing all the work and arriving at an answer, students entered their responses after clicking the "solve problem" button. The study was structured to give students two chances to solve the problem so that if they got the problem wrong in the first try, they could reconsider each step they had taken and try again. This allowed me to have an opportunity see how students revised their steps to answer the problem correctly on the second try. The second chance also allowed students who might have made a simple arithmetic error to backtrack and correct it.

RESULTS

Males scored about 20 points higher ($\overline{x} = 642.12$, $sd = 73.946$) than did females ($\overline{x} = 621.43$, $sd = 69.372$) on the paper-and-pencil SAT mathematics section, but a two-tailed t-test indicated that this difference was not statistically significant. Note that the average SAT mathematics score in this study is more than 100 points above the national SAT mathematics average of 515

(College Board, 2007). This higher achievement may be due to the fact that at the time of the study, most (84%) of the participants were enrolled in at least one Advanced Placement course (mainly AP Chemistry).

I conducted analyses of the amount of work shown on the paper version of the SAT mathematics section and analyses of students' problem solving on IMMEX: the number of steps students took to solve each problem correctly, the amount of time spent on each step and on each problem, and the number of attempts made at solving each problem. All of these analyses revealed differences in females' and males' performance.

Test Booklet Analysis

Each student's test book was coded by the work they showed and the number of steps they used in the mathematics sections. The student markings on the test booklets provided insights into the different ways that students approached the problems. Since students did not mark their test books for every problem, only the items where they showed their work could be coded for analysis. Note that students were not asked to show their work; they were asked to take the test and use the materials they were given as they would in a live testing situation. Table 9.2 shows the number of students who showed work by gender. Almost all students showed work on the test booklets. The mean in Table 9.2 indicates the average number of items for which students showed work.

Females tended more often than males to show their work, across each of the three sections of the SAT. Totaling the number of items in which females showed work across all three sections, females showed their work

TABLE 9.2 Average Number of Items with Work Shown on the Paper SAT by Gender

Showing Work	Gender	N	Mean number of items showing work	SD
Section 1	Female	69	14.22	5.67
	Male	48	11.38	6.05
Section 3	Female	67	14.60	5.51
	Male	48	12.77	6.60
Section 6	Female	69	4.93	2.76
	Male	47	4.19	2.65

Note: In sections 1 and 3, the total number of items was 25 per section, whereas section 6 consisted of 10 items. The total number of SAT math items is thus 60.

on almost 34 of the 60 items on average, while males showed their work on 28 items. Although two-tailed t-tests indicated these differences were not statistically significant ($p > 0.05$), they were illustrative of how time might be spent during timed high-stakes test taking, and might be a contributing factor to the gender gap in mathematics achievement on such assessments.

IMMEX: Number of Steps Taken

Using the data IMMEX collected, the number and types of steps the students took to solve each problem were analyzed. In general, the fewer steps a student had taken in attempting to solve a problem, the more likely it was that the student had solved the problem correctly—as students who unsuccessfully completed a problem tended to take more irrelevant steps that were not helpful to solving the problem. In addition, the number of steps taken differed between males and females. On average, males took two fewer steps than females did to solve a problem: Males averaged about four steps ($\bar{x} = 3.91$, sd = 2.06), whereas females averaged about six steps ($\bar{x} = 6.12$, sd = 1.77). Even with these differences, females attempted more IMMEX items, took longer to solve each problem, and answered more problems correctly than males did; the reason for the higher success of females on these items appears to be that they verified their steps, not that they were inefficiently taking extra steps. On the paper-and-pencil SAT mathematics sections, however, females scored lower than did the males. The IMMEX test had no time constraints, so it appeared that females performed better in the untimed situation than males did. One could also argue that students had 'clues' from the submenus of the problem space; however, students would need to get to the appropriate top menus to proceed.

Informal interviews with the participants suggested that the females liked to be sure of their answers and would use any available information to verify them. They wanted their answers to be correct on the first attempt, and they took more steps and more time to ensure correctness. One female indicated, "I don't like getting a wrong answer. I want to make sure I get it right and look at my work very carefully. If I get it wrong the first time, I tend to start over from the beginning." The interviews suggested that males, on the other hand, tended to be inclined toward an answer and would select it, knowing they had a second chance in IMMEX if they got it wrong. This method resulted in fewer steps and less time taken per problem. As one male said, "It's not like it counts for anything [if you get it wrong the first time], so guessing seems ok. And it means I get done faster."

IMMEX: Amount of Time Spent per Step and per Problem

I also analyzed time spent per step and per problem. Females tended to take four seconds more per step ($\bar{x} = 21$ seconds, $sd = 21$) than did males ($\bar{x} = 17$ seconds, $sd = 18$). Given that females also took more steps per problem, on average females took about 55 seconds more per problem ($\bar{x} = 2{:}06$ steps, $sd = 2{:}51$) than males ($\bar{x} = 1{:}11$ steps, $sd = 1{:}45$). Two-tailed t-tests indicated that these differences were statistically significant ($p < 0.01$). This difference in time spent per step, coupled with the fact that females took more steps than males did, may well help to explain why there was the general finding that females tended to score lower on standardized mathematics tests under timed conditions. Females might not be able to complete as many problems in the allocated time for the test, and the time constraint on the paper version of the SAT posed a larger challenge to them than to males given their preferred methods of solving problems.

I also analyzed differences in time spent per step and per problem according to whether the problem was solved correctly or incorrectly. Problems that a student answered incorrectly on both attempts were considered to be incorrect. A t-test showed that the time differences on problems that students answered correctly versus those answered incorrectly were statistically significant, with the time per step and per problem en route to correctly solved problems each measuring significantly less time than the time spent on problems that were not solved correctly. On average, correctly solved problems took students four seconds less per step ($\bar{x} = 16$ seconds, $sd = 28$) than did unsolved problems ($\bar{x} = 20$ seconds, $sd = 32$). The total time students spent on solved problems ($\bar{x} = 1$ minute:34 seconds, $sd = 1{:}41$) was 30 seconds less than the time they spent on problems they did not answer correctly ($\bar{x} = 2$ minute:04 seconds, $sd = 2{:}30$). This difference was statistically significant ($p < 0.05$). In summary, it appeared that problems students could not solve required more time because more time was spent searching the problem space, possibly looking for clues to help solve the problem. Presumably the approach for these problems was not as clear-cut as it was for the problems students did solve correctly.

IMMEX: Number of Attempts Made

Students were given two opportunities to solve each IMMEX problem, so I could document the changes they made in their steps from the first to the second attempt. The majority (61%) of students solved the problems correctly on the first try, and an additional 23% answered the problems correctly on their second try. The steps students took on the second tries for both correct and incorrect answers were analyzed. Students who correctly solved a problem on the second try demonstrated an orderly process in which they deliberately retraced their steps to verify their answers and more systematically regulated their steps, indicating that these students knew what they were doing but had made a small error in computation at some point. Students who did not solve the problem correctly on the second try, on the other hand, showed less organization and planning in their process.

Observing the processes used by the participants gives an inside view of how they regulated their learning as they solved problems and suggests reasons for the performance differences documented between females and males. The amount of time and number of steps to solve each problem varied between males and females, with females taking extra steps to verify their answers and therefore taking more overall steps per problem than males. This verification process resulted in females spending more time on each problem than did males. The extra steps and time, however, paid off in that the females performed slightly better overall than the males on the IMMEX problems in the untimed condition.

LIMITATIONS OF THE STUDY

This study is not generalizable to all high school students, since the final sample of students who took both the paper SAT and completed the online IMMEX problems were mainly high-achieving students in Advanced Placement courses who achieved scores much higher than average SAT scores. However, it can still be seen that even among high-performers, females demonstrated the need to use an extra step to verify their mathematics work before choosing a final answer, and it is not unreasonable to assume that this would also be found within the larger female population.

Additionally, this study does not purport to analyze more complex mathematics problems, but merely those that are part of the high-stakes SAT assessment arena. The reason for choosing IMMEX was because many students do not show their work (even when asked) and have difficulty articulating their processes and their work steps in talk-aloud protocols. Using technology in this study allowed me to look at the time per step and per problem, as well as the number and types of steps. These analyses could not have been conducted without the technology.

Given that that the problem space of IMMEX was constructed to encompass many, but not all, ways students could approach the mathematics items, the process for how students would choose their steps was somewhat artificial. However, the students were able to navigate the problem space and solve problems despite this limitation.

Last, the use of two attempts to correctly solve each problem provided students with some feedback, in that if they would know that the first answer was wrong, and that they could have another attempt at solving the item. It is possible that the shorter time males took compared to females in correcting their mistakes may only be typical in this setting and not representative of what would happen if no feedback were provided. However, using this option provided verification of males' higher risk-taking behavior compared to females' when choosing a final answer, since females took more steps and more time to trace their steps.

DISCUSSION

Males seemed more able than females to solve mathematics problems efficiently. Males tended to collapse or skip steps more often than did females, thus spending less time and taking fewer steps per problem. IMMEX analyses confirmed that females took more steps to validate their correct responses than did males, which, under timed conditions, may limit the number of items to which females could respond, and thus may result in females correctly solving fewer problems than males. Females' lower scores on the mathematics sections of the timed paper-and-pencil SAT seemed to be a result of the pressure to complete in a certain time.

In the untimed IMMEX situation, females outperformed males on the same items that they had been presented with in the timed tasks. Females tended to follow algorithms, verifying their steps and the answer before completing a problem. An informal talk with a few students revealed greater frustration among females than among males with getting a wrong answer on the first try. Males felt that in the IMMEX system they still had a second chance and were consequently less concerned. Females may have been more concerned because they had already planned their work carefully and therefore would have to be even more meticulous to find out why their answers were incorrect.

The findings of this study show the importance of understanding when solving mathematics problems. Tracing students' steps illuminated some of the possible reasons that females' performance on high-stakes mathematics tests are typically lower than males' scores: mainly, their need to verify their answer as correct before proceeding to the next problem. Finding out what students know no longer needs to be surmised only from final answers and scores, as the use of IMMEX in this study demonstrates. Researchers can and must continue to probe the processes that students employ and the knowledge they bring to bear when confronted with a mathematics problem. The more deeply educators and researchers can analyze student thinking, the better the measures obtained of students' competence, knowledge, and abilities—and thus there can be better design of tools and practices for effective instruction and assessment.

It appears that current high-stakes assessments in timed situations are under-predicting what females are able to do in mathematics. This study shows that in untimed situations without multiple-choice options, females perform slightly better than males on a subset of the same items they took in a simulated high-stakes situation. From this work, it appears that females and males have different modes of approaching and thinking about mathematics, and yet these differences are not accommodated in current assessment techniques. If assessments continue along their current path of being timed, the gender gap will continue to exist. Females are obviously showing strength in the way they can accurately verify their work and solve problems correctly, and should be provided with high-stakes testing opportunities that better reflect their actual performance. In other words, in untimed high-stakes testing situations, it may be possible to find females performing higher than they currently do with time-restricted assessments.

The fact that students had a second attempt to solve a problem provided insight into how students planned and reflected on their work. An under-utilized strategy in instruction is having students plan their steps before they actually begin to solve a problem and then having them reflect on their plans and actions. In fact, a four-stage problem-solving process (Polya, 1945) is recommended because it emphasizes the steps students should take for successful problem solving: (1) understanding the problem, (2) devising a plan to solve the problem, (3) implementing a solution plan, and (4) reflecting on the problem. If mathematics teachers are currently not using these steps, I would encourage their use to empower students to regulate their own learning. Students can reflect on and review their previous work and, if necessary, plan a better set of steps for upcoming problems. This would allow for more purposeful action in problem solving.

NOTES

1. The SAT, once called the Scholastic Aptitude Test, is also now known as the SAT® Reasoning Test. At the time this study was conducted (in the 2000–2001 academic year), the SAT was also known as the SAT®I.
2. The SAT®I was used in this study. It differs from the currently administered SAT® Reasoning test, which consists of three scores, each with a minimum score of 200 and a maximum of 800: critical reading, mathematics and writing. The SAT® Reasoning Test replaced the SAT®I in 2005.
3. AP is dedicated to providing high school students opportunities to enroll in college level courses while in high school. Historically, students in AP courses are the highest achieving on standardized assessments and most advanced in terms of course-taking and grade point average.

REFERENCES

American Association of University Women. (1992). *The AAUW report: How schools shortchange girls*. Researched by the Wellesley College Center for Research on Women. Washington, DC: Author.

American Association of University Women. (1998). *Gender gaps: Where schools still fail our children*. Washington, DC: Author.

Angoff, W. H. (1974). *Criterion-referencing, norm-referencing, and the SAT* (RR-74-01). Princeton, NJ: Educational Testing Service.

Atkinson, R. C. (2001). Achievement versus aptitude tests in college admissions. *Issues in Science and Technology 18*(2). Retrieved March 2, 2008 from http://works.bepress.com/richard_atkinson/28

Barba, R. H. (1990). Problem solving pointers. *Science Teacher, 57*(7), 32–35.

Bennett, R. E., Morely, M., & Quardt, D. (1998). *Three response types for broadening the conception of mathematical problem solving in computerized-adaptive tests*. (RR 98-45-ONR). Princeton, NJ: Educational Testing Service.

Berardi-Coletta, B., Buyer, L.S., Dominowski, R. L., & Relinger, E. R. (1995). Meta-cognition and problem solving: A process-oriented approach. *Journal of Experimental Psychology: Learning, Memory, and Cognition, 21*, 205–223.

Bozick, R., & Ingels, S. J. (2008). *Mathematics coursetaking and achievement at the end of high school: Evidence from the Education Longitudinal Study of 2002 (ELS:2002)* (NCES 2008-319). Washington, DC: National Center for Education Statistics, Institute of Education Sciences, U.S. Department of Education.

Braswell, J. S., Lutkus, A. D., Grigg, W. S., Santapau, S. L., Tay-lim, B. S. H., & Johnson, M. S. (2001). *The nation's report card: Mathematics 2000*. Washington, DC: National Center for Education Statistics.

Brown, A. L. (1978). Knowing when, where, and how to remember: A problem of metacognition. In R. Glaser (Ed.), *Advances in instructional psychology*, Vol. 1 (pp. 77–165). Mahwah, NJ: Lawrence Erlbaum.

Chi, M. T. H., & Glaser, R. (1985). *Problem-solving ability*. Report no. LRDC-1985/6. Pittsburgh, PA: University of Pittsburgh Learning Research and Development Center.

Coley, R. (2001). *Differences in the gender gap: Comparisons across racial/ethnic groups in education and work*. Princeton, NJ: Educational Testing Service, Policy Information Report.

College Board. (1997). *10 real SATs*. Forrester Center, WV: Author.

College Board. (2007). *2007 college-bound seniors total group profile report*. New York, NY: Author.

Dalton, B., Ingels, S. J., Downing, J., & Bozick, R. (2007). *Advanced mathematics and science coursetaking in the spring high school senior classes of 1982, 1992, and 2004.* (NCES 2007-312). Washington, DC: National Center for Education Statistics, U.S. Department of Education.

Dorans, N. J., & Lawrence, I. M. (1987). *The internal construct validity of the SAT* (RR-87-36). Princeton, NJ: Educational Testing Service.

Dwyer, C., & Johnson, L. (1997). Grades, accomplishments, and correlates. In W. Willingham & N. Cole (Eds.), *Gender and fair assessment* (pp. 127–156). Mahwah, NJ: Erlbaum.

Ericsson, K. A., & Simon, H. A. (1993). *Protocol analysis: Verbal reports as data* (Rev. ed.). Cambridge, MA: MIT Press.

Ericsson, K. A. (2004). Deliberate practice and the acquisition and maintenance of expert performance in medicine and related domains. *Academic Medicine, 79*(10), 70–81.

Gallagher, A. M. (1990). *Sex differences in the performance of high-scoring examinees on the SAT-M* (RR 90-27). Princeton, NJ: Educational Testing Service.

Gallagher, A. M. (1992). *Strategy use on multiple-choice and free-response items: An analysis of sex differences among high-scoring examinees on the SAT-M* (RR 92-33). Princeton, NJ: Educational Testing Service.

Gallagher, A. M., & De Lisi, R. (1994). Gender differences in the Scholastic Aptitude Test-Mathematics problem solving among high-ability students. *Journal of Educational Psychology, 86*(2). 204–211.

Gallagher, A. M., DeLisi, R., Holst, P. C., McGillicuddy-DeLisi, A. V., Morely, M., & Cahalan, C. (2000) Gender differences in advanced mathematical problem solving. *Journal of Experimental Child Psychology, 75,* 165–190.

Gallagher, A. M., & Kaufman, J. C. (2005). *Gender differences in mathematics.* New York: Cambridge University Press.

Geary, D., Saults, S. J., Liu, F., & Hoard, M. K. (2000) Sex differences in spatial cognition, computational fluency and arithmetical reasoning. *Journal of Experimental Child Psychology, 77,* 337–353.

Geiser, S., & Studley, R. (2001). *UC and the SAT: Predictive validity and differential impact of the SAT I and SAT II at the University of California.* Oakland, CA: University of California Office of the President.

Gentner, D., & Stevens, A. L. (1983). *Mental models.* Hillsdale, NJ: Erlbaum.

Glaser, R. (1990). Toward new models for assessment. *International Journal of Educational Research, 14,* 475–483.

Glaser, R. (1996). Changing the agency for learning: Acquiring expert performance. In A. Ericcson (Ed.), *The road to excellence: The acquisition of expert performance in the arts and sciences, sports, and games* (pp. 303–311). Mahwah, NJ: Erlbaum.

Grigg, W., Donahue, P., & Dion, G. (2007). *The nation's report card: 12th-Grade reading and mathematics 2005* (NCES 2007.468). U.S. Department of Education, National Center for Education Statistics. Washington, D.C.: U.S. Government Printing Office.

Halford, G. S. (1993). *Children's understanding. The development of mental models.* Hillsdale, NJ: Erlbaum.

Hannafin, M. J., Hill, J. R., Oliver, K., Glazer, E., & Sharma, P. (2003). Cognitive and learning factors in web-based distance learning environments. In M. G. Moore & W. G. Anderson (Eds.), *Handbook of distance education* (pp. 245–260). Mahwah, NJ: Erlbaum.

Hartley, K., & Bendixen, L. D. (2001). Educational research in the Internet age: Examining the role of individual characteristics. *Educational Researcher, 30*(9), 22–26.

Hyde J. S., & Linn, M. C. (2006). Gender similarities in mathematics and science. *Science, 314,* 599–600.

Inquiry Synthesis Project, Center for Science Education, Education Development Center, Inc. (EDC) (2006, April). *Technical report 2: Conceptualizing inquiry sci-*

ence instruction. Retrieved December 13, 2007 from http://cse.edc.org/products/inquirysynth/pdfs/technicalReport2.pdf

Katz, I. R., Friedman, D. E., Bennett, R. E., & Berger, A. E. (1996). *Differences in strategies used to solve stem-equivalent constructed-response and multiple-choice SAT-mathematics items* (RR-96-20). Princeton, NJ: Educational Testing Service.

Kenney-Benson, G. A., Pomerantz, E. M., Ryan, A. M., & Patrick, H. (2006). Sex differences in math performance: The role of children's approach to schoolwork. *Developmental Psychology, 42*(1), 11–26.

Lave, J., Smith, S., & Butler, M. (1988). Problem solving as everyday practice. In R. I. Charles & E. A. Silver (Eds.), *The teaching and assessing of mathematical problem solving* (pp. 61–81). Reston, VA: National Council of Teachers of Mathematics.

Lemke, M., Sen, A., Pahlke, E., Partelow, L., Miller, D., Williams, T., Kastberg, D., & Jocelyn, L. (2004). *International outcomes of learning in mathematics literacy and problem solving: PISA 2003 results from the U.S. perspective.* (NCES 2005–003). Washington, DC: U.S. Department of Education, National Center for Education Statistics.

Marshall, S. P. (1995). *Schemas in problem solving.* New York: Cambridge University Press.

Mislevy, R. J., Yamamoto, K., & Anacker, S. (1991). *Toward test theory for assessing student understanding* (RR 91-32-ONR). Princeton, NJ: Educational Testing Service.

Mullis, I. V. S., Martin, M. O., Fierros, E. G., Goldberg, A. L., & Stemler, S. E. (2000). *Gender differences in achievement.* Chestnut Hill, MA: TIMMS International Study Center.

Mullis, I. V. S., Martin, M. O., Beaton, A. E., Gonzalez, E. J., Kelly, D. L., & Smith, T. A. (1998). *Mathematics and science achievement in the final year of secondary school: IEA's Third International Mathematics and Science Study.* Chestnut Hill, MA: TIMSS International Study Center, Boston College.

National Academy of Sciences, National Academy of Engineering, Institute of Medicine. (2007). *Rising Above the Gathering Storm: Energizing and Employing America for a Brighter Economic Future.* Washington, DC: National Academies Press. http://www.nap.edu/catalog.php?record_id=11463

National Research Council (NRC). (2006). *America's lab report: Investigations in high school science.* Washington, DC: National Academy Press.

Newell, A., & Simon, H. A. (1972). *Human problem solving.* Englewood Cliffs, NJ: Prentice Hall.

Noddings, N. (1988). Preparing teachers to teach mathematical problem solving. In R. I. Charles & E. A. Silver (Eds), *The teaching and assessing of mathematical problem solving,* vol. 3 (pp. 244–258), Reston, VA: The National Council of Teachers and Mathematics.

Perie, M., Grigg, W. S., & Dion, G. S. (2005). *The nation's report card: Mathematics 2005.* Washington, DC: National Center for Education Statistics.

Perkins, R., Kleiner, B., Roey, S., & Brown, J. (2004). *The high school transcript study: A decade of change in curricula and achievement, 1990–2000* (NCES 2004-455). National Center for Education Statistics, Institute of Education Sciences, U.S.

Department of Education. Washington, DC: U.S. Government Printing Office.

Petrides, L. A. (2002). Web-based technologies for distributed (or distance) learning: Creating learner-centered educational experiences in the higher education classroom. *International Journal of Instructional Media, 29*(1), 69–77.

Polya, G. (1945). *How to solve it: A new aspect of mathematical method.* Princeton, NJ: Princeton University Press.

Pomerantz, E. M., Altermatt, E. R., & Saxon, J. L. (2002). Making the grade but feeling distressed: Gender differences in academic performance and internal distress. *Journal of Educational Psychology, 94,* 396–404.

Santapau, S. L. (2001). *The nation's report card: Mathematics highlights 2000.* Washington, DC: National Center for Education Statistics.

Schank, R. C. (1990). Case-based teaching: Four experiences in educational software design. *Interactive Learning Environments, 1*(4), 31–53.

Schoenfeld, A. H. (1988). Problem solving in context(s). In R. I. Charles & E. A. Silver (Eds.), *The teaching and assessing of mathematical problem solving* (pp. 82–92). Reston, VA: National Council of Teachers of Mathematics.

Shapley, P. (2000). On-line education to develop complex reasoning skills in organic chemistry. *Journal of Asynchronous Learning Networks, 4*(2). Retrieved January 4, 2008, from http://www.sloan-c.org/publications/jaln/v4n2/v4n2_shapley.asp

Siegler, R. S. (1988). Individual differences in strategy choices: Good students, not-so good students, and perfectionists. *Child Development, 59,* 833–851.

Soller, A., & Stevens, R. (2007). *Applications of stochastic analyses for collaborative learning and cognitive assessment* (IDA Document D-3421). Alexandria, VA: Institute for Defense Analyses.

Song, L., & Hill, J. R. (2007). A conceptual model for understanding self-directed learning in online environments. *Journal of Interactive Online Learning, 6*(1). Retrieved March 12, 2008 from http://ncolr.org/jiol/issues/viewarticle.cfm?volID=6&IssueID=19&ArticleID=98

Stevens, R. H. (1991). Search path mapping: A versatile approach for visualizing problem-solving behavior. *Academic Medicine, 66*(9), S72–S75.

Stevens, R. H., & Casillas, A. (2006). Artificial neural networks. In D. M. Williamson, I. I. Bejar, & R. J. Mislevy (Eds.), *Automated scoring of complex tasks in computer based testing* (pp. 259–312). Mahwah, NJ: Erlbaum.

Stevens, R. H., & Palacio-Cayetano, J. (2003). Design and performance frameworks for constructing problem-solving simulations. *Cell Biology Education, 2*(3), 162–179.

Stevens, R. H., & Soller, A. (2005). Machine learning models of problem space navigation: The influence of gender. *ComSIS, 2*(2), 83–98.

Stevens, R., Soller, A., Cooper, M., & Sprang, M. (2004). Modeling the development of problem solving skills in chemistry with a web-based tutor. In J. C. Lester, R. M. Vicari, & F. Paraguaca (Eds.), *Intelligent tutoring systems* (pp. 580–591). Berlin Heidelberg, Germany: Springer-Verlag.

Tatsuoka, K. K. (1990). Toward an integration of item-response theory and cognitive error diagnosis. In N. Frederiksen, R. Glaser, A. Lesgold, & M. G. Shafto

(Eds.), *Diagnostic monitoring of skill and knowledge acquisition* (pp. 453–488). Hillsdale, NJ: Erlbaum.

Tatsuoka, K. K. (1995). Architecture of knowledge structures and cognitive diagnosis: A statistical pattern recognition and classification approach. In P.D. Nichols, S. F. Chipman & R. L. Brennan (Eds.), *Cognitively diagnostic assignment* (pp.327–359). Hillsdale, NJ : Erlbaum.

U.S. Department of Education (1997). *The condition of education 1997* (NCES 97-388). National Center for Education Statistics, U.S. Department of Education, Washington, DC: U.S. Government Printing Office.

U.S. Department of Education, Institute of Education Sciences, National Center for Education Statistics, National Assessment of Educational Progress (NAEP) (n.d.). *1990–2007 Mathematics Assessments.* Scores for male and female students increased in 2007 at both grades 4 and 8. Retrieved May 8, 2009 from http://nationsreportcard.gov/math_2007/m0008.asp

U.S. Department of Education. (2007). *The nation's report card: 12th-grade reading and mathematics 2005* (NCES 2007.468). U.S. Department of Education, National Center for Education Statistics, Washington, DC: U.S. Government Printing Office.

Wainer, H. (1983). *An exploratory analysis of performance on the SAT* (RR-83-36). Princeton, NJ: Educational Testing Service.

CHAPTER 10

GENDER DIFFERENCES IN MATHEMATICS ACHIEVEMENT

Evidence from Regional and International Student Assessments

Xin Ma
University of Kentucky

BACKGROUND

Large-scale student assessments have long been considered a valid research tool for the purpose of cross-national comparisons (Robitaille & Travers, 1992). This chapter uses latest data from several regional and international student assessments to evaluate the current status of gender differences in mathematics achievement as well as change over time in gender differences. Mathematics achievement is defined as competence in mathematics tested through standardized paper-and-pencil instruments. In education, a student assessment is a procedure designed to obtain information about cognitive and affective outcomes of a group of learners for a variety of eval-

International Perspectives on Gender and Mathematics Education, pages 225–248
Copyright © 2010 by Information Age Publishing

uative purposes, one of which is to examine the equality issues (e.g., gender differences) in learning outcomes. All existing regional and international student assessments contain measures on learning outcomes that can be separated by gender. Therefore, it is appropriate to use regional and international student assessments to evaluate gender differences in learning outcomes. Given that regional and international student assessments usually draw nationally representative samples for data analysis, they provide excellent opportunities to investigate gender differences in learning outcomes at the national level, which then allows regional and international comparisons of gender differences in learning outcomes within a particular assessment.

In fact, the research literature has no lack of reviews of gender differences in learning outcomes from large-scale student assessments. The most influential student assessments on mathematics achievement come from the First and Second International Mathematics Study (FIMS and SIMS) and the Third International Mathematics and Science Study (TIMSS) conducted by the International Association for the Evaluation of Educational Achievement (IEA). In the 1964 FIMS with 12 educational systems, Husen (1967) reported evident differences in favor of boys in computation skills. However, without finding any significant gender differences within countries, Husen (1967) claimed that gender differences in favor of boys were a global phenomenon and locally might even favor girls within countries. In the 1970 FIMS with 19 educational systems, Comber and Keeves (1973) reported that boys outperformed girls at the age of 10 and 14 when considering all mathematics topics as a whole. Reanalyzing both 1964 and 1970 FIMS data, Steinkamp, Harnisch, Walberg, and Tsai (1985) concluded that gender differences in mathematics achievement were small but pervasive across countries.

Data from SIMS with 20 educational systems were also used to examine gender differences in mathematics achievement. Gender differences varied in both size and direction across educational systems in SIMS (Baker & Jones, 1993). Among students at the age of 8, out of 19 countries, three reported gender differences in favor of boys and four in favor of girls in arithmetic; four reported differences in favor of girls in algebra; five reported differences in favor of boys in geometry; five found differences in favor of boys in measurement; and three reported differences in favor of girls in statistics (see Robitaille & Garden, 1989). Within-country gender differences varied widely across countries and were smaller than gender differences among countries (Hanna 1989, 1994). Among students at the age of 14, out of 15 countries, three found differences in favor of boys and one in favor of girls in set theory; nine reported differences in favor of boys in number systems; four had differences in favor of boys in algebra; five reported differences in favor of boys in geometry; and seven found differences in favor of boys in functions (see Robitaille & Garden, 1989). At this age, gender differences increased compared with the age of 8, with girls becoming clearly less successful than boys (Hanna 1989, 1994).

Hanna (2003) summarized gender differences in mathematics achievement in TIMSS, based on initial reports by Mullis et al. (1997, 1998), Beaton et al. (1996), and Beaton and Robitaille (1999):

> The results of the TIMSS cross-national study, encompassing more than 40 countries and about half a million boys and girls, indicate that up to Grade 8 there are few significant gender differences in achievement. The results also show that at the level of advanced mathematics (in the last grade of secondary school), five out of the 16 participating countries provide conditions that have led to an almost total disappearance of gender differences in achievement. (Hanna, 2003, p. 209)

After reviewing gender differences in mathematics achievement from IEA student assessments (FIMS to SIMS to TIMSS), Hanna (2003) claimed "the end of gender differences" (p. 209). Specifically, Hanna (2003) elaborated that:

> Gender differences in mathematics decreased considerably over the 30 years or so covered by these [IEA] studies and indeed are on the way to disappearing. Perhaps the most significant contribution of these international comparisons, in the context of gender studies, is to have revealed that several countries have in effect achieved gender equity in mathematics. This fact presents a challenge to those countries that have not yet done so. (pp. 209–210)

This chapter extends Hanna's (2003) synthesis of IEA student assessments to include other regional and international student assessments, in an attempt to provide a more general (and enhanced) picture of gender differences in mathematics achievement.

DATA SOURCES

Regional student assessments include the Latino Americano Laboratorio de Evaluacion de La Calidad de La Educacion (LAB) (focusing on Latin America), the Programme d'Analyse des Systèmes Educatifs des Pays de la CONFEMEN (PASEC) (focusing on Francophone Africa) (Bernard, 2006), and the Southern Africa Consortium for Monitoring Educational Quality (SACMEQ) (focusing on Southern and Eastern Africa). International student assessments include the Programme for International Student Assessment (PISA) and the Third International Mathematics and Science Study (TIMSS). Table 10.1 summarizes the survey characteristics of these regional

TABLE 10.1   A Glance at Regional and International Student Assessments

	Year	Country	Subject	Grade	Age	Student	School
Regional Student Assessment							
LAB	1997	12	Language, Mathematics	3	9	26000	1200
	1997	12	Language, Mathematics	4	10	26000	1200
PASEC	2004	8	Language, Mathematics	2	8	1200	110
	2004	8	Language, Mathematics	5	11	1200	110
SCAMEQ	2002	14	Language, Mathematics	6	13	42000	2300
International Student Assessment							
PISA	2000	32	Language, Mathematics, Science	8	15	229000	8500
	2003	40	Language, Mathematics, Science	8	15	250000	10000
TIMSS	1995	25	Mathematics, Science	4	9	94000	4200
	1995	39	Mathematics, Science	8	14	255000	5100
	1995	21	Mathematics, Science	12	18	53000	3000
	1999	38	Mathematics, Science	8	14	133000	5700
	2003	25	Mathematics, Science	4	9	124000	4600
	2003	46	Mathematics, Science	8	14	238000	7800

Note: Average age is reported. Student sample is rounded to the nearest thousand. School sample is rounded to the nearest hundred. In both regional and international student assessments, sample for each participating country is obtained through a stratified random sampling procedure (i.e., sample is nationally representative of each participating country).

and international student assessments. Some regional student assessments are not necessarily current. For example, data in LAB were collected 10 years ago. Nevertheless, it remains the most comprehensive student assessment in Latin America. Obviously, LAB provides more of a historical perspective on gender differences in mathematics achievement in that region.

Some cautions are necessary to avoid the misuse of regional and international student assessments. First and foremost, context is the key in using all regional and international student assessments. Although regional student assessments have nationally representative samples, countries are not randomly selected from any region. Because of this situation, what will be presented cannot be generalized to the whole region. Instead, what is to be discussed in this chapter is valid only for countries participating in each regional student assessment. Therefore, throughout this chapter, the term "region" (or "regional") is used for convenience of discussions but not for generalization of conclusions.

GENDER DIFFERENCES IN MATHEMATICS ACHIEVEMENT

For each regional or international student assessment, the three countries with the largest gender differences in mathematics achievement are identified. Selecting this small number of countries allows this analysis to highlight, in a concise and manageable way, countries and regions that face the greatest challenges. If there are fewer than three countries with gender differences, all countries are included.

Regional Results from LAB

LAB measured mathematics achievement among 3rd and 4th graders. Mathematics achievement was measured by standardized scores with a mean of 250 points and a standard deviation of 50 points. Table 10.2 presents gender differences in mathematics achievement. At the 3rd grade, four out of 12 participating countries demonstrated gender differences, all in favor of boys. One way to gauge the magnitude of gender differences is to convert them into a common metric that reports statistical results in *SD* units. Based on the classification in Rosenthal and Rosnow (1984), gender differences more than 0.50 *SD* can be considered large, between 0.30 and

TABLE 10.2   Gender Differences in Mathematics Achievement: Results from LAB

	Grade 3		Grade 4	
	Boys Better	Girls Better	Boys Better	Girls Better
Argentina	5*		8*	
Bolivia	1		6*	
Brazil	7*		8*	
Chile	3		5*	
Colombia	1		4	
Cuba		2		6*
Honduras		2		2
Mexico	1			2
Paraguay	7*		1	
Peru	5*		2	
Dominican Republic		4	1	
Venezuela	1		2	

* Gender differences statistically significant at the 95% confidence level.

0.50 *SD* can be considered moderate, and less than 0.30 *SD* can be considered small. These measures of small, moderate, and large effects were used in this chapter to describe gender differences.

Gender differences in favor of boys in mathematics achievement at the 3rd grade were all small in magnitude (five points in Argentina and Peru and seven points in Brazil and Paraguay). The male advantage did not advance much at the 4th grade: four out 12 participating countries demonstrated gender differences in favor of boys, as in the case of the 3rd grade. Gender differences did reach the moderate level in two countries (eight points in Argentina and Brazil), and these two countries happened to show gender differences at both grade levels. Although Bolivia and Chile showed small gender differences at the 4th grade only (six and five points, respectively), Paraguay and Peru, with small gender differences at the 3rd grade, did not carry the male advantage over to the 4th grade. On the other hand, girls in Cuba showed a small female advantage (six points).

Overall, the pattern that emerged 10 years ago from the participating Latin American countries about gender differences in mathematics achievement portrayed a picture in which the male advantage in (middle-grade) primary school was limited to a small number of participating countries (one third), but half of these countries experienced widened gender differences as students progressed through the middle grades of primary school. Argentina and Brazil appeared to be the gravity center for this male advantage among the participating Latin American countries. Although a small female advantage occurred as early as the 4th grade in Cuba, gender differences in favor of girls were still a rather isolated phenomenon at this early stage of schooling among the participating Latin American countries.

Regional Results from PASEC

PASEC administered a mathematics test to the 2nd and 4th graders in the participating Francophone African countries. The PASEC mathematics scale was from 0 to 100 points. At the 2nd grade, there were gender differences in mathematics achievement in five out of eight participating countries, four of which were in favor of boys (see Table 10.3). Nevertheless, all gender differences ranged from two to five points (within a scale of 0 to 100). Such a magnitude is not a major concern in any practical sense. It is reasonable to conclude that there were no appreciable gender differences in mathematics achievement among the 2nd graders in the participating Francophone African countries.

At the 5th grade, there were gender differences in mathematics achievement in four out of eight participating countries, all in favor of boys. However, ranging from two to four points, such a male advantage, again, is not

TABLE 10.3   Gender Differences in Mathematics Achievement: Results from PASEC

	Grade 2		Grade 5	
	Boys Better	Girls Better	Boys Better	Girls Better
Burkina Faso	0.3		3.0*	
Cameroon	1.0		0.8	
Chad	4.0*		0.5	
Côte d'Ivoire	2.1*		0.9	
Madagascar		2.5*	0.3	
Mali	2.9*		3.6*	
Niger	1.5		2.0*	
Senegal	5.2*		2.0*	

* Gender differences statistically significant at the 95% confidence level.
Source: Bernard (2006).

a cause for practical concern. Only two out of eight participating countries, Mali and Senegal, had gender differences at both grade levels. Overall, there is no immediate concern about gender differences in mathematics achievement at the primary school level among the participating Francophone African countries. The magnitude of gender differences is so small that the slight male advantage in some participating countries requires nothing but continuous monitoring.

Regional Results from SACMEQ

The SACMEQ mathematics achievement test had standardized scores with a mean of 500 points and a standard deviation of 100 points. Table 10.4 shows differences in mean mathematics achievement between boys and girls. Gender differences were present in six out of 14 participating countries. Five out of these six countries showed gender differences in favor of boys. A moderate male advantage was found in Tanzania (33 points). Other male advantages (in Kenya, Mozambique, Zambia, and Zanzibar) were small, ranging from 10 to 22 points. On the other hand, moderate gender differences in favor of girls (38 points) occurred in Seychelles.

Overall, SACMEQ indicates that there were no gender differences in mathematics achievement at the upper primary school level (the 6th grade) in the majority of the participating Southern and Eastern African countries. Although the few gender differences were almost all in favor of boys, the female advantage was the largest among all gender differences. Therefore, among the participating Southern and Eastern African countries, an interesting balance appeared to be at work in which the male advantage out-

TABLE 10.4   Gender Differences in Mathematics Achievement: Results from SACMEQ

	Difference in Average Achievement		Percent at Low Level	
	Boys Better	Girls Better	Boys	Girls
Botswana		9	42	43
Kenya	22*		32	34
Lesotho		3	43	55
Malawi	10		52	48
Mauritius		11	33	28
Mozambique	18*		53	35
Namibia	5		45	49
Seychelles		38*	37	29
South Africa		8	43	44
Swaziland	5		43	45
Tanzania	33*		38	46
Uganda	4		47	36
Zambia	10*		51	48
Zanzibar	14*		46	49

* Gender differences statistically significant at the 95% confidence level.
Source: Ross, Saito, Dolata, and Ikeda (2004).

numbered the female advantage but in a much smaller magnitude in most cases, whereas the female advantage claimed the largest magnitude.

SACMEQ also defined a low level of mathematics competence. Percents of boys and girls at this low level in each participating country are also presented in Table 10.4. For countries that did not report gender differences, it is still possible to observe gender differences at this low level in a couple of them. For example, 55% of girls were at the low level of mathematics competence in contrast to 43% of boys in Lesotho, whereas 47% of boys were at the low level of mathematics competence in contrast to 36% of girls in Uganda. On the other hand, countries that reported gender differences in favor of boys might actually have a female advantage at the low level of mathematics competence. The most typical case is Mozambique. Even though, on average, there was a male advantage, 53% of boys were actually at the low level of mathematics competence in contrast to 35% of girls, an obvious female advantage at this low level.

International Results from PISA

PISA defines mathematics achievement as mathematics literacy: the ability to formulate and solve mathematical problems in situations encountered in life (see Organization for Economic Cooperation and Development, 2001). The PISA mathematics literacy had standardized scores with a mean of 500 points and a standard deviation of 100 points. Table 10.5 shows gender differences in mathematics literacy at the 8th grade in PISA. In the case of PISA 2003, 28 out of 40 participating countries reported gender differences (this represents 70% of all participating countries), all but one in favor of boys. Gender differences in favor of boys were all small, ranging from 6 to 29 points. The small female advantage (15 points) occurred in Iceland. Regionally, the three countries with the largest gender differences came from Western Europe (Liechtenstein with 29 points) and East Asia (Korea with 23 points and Macao with 21 points).

Six performance levels were developed based on scores on mathematics literacy in PISA 2003. A comparison of performance by boys and girls at the lowest and highest levels is shown in Table 10.5. At the top end of the performance spectrum, all participating countries indicated that the male percent was larger than (at least as large as) the female percent at the highest level of performance. Therefore, participating countries that did not report gender differences could still have "local" gender disparities. At the bottom end, 22 out of the 27 participating countries with gender differences in favor of boys indicated that the female percent was larger than (at least as large as) the male percent at the lowest level of performance, and Iceland with gender differences in favor of girls had a higher percent of boys at this level. Therefore, the vast majority of participating countries with gender differences in favor of boys (girls in the case of Iceland) commonly had a smaller percent of boys (girls in the case of Iceland) at the lowest level of performance—this might be a necessary (but not sufficient) condition for gender differences to occur.

Change in Gender Differences: PISA Results

A comparison between PISA 2000 and 2003 results indicates that the male advantage in mathematics literacy was on the rise in proportion (from 47% of the participating countries in 2000 to 70% of the participating countries in 2003) (see Table 10.5). All but two participating countries with gender differences in 2000 continued to have gender differences in 2003. However, the male advantage remained small from 2000 to 2003. Compared with 2000, 2003 witnessed the first case of a female advantage. During this 3-year period, Iceland experienced the rise of a small female advantage in 2003 from a status of no gender differences in 2000. East Asia and Western Europe consistently "produced" participating countries with the largest gender differences in favor of boys among all participating countries. Interestingly, the first case of female advantage came also from Western Europe (i.e., Iceland).

TABLE 10.5   Gender Differences in Mathematics Achievement: Results from PISA 2003 and 2000

	Difference in Average Achievement				Percent in Performance Level (PISA 2003)			
	PISA 2003		PISA 2000		≤ Level 1		≥ Level 6	
	Boys Better	Girls Better	Boys Better	Girls Better	Boys	Girls	Boys	Girls
East Asia and the Pacific								
Australia	5		12		5	4	7	5
Hong Kong	4				5	3	13	8
Indonesia	3				49	52	0	0
Japan	8		8		5	4	11	6
Korea	23*		27*		2	3	10	6
Macao	21*				2	2	7	3
New Zealand	14*			3	5	5	8	5
Thailand		4			25	23	0	0
Central and Eastern Europe								
Czech Republic	15*		12*		4	6	7	4
Hungary	8*		7		8	8	3	2
Latvia	3		6		8	7	2	1
Poland	6		5		8	6	3	2
Russia	10*			2	11	11	2	1
Serbia	1				19	16	0	0
Slovak Republic	19*				6	7	4	2
Turkey	15*				26	29	3	2
North America and Western Europe								
Austria	8		27*		6	5	5	3
Belgium	8		6		7	7	11	7
Canada	11*		10*		3	2	8	4
Demark	17*		15*		4	6	5	3
Finland	7*		1		2	1	8	5
France	9*		14*		6	5	5	3
Germany	9*		15*		9	9	5	3
Greece	19*		7		16	19	1	0
Iceland		15*		5	6	3	4	4
Ireland	15*		13*		4	5	3	2
Italy	18*		8		13	14	3	1
Liechtenstein	29*		12*		5	5	11	4
Luxembourg	17*		15*		7	8	3	1
Netherlands	5		11		2	3	8	7
Norway	6		11*		7	7	4	2
Portugal	12*		19*		12	11	1	0

	Difference in Average Achievement				Percent in Performance Level (PISA 2003)			
	PISA 2003		PISA 2000		≤ Level 1		≥ Level 6	
	Boys Better	Girls Better	Boys Better	Girls Better	Boys	Girls	Boys	Girls
Spain	9*		18*		8	8	2	1
Sweden	7*		7		6	6	5	3
Switzerland	17*		14*		4	6	9	5
United Kingdom			8					
United States	6*		10		11	10	3	1
Latin America								
Brazil	16*		27*		51	55	1	0
Mexico	11*		11		36	39	0	0
Uruguay	12*				24	28	1	0
Arab States								
Tunisia	12*				48	54	0	0

* Gender differences statistically significant at the 95% confidence level.
Source: Organization for Economic Co-operation and Development (2001, 2004).

Some comparisons are made for the 31 participating countries with data from both 2000 and 2003. Specifically, six countries experienced increased gender differences in mathematics literacy with Liechtenstein reporting an increase larger than 10 points (17 points), and five countries experienced decreased gender differences with no country reporting a decrease larger than 10 points. During the same period, six countries remained free of gender differences, eight countries witnessed the appearance of gender differences (all but one in favor of boys), and two countries (Austria and Norway) witnessed the disappearance of gender differences. Countries with the appearance of gender differences larger than 10 points include New Zealand (14 points), Iceland (15 points in favor of girls), Italy (18 points), and Greece (19 points).

International Results from TIMSS

With a focus on mathematics and science education (closely related to school curricula in mathematics and science), TIMSS was a four-year cycle of assessments of international trends in mathematics and science achievement. TIMSS 1995 (often simply referred to as TIMSS) was the first cycle of data collection. TIMSS 1995 targeted three populations of students in the 4th, 8th, and 12th grades. TIMSS 1999 (often referred to as TIMSS-Repeat or TIMSS-

R) was the second cycle, measuring the progress of students in mathematics and science achievement at the 8th grade. TIMSS 2003 was the latest cycle, assessing mathematics and science achievement of students at the 4th and 8th grades. The TIMSS mathematics achievement had standardized scores with a mean of 500 points and a standard deviation of 100 points.

TIMSS 2003 has seen major historical changes in gender differences in mathematics achievement. Table 10.6 indicates that, at the 4th grade, gender differences in favor of girls began to appear, with three participating countries (Armenia, Moldova, and Philippines) showing small female advantages ranging from 11 to 12 points (equivalent to 12% of all participating countries). Though small in magnitude, these gender differences are the first occurrence of a female advantage in mathematics achievement at the 4th grade in the entire IEA history. Gender differences in favor of boys occurred in six participating countries (equivalent to 24% of all participating countries). The male advantage (ranging from 6 to 11 points) was quite similar in magnitude to the female advantage. As a matter of fact, gender

TABLE 10.6 Gender Differences in Mathematics Achievement: Results from TIMSS 2003

	Grade 4		Grade 8	
	Boys Better	Girls Better	Boys Better	Girls Better
East Asia and the Pacific				
Australia	3		13	
Hong Kong	0			2
Indonesia				1
Japan	4		3	
Korea			5	
Malaysia				8
New Zealand	0			3
Philippines		12*		13*
Singapore	8*			10*
Taiwan		1		7
South and West Asia				
Iran			8	9
Central Asia				
Armenia		12*		10*
Central and Eastern Europe				
Bulgaria				1
Estonia				2

	Grade 4		Grade 8	
	Boys Better	Girls Better	Boys Better	Girls Better
Hungary	3		7*	
Latvia		1		6
Lithuania	1			5
Macedonia				9*
Moldova		11*		10*
Romania				4
Russia	4			3
Serbia				7*
Slovak Republic			0	
Slovenia	5			3
North America and Western Europe				
Belgium (Flemish)	2		11*	
Cyprus	9*			16*
England		2	0	
Israel			8	
Italy	9*		6*	
Netherlands	6*		7	
Norway	5*			3
Scotland	11*			5
Sweden			1	
United States	8*		6*	
Latin America				
Chile			15*	
Sub-Saharan Africa				
Botswana			2	
Ghana			17*	
South Africa			3	
Arab States				
Bahrain				33*
Egypt				1
Jordan				27*
Lebanon			10*	
Morocco	6		12*	
Palestinian A. T.				8
Saudi Arabia			10	
Tunisia		5	24*	

* Gender differences statistically significant at the 95% confidence level.
Source: Mullis, Martin, Gonzalez, and Chrostowski (2004).

differences in favor of girls balanced gender differences in favor of boys, resulting in the lack of gender differences at the 4th grade in TIMSS 2003 (i.e., the overall average gender differences were negligible).

Regionally, three out of the four countries with the largest gender differences reported female advantages. Specifically, Armenia from Central Asia and Philippines from East Asia reported a female advantage of 12 points, Moldova from Central and Eastern Europe reported a female advantage of 11 points, and Scotland from Western Europe reported a male advantage of 11 points.

Two phenomena appeared at the 8th grade regarding gender differences in mathematics achievement (see Table 10.6). First, for the first time in the entire IEA history, gender differences in favor of girls are observed. In fact, among all participating countries, countries showing gender differences in favor of girls were the same in number as countries showing gender differences in favor of boys (9 versus 9 or 20% each). The female advantage was mostly small, ranging from seven to 27 points, with the exception of Bahrain, which demonstrated a moderate female advantage (33 points). On the other hand, the male advantage was also small, ranging from six to 24 points. The second phenomenon quite noticeable in Table 10.6 is that among the three countries (Bahrain, Jordan, and Tunisia) with the largest gender differences, two of them (Bahrain and Jordan) showed gender differences in favor of girls. All three countries in this category are Arab countries. Such gender differences indicate that when girls in Arab countries are provided with equal opportunity to learn, they may well achieve far better in mathematics achievement than their male counterparts who traditionally have had privileged access to formal education.

The two phenomena discussed above also highlight the cross-sectional comparison between the 4th and 8th grades in TIMSS 2003. Two things are worth emphasizing. First, both grades witnessed the first occurrence of female advantage in mathematics achievement in the history of IEA. Secondly, the largest gender differences did not display any regional pattern between the 4th and 8th grades.

Change in Gender Differences: TIMSS Results

With three cycles of assessments, TIMSS provides a good opportunity to examine the trend of gender differences in mathematics achievement. Comparisons can be made at the 4th grade with TIMSS 1995 and 2003 as well as at the 8th grade with TIMSS 1995, 1999, and 2003. Table 10.7 presents a comparison of gender differences at the 4th grade between TIMSS 1995 and 2003. Clearly, gender differences have gone through dramatic change over time among participating countries in TIMSS. At the 4th grade, if gender differences ever occurred in 1995, they occurred in favor of boys. This pattern disappeared at the 4th grade in 2003. Gender differences in

TABLE 10.7   Gender Differences in Mathematics Achievement: 4th Grade Results from TIMSS 2003 and 1995

	TIMSS 2003		TIMSS 1995	
	Boys Better	Girls Better	Boys Better	Girls Better
East Asia and the Pacific				
Australia	3		2	
Hong Kong	0			1
Japan	4		8*	
Korea			15*	
New Zealand	0			10
Philippines		12*		
Singapore	8*			10
Taiwan		1		
Thailand				11
South and West Asia				
Iran		8	9	
Central Asia				
Armenia		12*		
Central and Eastern Europe				
Czech Republic			3	
Hungary	3		5	
Latvia		1		9
Lithuania	1			
Moldova		11*		
Russia	4			
Slovenia	5			3
North America and Western Europe				
Austria			8	
Belgium (Flemish)	2			
Canada			3	
Cyprus	9*		8	
England		2	5	
Greece				2
Iceland			1	
Ireland				3
Israel				9
Italy	9*			
Netherlands	6*		15*	
Norway	5		5	

(continued)

TABLE 10.7 Gender Differences in Mathematics Achievement: 4th Grade Results from TIMSS 2003 and 1995 (continued)

	TIMSS 2003		TIMSS 1995	
	Boys Better	Girls Better	Boys Better	Girls Better
Portugal			4	
Scotland	11*		0	
United States	8*		2	
Arab States				
Morocco	6			
Tunisia		5		

* Gender differences statistically significant at the 95% confidence level.
Source: Mullis, Martin, Fierros, Goldberg, and Stemler (2000). Mullis, Martin, Gonzalez, and Chrostowski (2004).

favor of girls have matched up in magnitude (not in number yet) with gender differences in favor of boys. In less than a decade, the male advantage in mathematics achievement has partially disappeared from the 4th grade among participating countries in TIMSS.

Some specific analyses of change over time in gender differences in mathematics achievement are possible at the 4th grade because 15 countries took part in both TIMSS 1995 and 2003. Nine countries (Australia, England, Hong Kong, Hungary, Iran, Latvia, New Zealand, Norway, and Slovenia) remained free of gender differences in both 1995 and 2003. With gender differences in favor of boys in both 1995 and 2003, the male advantage decreased nine points in the Netherlands. During the same period, the male advantage disappeared in Japan (eight points in 1995) but appeared in Cyprus (nine points in 2003), Scotland (11 points in 2003), Singapore (eight points in 2003), and the United States (eight points in 2003). The overall trend can best be described as a stable maintenance, with 10 out of 15 countries keeping their status of gender differences (either present or absent).

Table 10.8 presents a comparison of gender differences at the 8th grade among TIMSS 1995, 1999, and 2003. Similar to the case of the 4th grade, gender differences in mathematics achievement have experienced dramatic change over time at the 8th grade level among participating countries in TIMSS. In 1995, if gender differences ever occurred at the 8th grade, they occurred in favor of boys. In 1999, this pattern remained at the 8th grade. In 2003, this pattern disappeared at the 8th grade. In both number and magnitude, gender differences in favor of girls are matching up with gender differences in favor of boys. In less than a decade, the male advantage in mathematics achievement has disappeared at the 8th grade.

TABLE 10.8 Gender Differences in Mathematics Achievement: 8th Grade Results from TIMSS 2003, 1999, and 1995

	TIMSS 2003		TIMSS 1999		TIMSS 1995	
	Boys Better	Girls Better	Boys Better	Girls Better	Boys Better	Girls Better
East Asia and the Pacific						
Australia	13		2			5
Hong Kong		2		2	20	
Indonesia		1	5			
Japan	3			8	9*	
Korea	5		5		17*	
Malaysia		8	5			
New Zealand		3		7	9	
Philippines		13*		15		
Singapore		10*	2			2
Taiwan		7	4			
Thailand				4		9
South and West Asia						
Iran		9	24*		13*	
Central Asia						
Armenia		10*				
Central and Eastern Europe						
Bulgaria		1	0			
Czech Republic			17*		11	
Estonia		2				
Hungary	7*		6		0	
Latvia		6	5		4	
Lithuania		5	3			1
Macedonia		9*	0			
Moldova		10*	3			
Romania		4		5	3	
Russia		3	1			1
Serbia		7*				
Slovak Republic	0		5			4
Slovenia		3	1		8	
Turkey		2				
North America and Western Europe						
Austria					8	
Belgium (Flemish)	11*			4		4
Belgium (French)					6	
Canada			3			4

(continued)

TABLE 10.8   Gender Differences in Mathematics Achievement: 8th Grade Results from TIMSS 2003, 1999, and 1995 (continued)

	TIMSS 2003		TIMSS 1999		TIMSS 1995	
	Boys Better	Girls Better	Boys Better	Girls Better	Boys Better	Girls Better
Cyprus		16*		4		3
Demark					17*	
England	0		19		4	
Finland			3			
France					5	
Germany					3	
Greece					12*	
Iceland					2	
Ireland					14	
Israel	8		16*		29*	
Italy	6*		9			
Netherlands	7		5		8	
Norway		3				4
Portugal					11*	
Scotland		5			16	
Spain					10*	
Sweden	1				2	
Switzerland					5	
United States	6*		7		5	
Latin and South America						
Chile	15*		9			
Colombia					2	
Sub-Saharan Africa						
Botswana	2					
Ghana	17*					
South Africa	3		16		11	
Arab States						
Bahrain		33*				
Egypt		1				
Jordan		27*		7		
Lebanon	10*					
Morocco	12*		17			
Palestinian A. T.		8				
Saudi Arabia	10					
Tunisia	24*		25*			

* Gender differences statistically significant at the 95% confidence level.

Source: Mullis, Martin, Fierros, Goldberg, and Stemler (2000); Mullis, Martin, Gonzalez, and Chrostowski (2004); Mullis, Martin, Gonzalez, et al. (2000).

Some specific analyses of change over time in gender differences in mathematics achievement are also possible at the 8th grade with 21 participating countries in all three cycles (TIMSS 1995, 1999, and 2003). More than half of them remained free of gender differences across 1995, 1999, and 2003. Japan, Korea, Iran, and Israel come into one category. All of them experienced gender differences in favor of boys in 1995 but saw the disappearance of gender differences in 2003 (gender differences disappeared in Japan and Korea even earlier in 1999). Cyprus and Singapore witnessed the appearance of gender difference in favor of girls in 2003, although neither experienced gender differences in either 1995 or 1999. The female advantage was small in both countries (16 points in Cyprus and 10 points in Singapore). On the other hand, Belgium (Flemish-speaking), Hungary, and the United States witnessed the appearance of gender differences in favor of boys after being free of gender differences in both 1995 and 1999. The male advantage was small in all three countries (11 points in Belgium, seven points in Hungary, and six points in United States). Therefore, based on analysis of the participating countries with data from all three cycles, the overall trend can be characterized as a continuing absence of gender differences, with 12 out of 21 countries free of gender differences throughout 1995, 1999, and 2003 and with four more countries joining them in 2003.

UNDERSTANDING CHALLENGES

Trends of Gender Differences in Mathematics Achievement

Hanna (2003, p. 209) saw "the end of gender differences" in mathematics achievement based on a review of IEA student assessments (FIMS, SIMS, and TIMSS). The current review of recent regional student assessments (LAB, PASEC, and SACMEQ) has revealed two universal phenomena across all three regions. Firstly, gender differences in mathematics achievement are not only small in magnitude but also limited to a small number of participating countries. Secondly, there is no male predomination in mathematics achievement even though there are more gender differences in favor of boys than girls. Specifically, in the presence of gender differences in favor of boys, Cuba showed a small female advantage among participating Latin American countries in LAB; Madagascar showed a small female advantage among participating Francophone African countries in PASEC; and Seychelles showed a moderate female advantage among participating Southern and Eastern African countries in SACMEQ. It appears that these two phenomena are better characteristics of gender differences from a regional perspective than a claim of an end of gender differences.

What characterizes international student assessments is a dramatic change over time in gender differences in mathematics achievement. This claim is particularly true with TIMSS (1995, 1999, and 2003). For example, an examination of statistically significant gender differences in Table 10.8 shows that in 1995, if gender differences ever occurred at the 8th grade, they occurred in favor of boys. In 1999, this phenomenon remained at the 8th grade. In 2003, this phenomenon disappeared at the 8th grade. In both number and magnitude, gender differences in favor of girls matched up with gender differences in favor of boys in TIMSS 2003. Although PISA is certainly not as dramatic as TIMSS, a similar pattern occurred on a much smaller scale. In 2000, if gender differences ever occurred, they occurred in favor of boys. In 2003, this phenomenon disappeared. Iceland showed a small female advantage in PISA 2003. Referring to Hanna's (2003) claim of an end of gender differences in mathematics achievement, the current review of international student assessments has found one thing for certain from a global perspective—there is no longer male predominance in mathematics achievement. In other words, girls have begun to catch up with boys in mathematics achievement. This trend is much more evident in school curriculum related mathematics achievement (as measured in TIMSS) than survival skills related mathematics achievement (as measured in PISA).

Complexity of Gender Differences in Mathematics Achievement

Some regional and international student assessments have developed indices as additional ways to investigate outcomes of student learning. These indices have also provided additional ways to examine gender differences in learning outcomes. In fact, these indices have served to expose the complexity of gender differences. Gender differences in mathematics achievement in PISA 2003 indicate that all participating countries reported that the male percent was larger than (at least as large as) the female percent at the highest level of performance, indicating that participating countries that did not report gender differences could still have "local" gender disparities.

These scenarios suggest that the complexity of gender differences in learning outcomes constitutes a challenge for the understanding of gender equity for educational policy and practice. Countries without gender differences may not be entirely free of gender inequality, and countries with gender differences may still have some gender equality to celebrate. Overall, they call for a detailed analysis of gender differences, level by level. In other words, there ought to be different "analytical units" of gender differences. In this chapter, all existing indices provide opportunity to use competence level or ability level as a unit. Other units, such as attitudinal attribute (level) and socioeconomic background (level), can all be analyzed for gender differences (e.g., gender differences may vary according to different categories of socioeconomic background). The macro perspective of gender differences calls for, say, school district, geographic region, type of community, and type of school as units of analysis, whereas the micro perspective of gender differences calls for, say, family background, individual (cognitive and affective) attribute, and individual learning path (e.g., mathematics coursework) as units of analysis. Fortunately, regional and international student assessments are in a good position to provide meaningful units at both perspectives to address the complexity of gender differences. Each country should take full advantage of its participation in regional and international student assessments to analyze its data from these perspectives to gain a full understanding of the issue of gender equity in mathematics education (see more discussion in the next section).

DEVELOPING STRATEGIES

Policy initiatives for improving mathematics achievement of both genders are derived in this section to highlight the immediate major challenges that the international community faces in reducing gender differences in mathematics achievement and to recommend a course of action that the international community may take to improve gender equality in mathematics education.

Global Effort to Continue to Tackle Female Underachievement in Mathematics

This is not a new policy initiative but one that has been implemented for decades in response to female underachievement in mathematics. The implementation of this policy initiative has been quite comprehensive, including examining the nature of mathematics curricula and textbooks for gender biases (see Boaler, 1994) and scrutinizing the classroom practice of mathematics teachers for gender stereotypes (see Fennema, Peterson, Carpenter & Lubinski, 1990). The fruit of this effort has begun to appear with some female breakthroughs in mathematics achievement in both regional and international student assessments. The male advantage in mathematics achievement, however, continues to exist, even though on a much smaller scale. From the perspective of large-scale student assessments, to continue to tackle female underachievement in mathematics as a global effort requires a regular monitoring of gender differences in mathematics achievement at the regional or international level. This recommendation segues well into the next strategic recommendation.

National Effort to Keep Track of Gender Differences in Mathematics Achievement

One critical component in the continuing effort to achieve gender equity in mathematics achievement, particularly relevant to large-scale student assessments, is to regularly monitor gender differences in mathematics achievement both nationally and internationally. This is why continuing participation of countries from all regions of the world in international student assessments such as PISA and TIMSS is critically important and beneficial to each and every participating country. The longitudinal aspect of these assessments charts a historical pattern for each participating country to evaluate its effort in achieving gender equity. This aspect thus allows each participating country to predict both size and direction of gender differences in its own educational system. Meanwhile, the international aspect of these assessments generates a comparative context for each participating country to put gender differences in its own educational system into various global perspectives. This aspect thus allows each participating country to prioritize critical issues of gender differences. Policy changes as a function of results of major international student assessments depend primarily on prediction and prioritization in most countries. Indeed, prediction and prioritization become critically important, particularly when a participating country's educational system faces multiple urgent issues of educational policy and practice competing for limited resources.

Although general large-scale international student assessments such as PISA and TIMSS may not be the best means to pinpoint the origin of gender differences in mathematics achievement, they are so far the best means to observe the evolution of gender differences in mathematics achievement. Again, the ability to predict and to prioritize represents important strategies in the continuing effort to achieve gender equity in education, and large-scale student assessments provide unparalleled opportunities to each participating country to use its results to update gender differences in learning outcomes from both historical and comparative perspectives.

REFERENCES

Baker, D. P., & Jones, D. P. (1993). Creating gender equity: Cross-national gender stratification and mathematical performance. *Sociology of Education, 66,* 91–103.

Beaton, A. E., Mullis, I. V., M. O. Martin, E. J., Gonzalez, E. J., Kelly, D. L., & Smith, T. A. (1996). *Mathematics achievement in the middle school years: IEA's Third International Mathematics and Science Study.* Boston, MA: TIMSS International Study Center, Boston College.

Beaton, A. E., & Robitaille, D. F. (1999). An overview of the Third International Mathematics and Science Study. In G. Kaiser, E. Luna, & I. Huntley (Eds), *International comparisons in mathematics education* (pp. 30–47). Philadelphia, PA: Falmer.

Bernard, J. (2006). *Gender and primary school achievement in Francophone Africa: Analysis based on PASEC data.* Washington, DC: World Bank.

Boaler, J. (1994). When do girls prefer football to fashion? An analysis of female underachievement in relation to "realistic" mathematics context. *British Educational Research Journal, 20,* 551–564.

Comber, L. C., & Keeves, J. P. (1973). *Science education in nineteen countries.* New York: Halsted.

Fennema, E., Peterson, P. L., Carpenter, T. P., & Lubinski, C. A. (1990). Teachers' attributions and beliefs about girls, boys, and mathematics. *Educational Studies in Mathematics, 21,* 55–69.

Hanna, G. (1989). Mathematics achievement of girls and boys in grade eight: Results from twenty countries. *Educational Studies in Mathematics, 20,* 225–232.

Hanna, G. (1994). Cross-cultural gender differences in mathematics achievement. *International Journal of Educational Research, 21,* 417–426.

Hanna, G. (2003). Reaching gender equity in mathematics education. *Educational Forum, 67,* 203–214.

Husen, T. (1967). *International study of achievement in mathematics: A comparison of twelve countries.* Stockholm, Sweden: Almqvist & Wiksell.

Mullis, I. V., Martin, M. O., Beaton, A. E., Gonzalez, E. J., Kelly, D. L., & Smith, T. A. (1997). *Mathematics and science achievement in the primary school years: IEA's Third International Mathematics and Science Study.* Boston, MA: Center for the Study of Testing, Evaluation, and Educational Policy, Boston College.

Mullis, I. V., Martin, M. O., Beaton, A. E., Gonzalez, E. J., Kelly, D. L., & Smith, T. A. (1998). *Mathematics and science achievement in the final year of secondary school: IEA's Third International Mathematics and Science Study.* Boston, MA: Center for the Study of Testing, Evaluation, and Educational Policy, Boston College.

Mullis, I. V. S., Martin, M. O., Fierros, E. G., Goldberg, A. L., & Stemler, S. E. (2000). *Gender differences in achievement: IEA's Third International Mathematics and Science Study (TIMSS).* Chestnut Hill, MA: Boston College.

Mullis, I. V. S., Martin, M. O., Gonzalez, E. J., & Chrostowski, S. J. (2004). *TIMSS 2003 international mathematics report: Findings from IEA's Trends in International Mathematics and Science Study at the fourth and eighth grades.* Chestnut Hill, MA: Boston College.

Mullis, I. V. S., Martin, M. O., Gonzalez, E. J., Gregory, K. D., Garden, R. A., O'Connor, K. M., Chrostowski, S. J., & Smith, T. A. (2000). *TIMSS 1999 international mathematics report: Findings from IEA's Repeat of the Third International Mathematics and Science Study at the eighth grade.* Chestnut Hill, MA: Boston College.

Organization for Economic Cooperation and Development. (2001). *Knowledge and skills for life: First results from PISA 2000.* Paris: Author.

Organization for Economic Cooperation and Development. (2004). *Learning for tomorrow's world: First results from PISA 2003.* Paris: Author.

Robitaille, D. F., & Garden, R. A. (Eds.) (1989). *The IEA study of mathematics II: Contexts and outcomes of school mathematics.* Oxford: Pergamon.

Robitaille, D. F., & Travers, K. (1992). International studies in mathematics achievement. In D. A. Grouws (Ed.), *Handbook of research on mathematics teaching and learning* (pp. 687–709). New York: Macmillan.

Rosenthal, R., & Rosnow, R. L. (1984). *Essentials of behavioral research: Methods and data analysis.* New York: McGraw-Hill.

Ross, K., Saito, M., Dolata, S., & Ikeda, M. (2004). *SACMEQ data archive.* Paris: IIEP.

Steinkamp, M. W., Harnisch, D. L., Walberg, H. J., & Tsai, S. (1985). Cross-national gender differences in mathematics attitude and achievement among 13 year-olds. *Journal of Mathematical Behaviour, 4,* 259–277.

CHAPTER 11

MATHEMATICS ACHIEVEMENT IN ICELANDIC PLAYSCHOOLS

Examining When Gender Differences Emerge

Olof Bjorg Steinthorsdottir
University of North Carolina, Chapel Hill

Kimberly Dadisman
University of Chicago

and Dylan L. Robertson
Chicago Public Schools

Kristjana Steinthorsdottir
Krikaskoli Playschool, Mosfellsbaer, Iceland

INTRODUCTION

In Iceland, 10th-grade girls have been scoring higher on the Icelandic National Mathematics Testing Program over the last 10 years (Olafsson, Halldorsson, Skulason, & Bjornsson, 2007). Icelandic girls also scored significantly higher than boys in mathematics in the Programme for International Assessment (PISA) 2003 (OECD, 2003; Steinthorsdottir & Sriraman,

International Perspectives on Gender and Mathematics Education, pages 249–261

2008). The unique achievement in mathematics for girls in Iceland raises, among others, the question of when gender differences in mathematics achievement emerge. The study presented in this chapter examines girls' and boys' strategy use and achievement when solving problems centered on numeracy in one Icelandic playschool.

BACKGROUND

Context of the Study

If translated from Icelandic to English, the name of the schools that serve young children, ages 18 months to 5 years, is *playschool*. It is noteworthy that in Iceland, 5-year-old children are typically in playschool while in many other countries, children of this age have already entered elementary or primary school. Playschools are public, and more than 90% of 3- to 5-year-old children are enrolled in playschools. A certified playschool teacher has to hold a bachelor's degree in early childhood education from an accredited higher education institute. Playschool teachers are licensed to teach children from birth to first grade. The Icelandic national curriculum sets goals for playschools that emphasize creating learning environments that facilitate children's general development, health, and well-being through physical activities, literacy, art, nature, environment, and culture. Despite the emphasis on cognitive development in the national curriculum and research suggesting that young children are capable of and could benefit from activities that stress number skills and mathematical concepts, the term "mathematics" does not appear in the Icelandic national standards for playschools.

Research has reported that most 5-year-old children show varying degrees of number sense (Baroody, Lai, & Mix, 2006; Clements & Sarama, 2007). Traditionally, in Iceland, numeracy has not been included as a structured activity in playschools but rather it is discussed when opportunities arise during the children's free play. Research has shown that the frequency of children using numbers or number skills during play is relatively small. For example, one study of New Zealand children demonstrated that, during 70 hours of observations, 4-year-old preschoolers used number skills only 1.6% of the time during free play (Young-Loveridge, Carr, & Peters, 1995). Similar findings have been reported for preschool children in England. Gifford (1995, 2005) documented no observed instances of children using number skills during free play during a two-week period of preschool. These results suggest that preschoolers engage in few activities during free play that solicit the use and discussion of mathematical concepts and skills needed to facilitate learning numeracy. Gifford (2005) said this very elegantly by stating that the "laissez-faire approach to children learning in maths in the 'secret garden of play' does not work" (p. 2).

In this chapter, data from the pretest and the posttest from a four-year project on numeracy in one Icelandic playschool, fictitiously called Brekkukot Playschool, are presented. The larger study had three overarching goals. The first goal was to build on the Icelandic literature about girls' and boys' understanding of numeracy, to look for gender differences in relation to numeracy and mathematics problem-solving strategies in playschool. To our knowledge, no systematic investigations of mathematics and numeracy in Icelandic playschools have been undertaken. The second goal was to strengthen understanding about the role that teachers in playschools can have in developing children's understanding of mathematical concepts and skills that are needed to facilitate children's learning of numeracy. Lastly, the third goal was to provide evidence and arguments for the increasing importance of the role that playschools can play in developing children's mathematical competencies in Iceland. As mentioned previously, to date, the Icelandic national curriculum has not formalized goals related to promoting and fostering numeracy in playschools.

Gender Differences and Age

Much of the existing international literature suggests that gender differences in mathematics achievement typically emerge fairly late in students' careers. For example, in a meta-analysis, Hyde, Fennema, and Lamon (1990) reported that gender differences in mathematical problem solving did not generally exist in elementary and middle schools but frequently existed in high schools and college. Similarly, in the United States, Muller (1998) and Leahey and Guo (2001) found no gender differences in mathematics among middle school students and that differences emerged as students progressed through high school. A study from Iceland reported no gender differences in students' achievement in relationship to proportional reasoning, but significant gender differences were found in students' use of strategies (Steinthorsdottir & Sriraman, 2007).

The Programme for International Assessment (PISA), a survey of the knowledge and skills of approximately 400,000 15-year-old students (10th grade), in 2006 documented that males significantly outperformed females in terms of mathematics achievement in 35 of the 57 participating countries (OECD, 2007). Similar gender differences in mathematics achievement showing that males performed ahead of females were found in more than half of the participating countries in the 2000 and 2003 surveys (OECD, 2000, 2003). In PISA 2003, Iceland was the only country that reported a significant gender difference in favor of girls. That was not the case in PISA 2006; no gender differences were found in Iceland (OECD 2003, 2007).

The Trends in International Mathematics and Science Study 2003 study (TIMSS 2003) of fourth-grade (26 countries) and eighth-grade (46 countries) students documented that overall there were no gender differences in mathematics achievement. However, for some countries, there were notable performance differences; some favored girls while others favored boys. For example, in fourth grade, four countries showed significant gender differences in mathematics favoring girls, while seven countries had significant gender differences favoring boys. Similarly, in eighth grade, nine countries demonstrated gender differences in favor of girls, whereas 11 countries showed gender differences in favor of boys (Martin, Mullis, Gonzalez, & Chrostowski, 2004).

The TIMSS 2007 study reported a change in the gender dynamics. Similar to TIMSS 2003, there were no overall gender differences in mathematics achievement among fourth-grade students (37 countries). Data from individual countries in the fourth-grade study showed significant gender differences in favor of girls in eight countries and gender differences in favor of boys in 12 countries. The overall scores for eighth-grade students (50 countries) showed significant differences in favor of girls. Sixteen countries reported significant gender differences in favor of girls, compared to eight countries reporting significant differences in favor of boys. It is interesting that the largest differences in favor of girls were found in Middle Eastern countries (Mullis, Martin, & Foy, 2008). Iceland has not participated in the TIMSS studies.

During the years of compulsory education in Iceland (grades 1 through 10), the Icelandic National Mathematics Testing Program, a survey assessing mathematics performance, is administered in the 4th, 7th, and 10th grades. As mentioned earlier, in PISA 2003, a significant gender difference in mathematical achievement in favor of girls was reported in Iceland, contradicting the findings in numerous other countries that tended to show differences favoring boys. Some researchers questioned whether the PISA results were reliable for Iceland in the light of PISA 2006 in which no gender differences were found in Iceland. Olafsson, Halldorsson, and Bjornsson (2006) examined 10th-grade scores from the Icelandic National Mathematics program from 1996 to 2006 and found that gender differences in mathematics did exist favoring girls during most of the years studied. Sometimes the differences were found in urban areas, while other times in rural areas. Nonetheless, the Icelandic findings for 2003 mirrored the PISA 2003 and provided robust evidence that girls were outperforming boys in mathematics at the end of 10th grade.

In another study, the Educational Testing Institute in Iceland investigated gender differences in mathematics using data from the Icelandic National

Mathematics Testing Program about fourth- and seventh-grade students. Test scores from more than 65,000 students from 1996 to 2003 were analyzed. The findings from this study suggested that girls' overall achievement in mathematics is higher in each grade level, but that by seventh grade, the gender differences in favor of girls become larger (Skulason, Bjornsson, & Gunnarsson, 2004). Moreover, by 10th grade, girls were persistently outperforming boys in mathematics (S. Skulason, personal communication, November 8, 2008).

The literature on gender differences in mathematics in preschool and the early school years has been less prevalent and consistent. A growing number of international studies have reported gender differences in relation to the development of mathematical competencies and/or problem-solving strategies in the beginning school years. For example, in a large representative sample of third- and fourth-grade German students, boys were found to outperform girls in general mathematics achievement with differences attenuated somewhat by math content domains (Winkelmann, Heuvel-Panhuizen, & Robitzsch, 2008). In another study in the United States, Fennema, Carpenter, Jacobs, Franke, and Levi (1998) showed that although girls and boys were comparable in the proportion of correct solutions to math problems in the first through third grades, girls tended to use more concrete strategies, such as modeling and counting, whereas boys tended to use more abstract strategies, such as invented algorithms. In a third study, also from the United States, Carr and Jessup (1997) documented significant gender differences in mathematics problem-solving strategies used by first-grade children, reporting that boys used more complex strategies than girls, although the study revealed no significant differences between boys and girls in terms of the number of correct answers.

Few studies suggest that gender differences in mathematics may be rooted in phenomena prior to formal education or in the early years of schooling. For example, studies in the United States of gender differences in spatial abilities show that boys are superior to girls on mental rotation tasks beginning at the age of 4 years, 6 months (Levine, Huttenlocher, Taylor, & Langrock, 1999). Similarly, Levine, Vasilyeva, Lourenco, Newcombe, and Huttenlocher (2005) demonstrated that middle- and high-socioeconomic status (SES) boys in second and third grade in the United States score higher than girls on two spatial ability tasks. An increasing number of studies show that gender difference in mathematics can emerge as early as the first years of schooling or even prior to formal education. For example, descriptive results from the Early Childhood Longitudinal Study in the U.S., Kindergarten Class of 1998–99 (ECLS-K) showed that gender differences in mathematics achievement favoring boys emerge as early as first grade (Penner & Paret, 2007; Rathbun, West, & Germino-Hausken, 2004). On the other hand, findings have been reported from Finland showing that gender differences in numeracy, in favor of girls, can begin as early as the age of four (Aunio, Aubrey, Godrey, Pan, & Liu, 2008; Aunio, Hautamäki, Heiskari, & Van Luit, 2006). These studies imply that gender differences in mathematics can exist in preschool, but whether the differences are in favor of girls or boys is inconsistent.

In this chapter, we will examine whether playschool students' mathematics achievement and mathematics problem-solving strategies around numeracy differed for boys and girls. Using the pretest and posttest from the abovementioned project in Iceland, we examine variation in playschool children's number skills and strategies in relation to solving easy, moderate, and complex mathematical problems.

The numeracy work in Brekkukot Playschool was based on two distinctive theoretical frameworks that guided all the instruction and problem-solving activities. The word problems used for problem solving were based on a Cognitively Guided Instruction (CGI) framework (Carpenter, Fennema, Franke, Levi, & Empson, 1999), and the number games and the number activities stemmed from a Math Recovery framework from Australia (Wright, Stanger, Stafford, & Martland, 2006). During the school year, children engaged for about one hour in structured activities designed to foster numeracy and the acquisition of mathematical concepts four days a week: two times a week, children worked on word problems based on the CGI framework; and two times a week, the children worked on number games and activities, based on the Math Recovery framework. Two licensed playschool teachers were responsible for the numeracy work at Brekkukot Playschool. The teachers divided the work between them, with one teacher conducting problem-solving activities and the other carrying out number games and activities. Both teachers participated for all four years of the project.

METHOD

Sample

The overall sample consisted of 55 playschool children (26 boys and 29 girls) in their last year of one playschool in the metropolitan area of Reykjavik, Iceland. Our sample was one of convenience. Parental consent and participant assent were obtained from all participating students. The students in this study were all Caucasians from low- to high-middle-class homes. Unfortunately, we were not able to obtain the educational status of the parents, but anecdotally, we know that the parents' education attainment varied from high school diplomas to graduate degrees. Sample characteristics of the participating students largely mirrored those of the general playschool population and were reflective of the homogeneous nature of Iceland. The mean age of participants at the time of the posttest was 5.17 years (SD = 3.6 months).

Data were collected from four different cohorts of children over a period of four years beginning in the fall of 2002. Children were surveyed at the start of their last playschool year and again at the end of the school year. The school, classroom, and learning environments were similar for children in all cohorts in the study who all had the same playschool instructors. With the exception of gender (e.g., more boys in cohort 1 and fewer boys in cohorts 3 and 4), we found no differences within or across cohorts in terms of environment, background characteristics, and strategy use. Given the similarities across cohorts and that all children had the same playschool teachers, for the analysis presented below, the four cohorts of children are treated as one sample.

Measures and Procedures

Following established protocols, group administration procedures were used to survey children in the classroom using a CGI framework. The pre- and posttests were conducted as individual interviews in which the teacher read a problem to the students and asked them to solve it and think out loud. All pre- and posttests were videotaped and coded by the same two researchers. Prior work with children in kindergarten has investigated students' understanding of numeracy and the four operations (addition, subtraction, multiplication, division) by looking at strategy use and the type of problem that could successfully be solved (Carpenter et al., 1999). Carpenter and his colleagues developed a CGI framework that categorizes addition, subtraction, multiplication, and division problems according to the semantics of the problem. The researchers' work demonstrated that problem types affected students' strategy use (Carpenter et al., 1999; Carpenter, Ansell, Franke, Fennema, & Weisbeck, 1993). The CGI Kindergarten study investigated students' strategy use solving nine problems related to addition, subtraction, multiplication, and division in kindergarten. We adopted and modified the framework to include six of the original nine problems used in the Carpenter study (see Table 11.1). We eliminated three problems because of the length of the test and the time it took the children to solve all nine problems. The six problems were categorized based on the difficulty level of the problem: easy, Cronbach's α = .79; moderate, α = .76; and difficult. Following established protocols, problems using addition and subtraction were categorized as easy, problems using multiplication and division were considered moderate, and non-routine problems were classified as difficult (Carpenter, 1985).

TABLE 11.1   Pre- and Posttest Interview Problems

CGI problem type	Order given	Difficulty level	Problem
Separate result unknown	2	1	Elisa had 13 cookies. She ate 6 of them. How many cookies does Elisa have left?
Join change unknown	3	1	Thor has 7 kronas. How many more kronas does he have to have so that he can buy a lollipop that costs 11 kronas?
Comparison??	5	1	Jon has 12 balloons. Anna has 7 balloons. How many more balloons does Jon have than Anna?
Measurement division	4	2	Kitty had 15 guppies. She put 3 guppies in each bag. How many bags did Kitty put the guppies in?
Partitive division	6	2	Karl baked 20 cupcakes. He put the cupcakes into 4 boxes so that the same number of cupcakes was in each box. How many cupcakes did Karl put in each box?
Non-routine	7	3	19 children are taking a mini-bus to the zoo. They will have to sit either 2 or 3 to a seat. The bus has 7 seats. How many children will have to sit 3 to a seat, and how many can sit 2 to a seat?

The six problems were also coded according to the three strategies that the children used to solve the problem documented in the CGI framework. These three strategies were direct modeling, counting, and derived facts (Carpenter et al., 1993; Carpenter et al., 1999). Direct modeling is a strategy in which children use counters or drawings to directly model the relationships described in the problem. A characteristic of the direct modeling strategy is that all quantities are created by the student. Counting is a strategy in which children do not use counters or drawings to model the problem but count up or back from a given number. A distinctive feature of the counting strategies is that the child is able to keep one quantity in his or her mind. Derived facts is a strategy in which children use facts and/or derive new facts to find the answer to the problem (Carpenter et al., 1999).

RESULTS

Sample Characteristics

As shown in Table 11.2, aggregating all four cohorts, the sample was equally distributed across girls and boys (52.7% versus 47.3%). There were

TABLE 11.2   Pretest: Descriptive Statistics of Math Achievement and Problem-solving Strategies for Icelandic 5-year-old Children in Their Last Year of Playschool

	Boys	Girls	Total	p-value
Total children, N	26	29	55	—
Cohort 1, n	12	6	18	—
Cohort 2, n	6	5	11	—
Cohort 3, n	3	8	11	—
Cohort 4, n	5	10	15	—
Mean number of correct answers on the pretest, total	0.92	1.31	1.13	.345
Mean number of correct answers, easy	0.58	0.62	0.60	.855
Mean number of correct answers, moderate	0.31	0.52	0.42	.242
Mean number of correct answers, difficult	0.04	0.17	0.11	.116
Mean number of valid strategies on the pretest, total	1.08	1.52	1.31	.335
Mean number of valid direct modeling strategies	1.08	1.48	1.29	.378
Mean number of valid counting strategies	—	—	—	—
Mean number of valid derived fact strategies	—	—	—	—
Mean number of invalid no strategy use	4.92	4.48	4.69	.335

Note: No students used counting or derived fact strategies on the pretest.

fewer girls than boys in cohorts 1 and 2 and more boys than girls in cohorts 3 and 4, although none of the comparisons were statistically significant. On average, on the pretest, the participants answered correctly about one out of the six questions. The low rate of correct answers was not unexpected, as there had been no emphasis placed on numeracy prior to this project. Given the children's lack of numeracy skills in a problem-solving situation and exposure to formalized problem-solving strategies, not surprisingly, most children used invalid or no strategies when attempting to solve the given math problems. None of the children in the study used counting or derived facts as a strategy on the pretest. On the pretest, we found no evidence of gender differences in mathematics achievement in terms of the number of correct answers, difficulty of the problem, number of valid strategies, or type of strategy.

Posttest Means

Table 11.3 shows descriptive statistics for math achievement and problem solving at the end of the children's last year of playschool. By the end of the playschool year, on average, children answered correctly more than

TABLE 11.3   Posttest: Descriptive Statistics of Math Achievement and Problem-Solving Strategies for Icelandic 5-year-old Children in Their Last Year of PlaySchool

	Boys	Girls	Total	*p*-value
Mean number of correct answers on the posttest, total	4.15	4.55	4.36	.469
Mean number of correct answers, easy	2.23	2.31	2.27	.781
Mean number of correct answers, moderate	1.56	1.62	1.55	.447
Mean number of correct answers, difficult	0.46	0.62	0.55	.245
Mean number of valid strategies on the posttest, total	4.35	4.76	4.56	.448
Mean number of valid direct modeling strategies	4.04	4.59	4.33	.284
Mean number of valid counting strategies	0.27	0.17	0.22	.395
Mean number of valid derived fact strategies	—	—	—	—
Mean number of invalid/no strategy use	1.65	1.24	1.44	.448

Note: No students used the derived fact strategy on the posttest.

four of the six questions, making an improvement of approximately three questions or 53.8 percentage points (moving from 18.9 % to 72.7%) from the beginning to the end of the playschool year. Similarly, children were much more likely to use valid strategies on the posttest than the children were on the pretest. For example, children used valid strategies when attempting to solve a problem more than 76% of the time on the posttest compared to only 21.8% of the time on the pretest.

Two additional analyses were performed to assess for gender differences. First, in primary regression analyses, we examined whether boys and girls differed on math achievement at the end of the playschool year above and beyond the influence of pretest scores. After controlling for pretest scores, we found that gender was unrelated to all achievement outcomes on the posttest. In secondary regression analyses, we examined whether the effect of the pretest scores on posttest achievement differed for boys and girls. In these analyses, a pretest by gender interaction term was added in the second step to the primary regression model and tested for significance. We found no evidence to suggest that the effect of the pretest on mathematics achievement at the end of the playschool year differed according to gender.

DISCUSSION

This study examined gender differences in mathematical performance for a relatively small sample of playschool boys and girls in Iceland. Prior international research on gender differences in mathematical performance and problem-solving strategies has been mixed, depending on the age of the children studied and the country of origin. For example, studies in the United States typically show that gender differences in mathematics emerge in middle or later grades of schooling and generally favor boys (Levine et al., 2005; Penner & Paret, 2007; Rathbun et al., 2004). On the other hand, international findings are more mixed, with some countries reporting boys outperforming girls and others reporting girls doing better than boys depending on age (Martin et al., 2004). For example, studies with children in Iceland have suggested that gender differences favor girls and begin to emerge around the fourth grade, becoming more pronounced by 10th grade (Skulason et al., 2004).

Previous research in Iceland did not examine gender differences in mathematics achievement with playschool-age children. The fact that we found no differences between boys and girls in terms of mathematical performance and problem-solving abilities among five-year-old children in playschool is significant for two reasons. First, given prior research on Icelandic children, our findings suggest that gender differences may not emerge until sometime after the beginning of formal primary schooling. Clearly, more studies of playschool children are needed to validate and confirm our findings. Nonetheless, our results add to the complicated and sometimes contradictory research findings relating children's mathematical performance to gender (Aunio, Aubrey, Godfrey, Pan, & Liu, 2008; Carr & Jessup, 1997). Second, our results indicate that children improved their mathematics achievement between the pre- and posttests. This suggests that quality mathematics instruction should be part of the national playschool curriculum.

Future research should continue to study playschools that are implementing math instruction (1) to examine possible gender differences to validate the data presented here and (2) to begin to develop a national standard/curriculum in mathematics for playschools. In addition, future research should try to pinpoint where the gender differences emerge in primary school. Finally, future research should examine mathematics classrooms from first through fifth grade and study what the classroom contributes to the emergent gender differences in favor of girls in Iceland.

REFERENCES

Aunio, P., Aubrey, C., Godrey, R., Pan, Y., & Liu, Y. (2008). Children's early numeracy in England, Finland, and People's Republic of China. *International Journal of Early Years in Education, 16*(3), 203–221.

Aunio, P., Hautamäki, J., Heiskari, P., & Van Luit, J. E. H. (2006). The early numeracy test in Finnish—Children's norms. *Scandinavian Journal of Psychology, 47*, 369–378.

Baroody, A. J., Lai, M., & Mix, K. S. (2006). The development of young children's early number and operation sense and its implications for early childhood education. In B. Spodek & O. N. Saracho (Eds.), *Handbook of research on the education of young children*, 2nd ed. (pp. 187–221). Mahwah, NJ: Lawrence Erlbaum Associates.

Carpenter, T. P. (1985). Learning to add and subtract: An exercise in problem solving. In E. A. Silver (Ed.), *Teaching and learning mathematical problem solving: Multiple research perspectives* (pp. 17–40). Hillsdale, NJ: Lawrence Erlbaum.

Carpenter, T. P., Ansell, E., Franke, M. L., Fennema, E., & Weisbeck, L. (1993). Models of problem solving: A study of kindergarten children's problem-solving processes. *Journal for Research in Mathematics Education, 24*(5), 428–441.

Carpenter, T.P., Fennema, E., Franke, M.L., Levi, L., & Empson, S.B. (1999). *Children's mathematics: Cognitively guided instruction.* Portsmouth, NH: Heinemann.

Carr, M., & Jessup, D. (1997). Gender differences in first grade mathematics strategy use: Social, metacognitive, and attribution influences. *Journal of Educational Psychology, 89*(2), 318–328.

Clements, D. H., & Sarama, J. (2007). Early childhood mathematics learning. In F. K. Lester (Ed.), *Second handbook of research on mathematics teaching and learning* (pp. 461–555). Charlotte, NC: Information Age Publishing.

Fennema, E., Carpenter, T. P., Jacobs, V., Franke, M. L., & Levi, L. (1998). A longitudinal study of gender differences in young children's mathematical thinking. *Educational Researcher, 27*(5), 6–11.

Gifford, S. (1995). Number in early childhood. *Early Child and Development and Care, 109*, 95–115.

Gifford, S. (2005). *Teaching Mathematics 3–5: Developing learning in the foundation stages.* Berkshire, England: Open University.

Hyde, J. S., Fennema, E., & Lamon, S. J. (1990). Gender differences in mathematics performance: A meta-analysis. *Psychological Bulletin, 107*, 139–155.

Leahey, E., & Guo, G. (2001). Gender differences in mathematical trajectories. *Social Forces, 80*(2), 713–732.

Levine, S. C., Huttenlocher, J., Taylor, A., & Langrock, A. (1999). Early sex differences in spatial skills. *Developmental Psychology, 35*, 940–949.

Levine, S. C., Vasilyeva, M., Lourenco. S. F., Newcombe, N. S., & Huttenlocher, J. (2005). Socioeconomic status modifies the sex differences in spatial skill. *Psychological Science, 16*, 841–845.

Martin, M. O., Mullis, I. V. S., Gonzalez, E. J., & Chrostowski, S. J. (2004). *TIMSS 2003 international science report: Findings from IEA's trends in international mathematics and science study at the fourth and eighth grades.* Chestnut Hill, MA : International Study Center, Boston College, Lynch School of Education.

Muller, C. (1998). Gender differences in parental involvement and adolescents' mathematics achievement. *Sociology of Education, 71*, 336–356.

Mullis, I. V. S., Martin, M. O., & Foy, P. (2008). *TIMSS 2007 international: Findings from IEA's Trends in International Mathematics and Science Study at the fourth and eighth grades.* Chestnut Hill, MA: TIMSS & PIRLS International Study Center.

Olafsson, R. F., Halldorsson, A. M., & Bjornsson, J. K. (2006). Gender and the urban-rural differences in mathematics and reading: An overview of PISA 2003 results in Iceland. In J. Mejding & A. Roe (Eds.), *Northern lights on PISA 2003—A reflection form Nordic countries* (pp. 185–198). Copenhagen: Nordic Council of Ministers.

Olafsson, R. F., Halldorsson, A. M., Skulason, S., & Bjornsson, J. K. (2007). *Kynjamunur í PISA og samræmdum prófum 10. bekkjar* [Gender differences in PISA and the Icelandic national mathematics test in 10th grade]. Reykjavik: Namsmatstofnun.

Organisation for Economic Co-operation and Development (OECD). (2000). *Literacy skills for the world of tomorrow–further results from PISA 2000.* Paris: Author.

Organisation for Economic Co-operation and Development (OECD). (2003). *Learning for tomorrow's world: First results from PISA 2003.* Paris: Author.

Organisation for Economic Co-operation and Development (OECD). (2007). *Science competencies for tomorrow's world.* Paris: Author.

Penner A. M., & Paret, M. (2007). Gender differences in mathematics achievement: Exploring the early grades and the extremes. *Social Science Research, 37,* 239–253.

Rathbun, A., West, J., Germino-Hausken, E. (2004). *From kindergarten through third grade: Children's beginning school experiences.* Washington, DC: National Center of Educational Statistics.

Skulason, S., Bjornsson, J., & Gunnarsson, F. (2004, October). *Kynjamunur í námsárangri: Kynjamunur á samræmdum prófum 1996-2003* [Gender differences in achievements: Gender differences in the Icelandic National Mathematics Testing Program, 1996–2003]. Paper presented at the annual conferences at the University of Iceland, School of Education, Reykjavik, Iceland.

Steinthorsdottir, O. B., & Sriraman, B. (2007). Gender and strategy use in proportional situations: An Icelandic study. *Nordic Studies in Mathematics Education, 12*(34), 25–56.

Steinthorsdottir, O. B., & Sriraman, B. (2008). Exploring gender factors related to PISA 2003 results in Iceland: A youth interview study. *ZDM: The International Journal on Mathematics Education, 40*(4), 591–600.

Winkelmann, H., van de Heuvel-Panhuizer, M., & Robitzsch, A. (2008). Gender differences in the mathematics achievement of German primary school students: Results from a German large-scale study. *ZDM: The International Journal of Mathematics Education, 40*(4), 601–616.

Wright, R. J., Stanger, G., Stafford, A. K., & Martland, J. (2006). *Teaching number in the classroom with 4–8 year olds.* Thousand Oaks, CA: Sage.

Young-Loveridge, J., Carr, M., & Peters, S. (1995). *Enhancing the mathematics of four year olds: The EMI-4s study.* Hamilton, New Zealand: University of Waikato.

CHAPTER 12

MATHEMATICS TEACHER EDUCATION AND GENDER EFFECTS

Sigrid Blömeke
Humboldt University of Berlin

Gabriele Kaiser
University of Hamburg

There is a growing body of research about gender effects on student achievement in mathematics. Even if a lot of controversy still exists about how to *explain* differences in the achievement between girls and boys, the discussion on gender equity can be regarded as a well-established research field. The situation is very different when it comes to teacher education. Only a few studies exist that include gender-specific data about future teachers' achievement during their education (see, for example, the recent *International Handbook of Mathematics Teachers Education,* in which neither gender aspects under a general perspective nor gender differences in achievement are discussed in special chapters; Sullivan & Wood, 2008)—not least because teacher education research is still a fairly new field itself and evaluating teachers does not have a long tradition.

With the present study we intend to offer first steps in order to close this research gap. We examine gender differences in the knowledge of future lower-

International Perspectives on Gender and Mathematics Education, pages 263–283

secondary mathematics teachers in Germany. The study design does not only allow us to analyze the knowledge of male and female teachers *at the end of their education* but also to *compare different subgroups of lower-secondary teachers* (those trained as part of a primary program vs. those trained as part of an upper-secondary program) as well as different *cohorts* of future teachers (those at the beginning vs. those at the end of teacher education). Such a design allows the development of possible explanations if gender-related differences occur.

In the following, after a brief review of the state of research about gender-related differences in mathematical achievement, we describe the theoretical framework and the design of our study. The study described here is part of a larger cross-national assessment study called *Mathematics Teachers in the 21st Century (MT21)*, in which data about the effectiveness of mathematics teacher education in six countries is gathered (Blömeke, Kaiser & Lehmann, 2008; Schmidt et al., 2007; Schmidt, Blömeke, & Tatto, in press). We present our results about gender-related achievement differences based on the German data.

GENDER-RELATED DIFFERENCES IN MATHEMATICAL ACHIEVEMENT AND RESEARCH HYPOTHESES

For a long time, mathematics has been regarded as a male-dominated science. Two indicators continuously supported this view: Firstly, females were (and still are) strongly underrepresented in advanced mathematics courses in school as well as in science, technology, engineering, and mathematics professions (NRC, 1994). Even in countries with a long tradition of gender equity like Sweden, mathematics appears to have the strongest gender imbalance of almost all educational and professional fields from upper-secondary school on (Brandell, 2008). Secondly, females showed lower achievement compared to their male counterparts on mathematics tests in higher school grades (Fan, Chen, & Matsumoto, 1997). Whereas Hyde, Fennema, and Lamon (1990) did not find gender differences on the primary and lower-secondary school level in their meta-analysis of 100 studies published in English-speaking journals, they found significant achievement differences in favor of boys on the upper-secondary school and college level.

Explanations for gender inequity point either to biological differences (mainly in early publications, e.g.. Benbow & Stanley, 1980, 1983) or to socio-psychological aspects. The latter perspective stresses that gender is a socially constructed feature that results in gender stereotyping, with masculinity related to logic and femininity to emotion (Brandell, 2008). In this perspective, females receive less support and encouragement from teachers and parents in fields like mathematics, and they have fewer formal and informal opportunities to learn mathematics (Kimball, 1989; Henrion, 1997). These approaches emphasize, based on the stereotyping of mathematics as a male domain, that teachers tend to overrate male students' mathematics capability, that they have higher expectations for male students, and more positive attitudes about male students (Li, 1999). However, it is important to notice that gender inequity in mathematics cannot completely be explained by fewer learning opportunities. In a large longitudinal study from grades 8 through 12, Stockdale (1995) controlled for differences in class-taking patterns as well as for support from teachers and parents. The hypothesis that achievement differences between females and males would disappear was not supported by the data. Boys still scored significantly higher on mathematics tests than girls.

Some trends indicate changes in the gender-related achievement gaps, though. In most Western countries, the differences in mathematics achievement between boys and girls have decreased over the past decades (Burton, 2001; Forgasz & Leder, 2001). In the United States, for example, in a recent study Hyde, Lindberg, Linn, Ellis, and Williams (2008) did not find any relevant gender differences anymore from grades 2 through 11. Even at the upper-secondary school level the effect sizes were small (< 0.06). A review of studies about gender differences in mathematics in Australia and New Zealand published since 2004 mainly came to the same conclusion, although some inconsistencies had to be noticed with respect to content areas and grades (Vale & Bartholomew, 2008).

The main controversy left is the one about gender differences at the highest level of achievement. Vale and Bartholomew (2008) point to significantly lower proportions of females when it comes to this level in Australia. Correspondingly, Leder and Forgasz (2008) challenge the results of Hyde et al. (2008) and point out that their study did not include enough high-level cognitive tasks. According to them, it is at least questionable whether the study really shows that gender similarities characterize mathematics performance or whether the description given by Fennema (1974) thirty years ago is still valid; this study indicated that in primary school gender differences in mathematics could not be detected, although in upper-secondary school, male students outperformed females on cognitively demanding tasks.

International studies display varied results: In the PISA and TIMSS studies, in the majority of countries either no gender differences in mathematical achievement at primary and lower-secondary level were measured, or differences in favor of the boys occurred. However, there are a growing number of countries where the situation is reversed and females outperform male students. Students in a growing number of non-Western countries do not share a gender-specific perspective on mathematics anymore. A closer look at eight Asian-Pacific countries, for example, reveals no gender-related differences in six countries and significant differences in favor of girls in two countries (Dindyal, 2008). The same trend applies to many Arab countries. Girls and boys either did not show different levels of performance or girls outperformed boys (see Mullis, Martin, Gonzalez, & Chrostowski, 2004 for Jordan and Bahrein; Nasser & Birenbaum, 2005 for Israel, United Arab Emirates, and the Lebanon; OECD, 2007 for Qatar). The trend covers countries with very different levels of development and achievement.

Germany seems to represent a more typical Western country in this context. On the student level, there are still significant gender differences in mathematics achievement, and the differences are always in favor of boys (on the primary level, which covers grade 1 through 4 in Germany, see Pietsch & Krauthausen, 2006; Schwippert, Bos, & Lankes, 2003; and Winkelmann, van den Heuvel-Panhuizen, & Robitzsch, 2008; on the lower-secondary level, which comprises grade 5 through 10, see Schöps, Walter, Zimmer, & Prenzel, 2006 and Zimmer, Burba, & Rost, 2004; on the upper-secondary level, which covers grade 11 through 13, see Köller & Klieme, 2000 and Lehmann, Hunger, Ivanov, Gänsfuß, & Hoffman, 2004). These studies emphasize that differences in Germany are larger on the primary and lower-secondary level if the school type is controlled. Mathematics is required on these levels, and no further tracking takes place once students are sorted into three different types of school ("Hauptschule" which mainly prepares for blue-collar professions, "Realschule" which mainly prepares for white-collar professions, and the "Gymnasium" which mainly prepares for university studies) after grade 4. Students do not have choices which mathematics classes to take, and inequity fully manifests itself in achievement.

In the most recent German large-scale study by Winkelmann et al. (2008), evaluating the recently introduced standards for mathematics education with 10,000 students from grades three and four, boys outperformed girls and differences were higher in grade three than in grade four. However, substantial variation was found between the content domains and general mathematical achievement. In contrast to earlier findings from other studies, the girls' underperformance was highest in basic skills, and the advantage of boys in problem solving was relatively small. The boys' strongest advantage was on items requiring reproduction, whereas girls showed their strongest performance on items requiring drawing connections. The authors suggest that a possible explanation for these unexpected results could be inherent in the definition of problem solving in the German standards, which might differ from other studies. In any case, the results of this study make clear that studies of gender differences in mathematics achievement are strongly influenced by the measurement instruments and are not easily comparable.

On the upper-secondary level, German students have more choices in the academic type of school (the "Gymnasium") as well as in the vocational school system, which follows the "Hauptschule" and the "Realschule." Mathematics is still required, but now tracking within these types of schools takes place and students can decide whether they want to take mathematics courses on a basic or an advanced level. So, gender differences not only manifest themselves in achievement but also in the choice of courses. The proportion of female students opting for advanced courses in mathematics is significantly lower than the proportion of male students. Within those basic and advanced courses gender differences are less apparent. In the German TIMSS study for upper-secondary level, gender differences occurred neither in basic nor in advanced courses (Köller & Klieme, 2000). However, in a longitudinal study carried out in Baden-Württemberg and Hamburg, gender differences on the upper-secondary level are still in place even if the course level is controlled (Trautwein, Köller, Lehmann, & Lüdtke, 2007).

This heterogeneous situation continues in one of the rare studies on future teachers in mathematics carried out by Curdes, Jahnke-Klein, Lohfeld, and Pieper-Seier (2003) on a small-scale basis. Based on items of the TIMSS study they evaluated the mathematical knowledge of future mathematics teachers at the beginning of their university study and after several years of study. Only small gender differences in favor of the males were found at the beginning; the female future teachers even showed better results in self-developed proof items. The tests with more advanced future teachers showed greater differences between the males and females, although the differences in an additional test on spatial abilities did not yield significant differences. Due to the fact that no differentiation between different types of future teachers (e.g., according to their license level) were made, the results are difficult to interpret.

Based on these inconsistent results, we formulate our working hypotheses, taking into account the different forms of teacher education for lower-secondary mathematics teachers common at German universities. Future teachers can opt between two main forms of teacher education for the lower-secondary level: they can choose to be trained either in a combined primary and lower-secondary program or in a combined lower- and upper-secondary program. Significant differences exist in the proportion of males and females in these programs and in the levels of mathematics taught.

1. We hypothesize that more males opt for the secondary mathematics program, which is highly loaded with mathematics courses on the level of diploma students (equivalent to a master study in mathematics), whereas females prefer the combined primary and lower-secondary program, which is more orientated towards pedagogy.
2. Significant gender differences in favor of males exist in the future teachers' achievement on our test with respect to mathematics knowledge as well as to the knowledge of mathematics pedagogy if the program level is not controlled. Taking into account the uncertain state of research about gender differences when the course level is controlled, we have to leave it open whether the gender differences assumed would still occur if the program level is controlled.

THEORETICAL FRAMEWORK

The basis of our study, Mathematics Teaching in the 21st Century (MT21), is the concept of professional competence as it is defined in cognitive psychology. Competence in this tradition means to have the knowledge and the skills to solve job-related tasks. Weinert (1999, 2001) divides competence into cognitive abilities and skills on the one hand and into the motivational, volitional, and social willingness and skills to successfully and responsibly apply solutions in variable situations. In our study this means teachers' professional knowledge on the one hand and their professional beliefs and job motivation on the other hand.

What the structure of professional competence looks like was modeled according to existing theories of teacher knowledge and teacher beliefs. Regarding the knowledge facets, we followed Shulman's (1985) distinction of content knowledge, pedagogical content knowledge, and general pedagogical knowledge (see also Blömeke, 2002; Baumert & Kunter, 2006). This means in our case mathematics knowledge, knowledge of mathematics pedagogy, and general pedagogical knowledge. The two content-specific facets were further subdivided into the areas of number, algebra, functions, geometry, and data as well as into the cognitive demands algorithms, problem solving, and modeling in order to have a fine-grained heuristic for the development of items. In analyses of the German data, a distinction between levels of mathematical complexity was introduced as well—namely, school mathematics for lower and upper-secondary mathematics, school mathematics from a higher point of view, and university mathematics, relating to special German perspectives (Kirsch, 2004).

The pedagogical content knowledge facet was further subdivided as well, into typical teacher tasks: instructional planning knowledge, knowledge about student learning, and curricular knowledge. The instructional planning knowledge is needed before instruction begins. The mathematics content for students must be chosen appropriately, simplified, and prepared with the use of various representations (Krauthausen & Scherer, 2007; Vollrath, 2001). The knowledge about student learning is related to the learning processes that occur during instruction. The *MT21* study focuses on teacher–student interactions. Such knowledge includes classifying student answers, verbal or written, in relation to the tasks or questions that stimulated them, asking questions at different levels of complexity, identifying common misconceptions, providing feedback, and reacting with appropriate intervention strategies. Such knowledge is assigned a prominent position, since it has been found relevant to advanced student performance (Blum et al., 2004; Helmke, 2004; Klieme, Schümer, & Knoll, 2001). The curricular knowledge focuses on the development and progression of student competence in mathematics during the years of schooling. Teachers have to deal with questions like: What would it mean for later lesson units, for example, if a classic subject area of school mathematics in the lower secondary curriculum were removed or taught as a part of a different sequence?

Regarding beliefs, we followed Richardson's (1996, p. 102) definition of beliefs as "psychologically held understandings, premises, or propositions about the world that are felt to be true" and we distinguish between epistemological beliefs, instructional beliefs and more general beliefs about the role of schooling and teaching. In addition, we surveyed self-centered beliefs like job motivation and personality features.

A visualization of the different facets of future teachers' professional competence as we conceptualize it in *MT21* is displayed in Table 12.1.

The distinction of pedagogical content knowledge and content knowledge has been intensely debated, especially concerning whether a separation is possible and how to conceptualize it. In their fundamental discussion on the development of Shulman's theoretical approach, Graeber and Tirosh (2008) point out that to a certain extent the concept of pedagogical content knowledge is still elusive, clarified more often by lists of examples than by theoretical reflections. In their attempt to expand Shulman's work, Ball and Bass (2003) introduce the notion of mathematical knowledge for teaching that describes the tasks of teachers as to "unpack" the compressed abstract mathematical ideas within the teaching process (Hill, Ball, & Schilling, 2008). They discriminate further areas of knowledge such as common content knowledge and specialized content knowledge, knowledge of content and students and knowledge of content and teaching.

TABLE 12.1   Facets of Mathematics Teachers' Professional Competence

	Mathematics Knowledge	Mathematics Pedagogy Knowledge	General Pedagogy Knowledge	Beliefs
Content Areas	Number, Algebra, Functions, Geometry, Data		Education, Psychology, Sociology	Mathematics, Generic
Level of complexity	School mathematics for lower and upper-secondary schools, school mathematics from a higher point of view, and university mathematics			
Teacher Tasks		Curricular knowledge, Planning knowledge, Knowledge of enacting mathematics	Lesson planning, Classroom management, Assessment	Nature of mathematics, Mathematics teaching and learning, Schooling

We cannot go into further details at this point but want to emphasize that the conceptualization of pedagogical content knowledge of our study refers beyond others to the German tradition of "Didaktik" (amongst others, it is similar to the French tradition of "didactique") which has strong philosophical roots and develops models for teaching and learning processes in a deductive way which is nevertheless informed by empirical research. Teachers' content knowledge as related to the content to be taught plays traditionally a very important role in these approaches (Pepin, 1999).

STUDY DESIGN

Sample

Data collection for *MT21* was carried out in 2006 in six countries: in Bulgaria, Germany, Mexico, South Korea, Taiwan, and the USA. The selection of countries was based on three criteria using data from existing international comparisons:

1. The set of countries had to cover the main types of teacher education (one-step university programs or two-step programs with university studies first followed by extended practical experiences; specialized lower-secondary programs, combined primary and lower-secondary programs, or combined lower and upper-secondary programs; one-subject programs or multiple-subject programs; programs of different length).
2. Student achievement as shown in TIMSS and PISA had to vary across the selected countries.
3. The countries had to show at least a middle grade degree of industrialization in order to avoid serious bias through socioeconomic differences.

The number of countries had to be restricted to six countries in order to make optimal use of the data and insights garnered through a discourse-oriented kind of project work (see Schmidt et al., 1996).

Within countries, the sampling was done at the institutional level. The goal was to obtain a reasonably representative sample for each country, including the variation found across all teacher-education institutions in the country. Four criteria were applied to accomplish this goal: type, size, location, and reputation of teacher-education institutions (see Table 12.2). The total number of institutions sampled across the six countries was 34. Within each sampled institution, the goal was to survey all eligible future teachers. The total number of future teachers sampled was 2,628. To ensure that

TABLE 12.2 Sampling in MT21

Countries	Bulgaria	Germany	Mexico	Korea	Taiwan	US
Criteria for Selecting Institutions	Size	Type, Size, Location, Reputation	Location	Reputation	Type, Size, Reputation	Type, Size, Location, Reputation
No. of Institutions	3	4 (1st step) + 22 (2nd step)	5	4	5	12
Cohorts (1 = beg., 2 = mid, 3 = end)	1 + 3	1 + 2 + 3	1 + 3	1 + 3	1 + 2 + 3	1 + 2 + 3
Number of FTs	161	849	358	210	668	382
Program Levels	Lower + Upper Sec.	Element. + Lower Sec.; Lower + Upper Sec.	Lower Sec.	Lower + Upper Sec.	Lower + Upper Sec.	Element. + Lower Sec., Lower Sec., Lower + Upper Sec.
Length of ed. (years)	4	3.5 + 1.5, 4.5 + 2	4	4	4	4 (5)

samples were as representative as possible, the results were weighted according to the total number of lower-secondary teachers prepared in every institution and the proportion of missing responses. The future teachers sampled were subdivided into two or three cohorts respectively: students at the beginning of their program, mid-term students, and students at the end of teacher education. This was done to estimate what individuals actually gain in their teacher education program, at least as estimated by cohort longitudinal data (Keeves, 1992; Wiley & Wolfe, 1992).

The subset of data we used in the present chapter is the German data. Eight-hundred-forty-nine future lower-secondary mathematics teachers from four regions were part of the study. We selected all universities and surrounding state seminars in these regions (i.e., four universities and 22 surrounding state seminars). In Germany, teacher education consists of two consecutive steps. First, academic education in mathematics, mathematics pedagogy, and general pedagogy takes place at university. Second, a practical training step takes place at small institutions specialized on teacher education and run by the state. We aimed at a complete census at those institutions that took part in our study. Students in the second institution represent future teachers at the end of teacher education (cohort 3: "finishers"). Post hoc, we divided the university group into two cohorts, which we labeled as "beginners" (cohort 1) and "mid-term future teachers" (cohort 2). We have a response rate of 87% in the group of finishers and a response rate of 55% in the group of beginners, both of which we will analyze in this chapter. To make this non-random sample as representative as possible, the results were weighted according to license level (combined primary and lower secondary vs. secondary program), cohort, and institution future teachers belong to.

As mentioned, lower-secondary teachers in Germany are trained in two different programs. The combined primary and lower-secondary program usually lasts for 3.5 years at university, followed by 1.5 years practical training at the state seminars. These future teachers would have received roughly 35 to 40 credit points in mathematics. In contrast, lower-secondary teachers as part of a secondary program receive at least 60 credit points in mathematics during a 4.5-year-long university education, and in most federal states they also benefit from a longer practical training.

Whereas the entrance requirements formally are the same for the two programs (final secondary school-examination "Abitur"), self-selectivity, different ratios of applicants and study places available, and different drop-out rates lead to different characteristics of the two groups of future teachers. According to a *t-test* of mean differences, female finishers in the secondary program had a significantly better grade-point average (GPA) in their final school examination than their male counterparts, whereas there were no significant differences between females and males at the beginning of teacher education (see Table 12.3; a grade-point average of "1" is the best

TABLE 12.3 Gender-specific Grade Point Average in the Final Secondary School Examination ("Abitur")

	Elementary-Lower Secondary						Upper-Lower Secondary					
	Cohort 1: Beginners (n = 294)			Cohort 3: Finishers (n = 133)			Cohort 1: Beginners (n = 74)			Cohort 3: Finishers (n = 153)		
Female	Male	Diff.	Female	Male	Diff.	Female	Male	Diff.	Female	Male	Diff.
2.6	2.7	ns	2.4	2.5	ns	2.5	2.4	ns	1.9	2.1	$t = -2.61$*

Note: ns = not significant, * $p < .05$

TABLE 12.4 Gender-specific Proportion of Classes in Advanced Mathematics (in %)

	Elementary-Lower Secondary						Upper-Lower Secondary					
	Cohort 1: Beginners (n = 294)			Cohort 3: Finishers (n = 133)			Cohort 1: Beginners (n = 74)			Cohort 3: Finishers (n = 153)		
Female	Male	Diff.	Female	Male	Diff.	Female	Male	Diff.	Female	Male	Diff.
40.3	53.4	$\chi^2 = 3.86$*	50.6	76.8	$\chi^2 = 9.37$**	62.5	80.0	ns	82.8	85.3	ns

Note: ns = not significant, * $p < .05$, ** $p < .01$

mark and a GPA of "6" is the worst mark, with a GPA lower than 4.0 counting as failed). In contrast, according to a *chi-square test* of the frequency distribution, males in the combined primary and lower-secondary program had a course in advanced mathematics to a significantly larger extent than their female counterparts—and this holds at the beginning and even more at the end of teacher education (see Table 12.4).

Instruments

Test items were distributed over two forms in order to make the best use of future teachers' limited study time. Overall, the survey took 90 minutes with the testing part on mathematics and mathematics pedagogy lasting for 60 minutes. Due to the matrix design, the Rasch model (one-parameter Item-Response-Theory model) was used to obtain scale scores. The two dimensions—knowledge in mathematics and knowledge in mathematics pedagogy—were scaled simultaneously with no overlapping item loadings on the two factors (simple-structure modeling or between-item multidimensionality). With this procedure the conceptual overlap between mathematics and mathematics pedagogy becomes visible in the shared covariance, which is high: $r = 0.81$. The resulting parameters were transformed so that the mathematics pedagogy scale as the scale with the lower mean had a mean of 100.00 and a standard deviation of 20.00. For mathematics as the scale with the higher mean, the transformation led to a mean of 101.61 and a standard deviation of 23.25.

Copyright restrictions prevent the documentation of items used in the final test forms. However, Figure 12.1 is an example of an item included in the pool of items piloted addressing the topic of functions and very similar

For each of the statements below, indicate the largest set from which the exponents can come to make the statement true for all $x > 0$.

Check **one box** in each **row**.

	Natural numbers	Integers	Rational numbers	Real numbers
1. $x^a x^b = x^{a+b}$	☐	☐	☐	☐
2. $\dfrac{x^a}{x^b} = x^{a-b}$	☐	☐	☐	☐
3. $\dfrac{x^a}{x^a} = 1$	☐	☐	☐	☐
4. $\sqrt[q]{x^p} = x^{\frac{p}{q}}$, $q \neq 0$	☐	☐	☐	☐

Figure 12.1 Example of a *mathematics knowledge* item.

The following problem has proven very difficult for many students.

A rectangular shaped swimming pool has a paved walkway around it as shown.

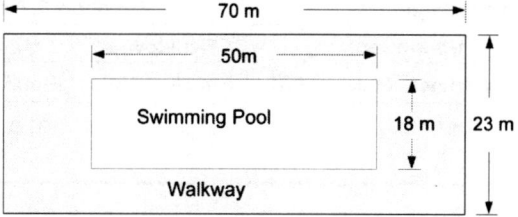

What is the area of the paved walkway?

What do you think is the *main reason* this particular item is so difficult for students?

Check **one** box.

1. Area measurement items are always difficult for students..................... ☐
2. There is no simple formula to solve this problem. .... ☐
3. They do not know that an area can be found as the difference of two other areas... ☐
4. They are confused by the difference between perimeter and area.......... ☐
5. There are too many numbers and calculations for them to deal with...... ☐
6. I don't know.. ☐

Figure 12.1 Example of a *mathematics pedagogy knowledge* item.

to items in the final test. An example of a mathematics pedagogy item is shown in Figure 12.2.

RESULTS

Almost two thirds of the future teachers in our sample are females (63% vs. 37%). As expected, the proportion of females is substantially higher in the less subject-matter focused combined elementary and lower-secondary teacher-education program (72% vs. 28% males) than in the combined upper- and lower-secondary program (44% vs. 56% males). The gender imbalance between the two programs applies both to beginners and finishers of teacher education (see Table 12.5). In both cases, the proportion of females is significantly lower at the end compared to the beginning of the program which might indicate a higher drop-out rate of female future teachers during teacher education.

Referring to the knowledge level of female and male future teachers, the female group shows significantly lower mathematics knowledge than the male group—and this holds at the beginning as well as at the end of teacher education (see Table 12.6, rows C1 and C3). The effect size at the

TABLE 12.5 Proportion of Female and Male Future Lower-Secondary Mathematics Teachers in Germany (in %)

Cohort 1: Beginners					Cohort 3: Finishers				
Elem-lowsec		Up-lowsec			Elem-lowsec		Up-lowsec		
Female	Male	Female	Male	Diff.	Female	Male	Female	Male	Diff.
75.2	24.8	54.1	45.9	$\chi^2 =$ 12.78***	57.9	42.1	37.9	62.1	$\chi^2 =$ 11.40**

$** p < .01, *** p < .001$

TABLE 12.6 Gender-Specific Achievement of Lower-Secondary Future Teachers at the Beginning (C1) and at the End of Teacher Education (C3)

	Mathematics Knowledge					Mathematics Pedagogy				
	Female	Male	Diff.	F	η^2	Female	Male	Diff.	F	η^2
C1	88.7	98.0	+9.3	13.64***	.04	91.9	96.1	+4.2	3.32ns	—
C3	107.0	113.9	+8.9	5.79*	.02	101.9	109.7	+7.8	11.38**	.04
Diff. C1–C3	+18.3	+15.9				+10.0	+13.6			

Note: Effect size η^2: 0.01 small effect, 0.06 medium effect, 0.14 large effect.
   ns = not significant, $* p < .05, ** p < .01, *** p < .001$

end is relatively small, though. It seems as if females gain *more* mathematics knowledge during teacher education than males, which is supported by the difference in score points between the beginning and the end of teacher education (see row "Diff. C1 – C3").

With respect to the knowledge of mathematics pedagogy, the difference between female and male future teachers at the beginning of teacher education is not significant. What is interesting, though, is that in this case *males* seem to gain more mathematics pedagogy knowledge than females during teacher education so that at the end there is a significant difference in favor of male future teachers.

The differences between males and females become even more visible if one takes background characteristics into account. The grade point average in the final secondary school examination and the kind of mathematics courses taken at upper-secondary school level strongly correlate with the knowledge in mathematics and mathematics pedagogy. If these variables are controlled, the gender effect does not increase at the end of teacher education but at the beginning: With respect to mathematics knowledge it increases from to $\eta^2 = .04$ to $\eta^2 = .06$ ($F = 31.43***$), and with respect to knowledge of mathematics pedagogy, the gender effect now becomes a significant effect: $\eta^2 = .02$ ($F = 8.28^{**}$).

While these results support our second hypothesis that significant gender differences in favor of males exist in the future teachers' achievement, we had to leave it an open question as to whether the differences still occur if we control the program. In fact, if one takes into account the program that offers different opportunities to learn mathematics and mathematics pedagogy, the picture changes remarkably. Then, the effect of gender on mathematics knowledge is less apparent. Controlling for the program ($F = 5.57^{*}$, $\eta^2 = .02$), males still have significantly more mathematics knowledge than females at the beginning of teacher education, but the effect size is already very small: $F = 11.23^{**}$, $\eta^2 = .03$. At the end of teacher education, the effect has completely disappeared. Only the program influences future teachers' knowledge of mathematics, and this to a large extent: $F = 67.08^{***}$, $\chi^2 = .19$. Once this feature is controlled, males and females do not differ significantly in their mathematics knowledge anymore: $F = 1.34$, ns. None of these results change in a relevant way if background characteristics are taken into account.

With respect to the knowledge of mathematics pedagogy, the gender effect again is even less relevant once we control for the program. In this case, there is no significant difference between males and females at the beginning of teacher education: $F = 0.25$, ns, whereas a small difference in the knowledge between future teachers of different programs exists: $F = 7.72^{**}$, $\eta^2 = .02$. The program effect at the end is—as one would expect—large: $F = 30.76^{***}$, $\eta^2 = .10$. In addition, a significant but only small gender effect exists: $F = 5.93^{*}$, $\eta^2 = .02$. Again, these results do not change in a relevant way when the background is controlled.

The analyses on our second hypothesis can be summarized in a regression analysis that demonstrates that even if there is a gender effect at single time points, its practical relevance is relatively low compared to other factors (see Table 12.7). With respect to mathematics knowledge, the variance explained by belonging to different programs is of only medium relevance at the beginning ($R^2 = .05$) but high at the end of teacher education ($R^2 = .20$); the variance explained by background is—as one would expect—high at the beginning ($R^2 = .13$) but of lower relevance at the end ($R^2 = .05$); the variance explained by gender is very low at both time points ($R^2 = .03$ at the beginning and $R^2 = .01$ at the end of teacher education). With respect to knowledge of mathematics pedagogy, the variance explained by belonging to different programs is $R^2 = .05$ at the beginning and $R^2 = .11$ at the end of teacher education; the variance explained by background is $R^2 = .07$ at the beginning and $R^2 = .06$ at the end; and the variance explained by gender is $R^2 = .00$ at the beginning and $R^2 = .03$ at the end of teacher education.

TABLE 12.7 Linear Regression of Knowledge of Mathematics and Mathematics Pedagogy on Program, Background and Gender

Variable	Mathematics Knowledge (C1/C3)			Mathematics Pedagogy (C1/C3)		
	B	T	p	B	T	p
Program	6.08/15.99	2.33/5.60	*/***	7.62/7.45	3.07/3.07	**/**
GPAS	−8.75/−8.10	−4.76/−3.49	***/**	−4.92/−8.94	−2.82/−4.54	**/***
AdvMath	9.84/7.52	4.68/2.49	***/*	7.82/2.30	3.91/0.89	***/ns
Gender	8.87/4.10	3.91/1.55	***/ns	2.81/6.94	1.30/3.08	ns/**

Note: GPAS: Grade point average in final secondary school-examination (inverse metric: "1" is the best mark, "4" the worst but still successful – therefore the negative sign)
AdvMath: Kind of mathematics course taken at upper-secondary level (0/1, 1 = advanced mathematics)
ns = not significant, * $p < .05$, ** $p < .01$, *** $p < .001$

CONCLUSION

The study *MT21* was carried out in six countries, beyond others in Germany, to examine the knowledge and the beliefs of future lower-secondary mathematics teachers and how they develop during teacher education in relation to opportunities to learn provided in different programs. The present study uses the German data in order to test two hypotheses about gender effects in teacher education: (1) that as a matter of self-selectivity, males prefer the combined lower and upper-secondary program which is heavily loaded with mathematics whereas females prefer the combined primary and lower-secondary program which is more orientated towards pedagogy; (2) that significant gender differences in favor of males exist in the future teachers' mathematics and mathematics pedagogy knowledge which was measured with a standardized test.

The analyses reveal that in fact significantly more males enroll in the secondary teacher-education program than females. The latter dominate instead the combined elementary and lower-secondary program. If one neglects this difference in enrollment numbers, male future teachers significantly outperform their female counterparts on the *MT21* mathematics and mathematics pedagogy tests, and this at the beginning as well as at the end of teacher education. Once the program is taken into account, the results shift. The gender difference is either not significant or it is of low practical relevance. So, it seems as if in teacher education the same mechanisms take place as on the student level: gender inequalities mainly manifest themselves in different choices, less in actual achievement when these choices are controlled.

It is plausible to attribute differences in choices to differences in attitudes. We do not have the possibility to test this hypothesis with our data, but we know from other studies that females hold more negative attitudes towards mathematics than males (Hyde et al., 1990; Nosek, Banaji, & Greenwald, 2002). On average, they have received less support and encouragement from teachers and parents while growing up and so, in a field like mathematics, they internalize a self-perception as inferior to boys (Kimball, 1989; Henrion, 1997). In contrast, pedagogy is often declared to be a "natural" choice of women. The split between mathematics and pedagogy would also explain why the gender difference on the mathematics pedagogy test is smaller than on the mathematics test.

What is interesting, though, is that it seems as if female future teachers are able to close part of the knowledge gap in mathematics during teacher education. This indicates higher learning gains like Lehmann (2006) stated for student achievement during lower-secondary schooling. These results from a true longitudinal study in the German federal state of Hamburg were confirmed by another true longitudinal study in Germany added to PISA 2003 with a lower-secondary student sample from grade 9. Even if girls entered a required mathematics class with a lower knowledge level than boys, they seemed to learn more during the school year so that gender-related differences decreased (Schöps et al., 2006). Although there is still a lot of evidence that gender differences actually increase during schooling in the long run, and although we have to be careful with too far-reaching conclusions based on our data set because *MT21* was a cross-sectional study, we have to point out that the differences in the achievement between males and females at the end of teacher education are lower than at the beginning.

Against this background, one would hope for the future that true longitudinal studies will be carried out also in the field of teacher education. Such studies could clarify a lot of open issues and provide much more insight into gender effects and how they develop. Cross-sectional data always have to live with the caveat that they cannot control for independent variables that might influence the outcome measures. Furthermore, it is not possible to decide about the chicken-and-egg causality dilemma based on this kind of data, which means that even if a study is designed very carefully, conclusions about the causes of effects are difficult to draw. In such future longitudinal studies, not only cognitive measures should be taken into account but also affective/motivational ones. They belong to teachers' competence from a conceptual point of view, and we have evidence that they are related to cognitive achievement from an empirical point of view.

REFERENCES

Ball, D. L., & Bass, H. (2003). Interweaving content and pedagogy in teaching and learning to teach: Knowing and using mathematics. In J. Boaler (Ed.), *Multiple perspectives on mathematics teaching and learning* (pp. 83–104). Westport, CT: Ablex Publishing.

Baumert, J., & Kunter, M. (2006). Stichwort: Professionelle Kompetenz von Lehrkräften. *Zeitschrift für Erziehungswissenschaft, 9,* 469–520.

Benbow, C. P., & Stanley, J. C. (1980). Sex differences in mathematical ability: Fact or artifact? *Science, 210*(4475), 1262–1264.

Benbow, C. P., & Stanley, J. C. (1983). Sex differences in mathematical reasoning ability: More facts. *Science, 222*(4627), 1029–1031.

Blömeke, S. (2002). *Universität und Lehrerausbildung.* Bad Heilbrunn: Klinkhardt.

Blömeke, S., Kaiser, G., & Lehmann, R. (Eds.). (2008). *Professionelle Kompetenz angehender Lehrerinnen und Lehrer. Wissen, Überzeugungen und Lerngelegenheiten deutscher Mathematikstudierender und -referendare – Erste Ergebnisse zur Wirksamkeit der Lehrerausbildung.* Münster: Waxmann.

Blum, W., Neubrand, M., Ehmke, T., Senkbeil, M., Jordan, A., Ulfig, F., & Carstensen, C.H. (2004). Mathematische Kompetenz. In M. Prenzel, J. Baumert, W. Blum, R. Lehmann, D. Leutner, M. Neubrand et al. (Eds.), *PISA 2003. Der Bildungsstand der Jugendlichen in Deutschland – Ergebnisse des zweiten internationalen Vergleichs* (pp. 47–92). Münster: Waxmann.

Brandell, G. (2008). Progress and stagnation of gender equity. Contradictory trends within mathematics research and education in Sweden. *ZDM—The International Journal on Mathematics Education, 40*(4), 659–672.

Burton, L. (2001). Fables: The tortoise? The hare? The mathematically underachieving male? In B. Atweh, H. Forgasz, & B. Nebres (Eds.), *Sociocultural research on mathematics education—An international perspective* (pp. 379–392). Mahwah, NJ: Lawrence Erlbaum Associates.

Curdes, B., Jahnke-Klein, S., Lohfeld, W., & Pieper-Seier, I. (2003). *Mathematikstudentinnen und –studenten – Studienerfahrungen und Zukunftsvorstellungen.* Wissenschaftliche Reihe NFFG 5. Norderstedt, BoD.

Dindyal, J. (2008). An overview of the gender factor in mathematics in TIMSS-2003 for the Asia-Pacific region. *ZDM – The International Journal on Mathematics Education, 40*(4), 993–1005.

Fan, X., Chen, M., & Matsumoto, A. R. (1997). Gender differences in mathematics achievement. Findings from the National Education Longitudinal Study of 1988. *Journal of Experimental Education, 65,* 229–242.

Fennema, E. (1974). Sex differences in mathematics achievement: A review. *Journal for Research in Mathematics Education, 5,* 126–139.

Forgasz, H., & Leder, G. (2001). "A+ for girls, B for boys": Changing perspectives on gender equity and mathematics. In B. Atweh, H. Forgasz, & B. Nebres (Eds.), *Sociocultural research on mathematics education—An international perspective* (pp. 347–366). Mahwah, NJ: Lawrence Erlbaum Associates.

Graeber, A., & Tirosh, D. (2008). Pedagogical content knowledge: Useful concept or elusive notion. In P. Sullivan, & T. Woods (Eds.), *Knowledge and beliefs in mathematics teaching and teaching development. The international handbook of mathematics teacher education*, Vol. 1 (pp. 117–132). Rotterdam: Sense Publisher.

Helmke, A. (2004). *Unterrichtsqualität: Erfassen, Bewerten, Verbessern* (3rd ed.). Seelze: Kallmeyersche Verlagsbuchhandlung.

Henrion, C. (1997). *Women in mathematics. The addition of difference*. Bloomington: Indiana University Press.

Hill, H. C., Ball, D. L., & Schilling, S. G. (2008). Unpacking pedagogical content knowledge: Conceptualising and measuring teachers' topic-specific knowledge of students. *Journal for Research in Mathematics Education, 39*(4), 372–400.

Hyde, J. S., Fennema, E., & Lamon, S. J. (1990). Gender differences in mathematics performance: A meta-analysis. *Psychological Bulletin, 107(2),* 139–155.

Hyde, J. S., Lindberg, S. M., Linn, M. C., Ellis, A. B., & Williams, C. C. (2008). Diversity: Gender similarities characterize math performance. *Science, 321*(5888), 494–495.

Keeves, J. P. (1992). *Methodology and measurement in international educational surveys.* The Hague, The Netherlands: IEA.

Kimball, M. M. (1989). A new perspective on women's math achievement. *Psychological Bulletin, 105*(22), 198–214.

Kirsch, A. (2004). *Mathematik wirklich verstehen* (4th ed.). Köln: Aulis Verlag Deubner

Klieme, E., Schümer, G., & Knoll, S. (2001). Mathematikunterricht in der Sekundarstufe I. „Aufgabenkultur" und Unterrichtsgestaltung. In Bundesministerium für Bildung und Forschung (Eds.), *TIMSS – Impulse für Schule und Unterricht* (pp. 43–57). Bonn: BMBF.

Köller, O., & Klieme, E. (2000). Geschlechtsdifferenzen in den mathematisch-naturwissenschaftlichen Leistungen. In J. Baumert, W. Bos, & R. Lehmann (Eds.), *TIMSS/ III: Dritte Internationale Mathematik und Naturwissenschaftsstudie – Mathematische und naturwissenschaftliche Bildung am Ende der Schullaufbahn. Bd. 2: Mathematische und physikalische Kompetenzen am Ende der gymnasialen Oberstufe* (pp. 373–404). Opladen: Leske + Budrich.

Krauthausen, G., & Scherer, P. (2007). *Einführung in die Mathematikdidaktik.* München: Elsevier.

Leder, G., & Forgasz, H. (2008). Mathematics education: New perspectives on gender. *ZDM—The International Journal on Mathematics Education, 40*(4), 513–518.

Lehmann, R., Hunger S., Ivanov, St., Gänsfuß, R., & Hoffmann, E. (2004). *Aspekte der Lernausgangslage und der Lernentwicklung – Klassenstufe 11 (LAU 11).* Hamburg: Behörde für Bildung und Sport.

Lehmann, R. (2006). Mädchen und Mathematik in der gymnasialen Sekundarstufe I – Ergebnisse einer Längsschnittstudie. In I. Hosenfeld, & F.-W. Schrader (Eds.), *Schulische Leistung. Grundlagen, Bedingungen, Perspektiven* (pp. 107–120). Münster: Waxmann.

Li, Q. (1999). Teachers' beliefs and gender differences in mathematics: A review. *Educational Research, 41*(1), 63–76.

Mullis, I. V. S., Martin, M. O., Gonzalez, E. J., & Chrostowski, S. J. (2004*). Findings from IEA's Trends in International Mathematics and Science Study at the fourth and eighth grades.* Chestnut Hill, MA: Boston College.

Nasser, F., & Birenbaum, M. (2005). Modelling mathematics achievement of Jewish and Arab eight graders in Israel. The effects of learner-related variables. *Educational Research and Evaluation, 11*(3), 277–302.

National Research Council (NRC), Committee on Women in Science and Engineering. (1994). *Women scientists and engineers employed in industry: Why so few?* Washington, DC: National Academy Press.

Nosek, B. A., Banaji, M. R., Greenwald, A. G. (2002). Math = male, me = female, therefore math ≠ me. *Journal of Personality and Social Psychology, 83*(1), 44–59.

OECD (2007). *PISA 2006 science competencies for tomorrow's world*. Vol. 1: Analysis. Paris: Author.

Pepin, B. (1999). Existing models of knowledge in teaching: Developing an understanding of the Anglo/American, the French and the German scene. In B. Hudson, F. Buchberger, P. Kansanen, & H. Seel (Eds.), *Didaktik/Fachdidaktik as science(-s) of the teaching profession?* (pp. 49–66). Umea: TNTEE Publication.

Pietsch, M., & Krauthausen, G. (2006). Mathematisches Grundverständnis von Kindern am Ende der vierten Jahrgangsstufe. In W. Bos, & M. Pietsch (Eds.), *KESS 4 – Kompetenzen und Einstellungen von Schülerinnen und Schülern am Ende der Jahrgangsstufe 4 in Hamburger Grundschulen* (pp. 143–163). Münster: Waxmann.

Richardson, V. (1996). The role of attitudes and beliefs in learning to teach. In J. Sikula, T. Buttery & E. Guyton (Eds.), *Handbook of research on teacher education* (2nd ed.) (pp. 102–119). New York: Macmillan.

Schmidt, W. H., Jorde, D., Cogan, L. S., Barrier, E., Gonzalo, I., Moser, U., et al. (1996). *Characterizing pedagogical flow: An investigation of mathematics and science teaching in six countries*. New York: Springer.

Schmidt, W. H., Tatto, M. T., Bankov, K., Blömeke, S., Cedillo, T., Cogan, L., et al. (2007). *The preparation gap: Teacher education for middle school mathematics in six countries – mathematics teaching in the 21st century (MT21)*. East Lansing, MI: MSU. Retrieved December 12, 2007 from http://usteds.msu.edu/related_research.asp

Schmidt, W. H., Blömeke, S., & Tatto, M. T. (in press). *Teacher preparation from an international perspective*. New York: Teachers College Press.

Schöps, K., Walter, O., Zimmer, K., & Prenzel, M. (2006). Disparitäten zwischen Jungen und Mädchen in der mathematischen Kompetenz. In M. Prenzel, J. Baumert, W. Blum, R. Lehmann, D. Leutner, M. Neubrand, et al. (Eds.), *PISA 2003. Untersuchungen zur Kompetenzentwicklung im Verlauf eines Schuljahres* (pp. 209–224). Münster: Waxmann.

Schwippert, K., Bos, W., & Lankes, E.-M. (2003). Heterogenität und Chancengleichheit am Ende der vierten Jahrgangsstufe im internationalen Vergleich. In W. Bos, E.-M. Lankes, M. Prenzel, K. Schwippert, G. Walther, & R. Valtin (Eds.), *Erste Ergebnisse aus IGLU. Schülerleistungen am Ende der vierten Jahrgangsstufe im internationalen Vergleich* (pp. 265–302). Münster: Waxmann.

Shulman, L. S. (1985). Paradigms and research programs in the study of teaching: a contemporary perspective. In M.C. Wittrock (Ed.), *Handbook of research on teaching* (3rd ed.) (pp. 3–36). New York: Macmillan.

Stockdale, S. (1995). *Gender differences in high-school mathematics achievement. An empirical application of the propensity score adjustment*. Los Angeles, CA: UCLA.

Sullivan, P., & Wood, T. (2008). *The handbook of mathematics teacher education.* Volume 1. Rotterdam, The Netherlands: Sense Publishers.

Trautwein, U., Köller, O., Lehmann, R. H., & Lüdtke, O. (Eds.) (2007). *Schulleistungen von Abiturienten. Regionale, schulformbezogene und soziale Disparitäten.* Münster: Waxmann.

Vale, C., & Bartholomew, H. (2008). In H. Forgasz et al., *Research in mathematics education in Australasia 2004–2007* (pp. 271–290). Rotterdam/Taipei: Sense Publishers.

Vollrath, H.-J. (2001). *Grundlagen des Mathematikunterrichts in der Sekundarstufe.* Heidelberg: Spektrum.

Weinert, F. E. (1999). *Konzepte der Kompetenz. Gutachten zum OECD-Projekt „Definition and Selection of Competencies: Theoretical and Conceptual Foundations (DeSeCo)".* Neuchatel, Schweiz: Bundesamt für Statistik.

Weinert, F. E. (2001). Concept of competence: A conceptual clarification. In D. S. Rychen, & L. H. Salganik (Eds.), *Defining and selecting key competencies* (pp. 45–66). Göttingen: Hogrefe.

Wiley, D. E., & Wolfe, R. G. (1992). Major survey design issues for the IEA Third International Mathematics and Science Study. *Prospects, 22*(3), 297–304.

Winkelmann, H., van den Heuvel-Panhuizen, M., & Robitzsch, A. (2008). Gender differences in the mathematics achievements of German primary school students: Results from a German large-scale study. *ZDM – The International Journal on Mathematics Education, 40*(4), 601–616.

Zimmer, K., Burba, D., & Rost, J. (2004). Kompetenzen von Jungen und Mädchen. In M. Prenzel, J. Baumert, W. Blum, R. Lehmann, D. Leutner, M. Neubrand, et al. (Eds.), *PISA 2003. Der zweite Vergleich der Länder in Deutschland – Was wissen und können die Jugendlichen?* (pp. 211–223). Münster: Waxmann.

SECTION III

HIGH ACHIEVERS

CHAPTER 13

DISCOVERING THE POTENTIAL OF GIFTED FEMALES IN MATHEMATICS

Kyeong-Hwa, Lee
Seoul National University

Eun-Jung, Lee
WISE Chungbuk Center

Seoung-Hey, Paik
Korea National University of Education

HeiSook, Lee
Ewha Woman's University, South Korea

INTRODUCTION

Research studies on gender differences in mathematics have been conducted in many developed countries since the 1970s. These nations have endeavored to achieve gender equity through various intervention programs, and some countries have even reported that gender differences in mathematics achievement are on the decrease. In most Asian and developing countries, however, the societies are fraught with gender inequities that make the research theme difficult to explore overtly. Similarly, although Korea has shown large gender differences on international mathematics as-

International Perspectives on Gender and Mathematics Education, pages 287–314
Copyright © 2010 by Information Age Publishing

sessments such as the Third International Mathematics and Science Study (TIMSS) and the Programme for International Student Assessment (PISA), there have been no research projects on mathematics and gender initiated by authorities. As a result, little research has been conducted on gender differences in mathematics education in Korea.

General Background of Korean Society

The emphasis on education, originating from "the Confucian culture which places a high value on education" was a major contributing factor to the rapid economic growth in Korea (Guo, 2005, p. 75). In addition, many people in Korea perceive education as a means of improving one's status in society. Growth in the economy and improvement in education levels have led to an increase in the number of educated women and have improved women's status in Korean society. Nevertheless, male dominance still exists in many areas of society.

Tables 13.1 and 13.2, for instance, show the numbers of males compared to females enrolled in institutions of higher education and in the workforce, respectively. As can be seen in Table 13.1, for the 10-year period between 1995 and 2005, the number of females enrolled in higher education institutions was less than the number of males every year. While there was an increase in the female enrollment ratio from 36.2 % to 39 % over the

TABLE 13.1   Numbers of Students Enrolled in Korean Higher Education Institutions by Gender

Year	Total population	Number of Students Enrolled in Higher Educational Institutions		Female Ratio (%)
		Male	Female	
1995	45,092,991	1,495,749	848,145	36.2
1996	45,524,681	1,601,484	940,175	37.0
1997	45,953,580	1,742,503	1,049,907	37.6
1998	46,286,503	1,841,641	1,109,185	37.6
1999	46,616,677	1,965,142	1,189,103	37.7
2000	47,008,111	2,076,787	1,286,762	38.3
2001	47,353,519	2,150,037	1,350,523	38.6
2002	47,615,132	2,185,896	1,391,551	38.9
2003	47,849,227	2,175,741	1,382,370	38.9
2004	48,082,163	2,163,361	1,391,754	39.0
2005	48,294,143	2,148,797	1,399,931	39.0

Source: MEST (Ministry of Education, Science and Technology), *Statistical Yearbook of Education*, various years.

TABLE 13.2 Economically Active Population and Labor Force Participation Rate by Gender

Year	Population 15 years old and over[a]	Economically Active Population[a]	Labor Force Participation rate (%) Total	Male	Female
1995	33,659	20,845	61.9	76.4	48.4
1996	34,274	21,288	62.1	76.2	48.9
1997	34,851	21,782	62.5	76.1	49.8
1998	35,347	21,428	60.6	75.1	47.1
1999	35,757	21,666	60.6	74.4	47.6
2000	36,186	22,134	61.2	74.4	48.8
2001	36,579	22,471	61.4	74.3	49.3
2002	36,963	22,921	62.0	75.0	49.8
2003	37,340	22,957	61.5	74.7	49.0
2004	37,717	23,417	62.1	75.0	49.9
2005	38,503	23,526	61.1	73.6	49.2

Source: KNSO, *Annual Report on Economically Active Population,* various years.
[a] Numbers are in 1000s of people

10-year period, the actual numbers of females enrolled in higher education never equaled the smallest number for their male counterparts.

The figures in Table 13.2 indicate that the labor force participation rates of men and women also differed in the 10-year period between 1995 and 2005. Women continued to have lower participation rates than men. Generally over the 10-year period, 70 % of men participated in the labor force from among the economically active population, while less than 50 % of eligible women were in the labor force.

Because of this established gender disparity in society, many Korean women have experienced disadvantage at some point in their lives. According to the report on the social statistics survey conducted by the Korea National Statistical Office (KNSO) in 2002, 69.1% of females have experienced some form of gender-based discrimination in the work place, and 72.4% in social life. In addition, in the *Human Development Report* (2005, cited in KNSO, 2008), Korea was ranked low in the Gender-related Development Index (GDI, 27th); and very low in the Gender Empowerment Measure (GEM, 59th) and the Equal Opportunity Index (54th).

Participation in Mathematics

To understand the diverse levels of participation in mathematics courses between girls and boys, the mathematics curriculum in Korea needs explanation. The school system in Korea is a single track of 6 years primary, 3 years lower secondary, 3 years upper secondary, and 4 years higher education. Heavy emphasis is placed on mathematics and science at all school levels (Guo, 2005), and mathematics is a compulsory subject for all students until the end of high school. Students, however, can take elective credit mathematics courses in their second year of high school based on whether they plan to specialize in the humanities or the natural sciences.

As shown in Figure 13.1, the percentages of male students who selected the elective courses of Practical Mathematics, Mathematics 1, Probability & Statistics, and Discrete Mathematics in 2005 were slightly higher than for female students. However, larger gender imbalances favoring males existed in the Mathematics 2 and Differentiation & Integration courses, acknowledged as the most difficult mathematics courses in Korea. Mendick (2003) pointed out a similar pattern of gender disparities favoring male participation in advanced mathematics courses in many other countries. In England, in fact, the percentage of males who selected advanced level mathematics was higher than that of females between 1995 and 2001: 65 % for males and 35 % for females in 1995; 63 % for males and 37 % for females in 2001 (Mendick, 2003). Forgasz, Leder, and Thomas (2003) also reported male dominance in selection of grade 12 mathematics courses in Victoria, Australia. They found that more males than females enrolled in all levels of mathematics courses—Further Mathematics, Mathematical Methods, and Specialist Mathematics—from 1994 to 1999 and that male participation rates increased as the level of difficulty grew higher.

Table 13.3 shows the percentages of male and female representation in enrollments of Korean students in mathematics-related higher education

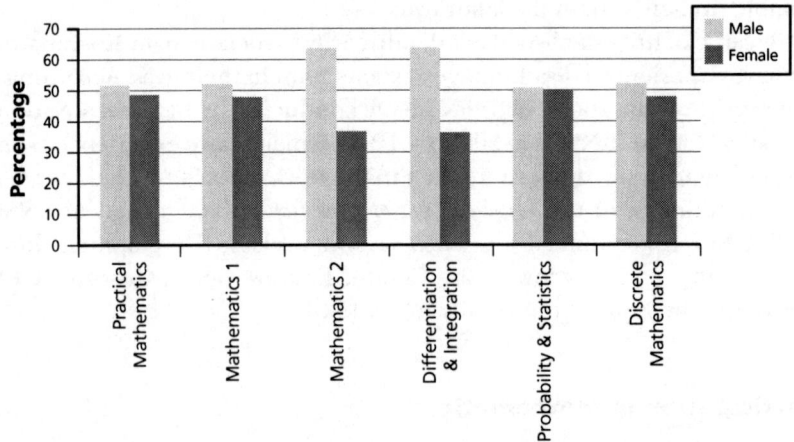

Figure 13.1 Male and female participation rates in elective courses in Korea (2005).

TABLE 13.3 Enrollment in Mathematics, Statistics, and Physics & Science by Gender in 2006

	University		Master's degree courses		Doctorate degree courses	
	Male (%)	Female (%)	Male (%)	Female (%)	Male (%)	Female (%)
Mathematics	53.10	46.90	54.70	45.30	58.60	41.40
Statistics	56.40	43.60	54.40	45.60	61.60	38.40
Physics & Science	72.70	27.30	75.60	24.40	81.20	18.80

Source: MEST (Ministry of Education, Science and Technology)

fields. More males than females participated in all three surveyed areas—mathematics, statistics, and physics and science—and the gender differences tended to increase as the education level became higher.

Achievement

In this section, gender differences in the performances of Korean students on international assessment programs—TIMSS 1995, 1999, and 2003 and PISA 2000, 2003, and 2006—are discussed. Comparisons over time are presented for the achievements of eighth graders on TIMSS and for 15-year-olds on PISA.

Table 13.4 presents the average scores of Korean students in eighth grade for TIMSS 1995, 1999, and 2003, and the gender differences between average scores. The achievement for boys was better than for girls on all three TIMSS assessments; the boys' mean score was higher than the girls' by 5 points in both 1999 and 2003 and by 17 points in 1995. While the boys' performance showed a slight improvement from 588 points in 1995 to 590 points in 1999, the girls' improvement was significantly better (571 points to 585 points). Therefore, the gender difference found in 1999 was 12 points less than in 1995 (Mullis, Martin, Gonzalez, & Chrostowski, 2004).

According to Organisation for Economic Co-operation and Development [OECD] (2001, 2004), Korea had statistically significant gender differences in favor of boys on both the PISA 2000 and 2003 mathematics assessments; the differences between the boys' mean scores and the girls' mean scores were approximately 27 points in 2000 and 23 points in 2003. These results placed Korea at the top of the 32 participating countries as the nation with the highest gender difference on PISA 2000 and as the nation with the second highest significant gender difference among the 40 participating countries on PISA 2003 (OECD, 2001, 2004). Surprisingly, however,

TABLE 13.4   Average Scores and Gender Differences of Eighth Graders in Korea (TIMSS)

TIMSS	Average Scores		Gender Difference (boys' score – girls' score)
	Boys	Girls	
1995	588	571	17
1999	590	585	5
2003	592	586	5

Source: Mullis, Martin, Gonzalez, & Chrostowski (2004, p. 50)

the gender difference on the PISA 2006 mathematics assessment dropped to nine points and was not statistically significant. While the males' mean score was unchanged at 552 in 2003 and 2006, the females' performance improved by 14 points from 529 in 2003 to 543 in 2006 (OECD, 2007).

PREVIOUS RESEARCH

Mathematics has been recognized as "a critical filter to further educational and career opportunities" (Leder, 1990, p.17). In the past, few opportunities were given to females in learning higher level mathematics, and women were not even allowed to enter university in many countries (Leder, 1992). Most women as well as men perceived mathematics as a male domain. Therefore, many females had a tendency to avoid higher-level mathematics courses and jobs in fields related to mathematics. Inferior performance by females was also found on international assessments such as the First International Mathematics Study (FIMS) (Hanna, 2003). Results from FIMS showed gender differences in favor of boys (13-year-olds) in 10 out of 12 participating countries and in boys' favor (ages 17–18) in all 10 participating countries (Hanna, 2003).

As gender differences in favor of males in both mathematics achievement and participation were recognized as problematic issues in mathematics education, many research studies have been conducted since the 1970s to uncover the factors that influence gender differences (Fennema, 2000; Leder, 1992). Research studies over the past three decades have "taught us that gender differences in mathematics achievement and participation are not due to biology, but to complex interactions among social and cultural factors, societal expectation, personal belief systems and confidence levels" (Kaiser & Rogers, 1995, p. 1).

The absence of sufficient evidence to support the view that biological variables cause gender differences in mathematics have also been noted by other researchers (e.g., Becker, 2001; Leder, 1992). According to Leder

(1992), gender differences pertained not only to learner-related variables such as cognitive and internal beliefs, but also to environmental variables such as school, home, and society.

Attitudes Toward Mathematics

There have been many studies on gender differences in student attitudes toward mathematics, and in most of them it was found that girls had less confidence in their mathematical ability than boys (Crombie et al., 2005; Fennema & Sherman, 1977; Geist & King, 2008; Joffe & Foxman, 1986). Geist and King (2008) also mentioned that girls tended to solve mathematical tasks with less confidence and felt fear about their performance while boys tended to feel more confident. Fennema and Sherman (1977), Joffe and Foxman (1986), and Meyer and Koehler (1990) pointed out that the difference in confidence levels between males and females in doing mathematics might be the cause of gender differences in mathematics achievement. Indeed, Ercikan, McCreith, and Lapointe (2005) found from their analysis of the Third International Mathematics and Science Study (TIMSS) that confidence in mathematics was one of the strongest predictors of achievement for girls from the U.S., but the relationship was much weaker for boys. It was noted, however, that this finding may be connected to the higher expressions of confidence by males even though their actual mathematics scores were lower than their female counterparts (Crombie et al., 2005). Forgasz, Leder, and Thomas (2003), however, reported changes in gender differences in attitudes toward mathematics. In Australia, compared to earlier studies, girls showed more confidence and pleasure learning mathematics than boys, and boys were more likely than girls to answer that they were not good at mathematics and that they gave up on difficult mathematical problems quickly.

Gender Differences in Mathematics Learning

Several researchers identified the differences in learning styles of females and males. While girls preferred cooperative activities in learning mathematics, boys preferred competitive activities (Chamdimba, 2003; Fennema, 2000; Geist & King, 2008). For example, Chamdimba (2003) found that during small-group activities in a mathematics class, girls tended to spend a lot of time on discussions with their peers and shared their ideas and mathematical knowledge rather than seeking a prompt attempt at a correct answer. On the other hand, boys had a tendency to concentrate on finding a solution with limited discussions with other group members.

Another established difference in learning style was that boys and girls approached mathematical problems differently. While girls tended to use rote skills by applying formulas, boys tried to develop new and creative techniques (Campbell & Evans, 1997; Gray, 1996; Leedy, LaLonde, & Runk, 2003). Hence, it may be that dissimilar traits in learning mathematics place girls at a disadvantage when solving more challenging problems and performing in a timed mathematics test (Campbell & Evans, 1997). Different children have different learning styles (Geist & King, 2008) and, likewise, children have different ways of learning mathematics. The differences between females and males need recognition, and appropriate curricula and teaching strategies for both must be developed and put into action in classrooms.

Parental Influence

The home environment strongly influences student success in mathematics learning (Bezuk, Whitehurst-Payne, & Aydelotte, 2000; Brew, 2003; Leder, 1992). Despite the importance of parental support, many research studies have reported that parents tended to treat their sons and daughters differently. Bhanot and Jovanovic (2005) asserted that parents who held gender stereotyped beliefs about mathematics were more inclined to underestimate their daughters' competency in mathematics and overestimate their sons' competency. Due to parents' differing beliefs about their children, parents of females believed that hard work is needed for their daughters to achieve high scores in mathematics (Eccles et al., 1985; Geist & King, 2008). What is more, parents had significantly lower expectations for their daughters' performance than their sons' (Wentzel, 1998). Also, generally fewer mothers than fathers assisted their children with mathematics homework, and accepted their daughters' poor performance because of their own low self-concepts about mathematics (Stallings, 1985). According to Geist and King (2008), different parental attitudes and expectations for girls and boys may shape their children's attitudes towards mathematics learning. Bhanot and Jovanovic (2005) also pointed out that student confidence in mathematics was more likely to depend on their parents' beliefs about their mathematical ability than on their actual test scores. Leedy, La-Londe, and Runk (2003) found that female students who participated in a regional mathematics contest showed little confidence in their ability and also believed that their mothers had low expectations for their success in mathematics. As indicated by these findings, it would appear that parents influence gender differences in mathematics. Therefore, intervention programs are needed not only for students but also for their parents to address gender stereotypical beliefs and encourage them to provide equal support to their daughters and their sons.

Gender Differences in Mathematics Achievement

Many recent studies indicated that gender differences in mathematics achievement seemed to have narrowed or were almost non-existent. Although the gender gap has decreased in regard to overall performance, the magnitude and direction of gender differences varied according to mathematics content and difficulty of items.

Hyde, Fennema, and Lamon (1990) conducted a meta-analysis of 100 studies and identified several patterns of gender differences according to the content of tests and the cognitive levels of test items. Females performed slightly better than males on computation problems, whereas males performed better on problem-solving tasks. Although males were superior to females in geometry, there were no gender differences in arithmetic and algebra. Similar results were found by Hanna (1989) and Cheung (1989). That is, boys were more successful in geometry and measurement, but there were no gender differences in arithmetic, algebra, probability, and statistics. Kaur (1990), however, reported that although boys were superior to girls in arithmetic, geometry, probability, and statistics, girls outperformed boys on algebra. Thus, findings on gender differences in mathematics topics are inconsistent. It is presumed that the content of tests alone does not affect gender differences in mathematics achievement, but rather the complex relationships among factors such as difficulty and content of items and students' age.

Some of the common research findings on gender differences in terms of mathematics content and difficulty of items suggest that males outperform females on problems that require spatial visualization skills (Barnes, 1997; Geist & King, 2008; Kaur, 1990; Shuard, 1986) and that boys are more successful than girls at higher cognitive level tasks such as comprehension and reasoning, whereas girls are better at lower levels such as manipulation and recalling (Cheung, 1989; Geist & King, 2008; Hyde, Fennema & Lamon, 1990; Kiamanesh, 2006; Shuard, 1986).

PURPOSE OF RESEARCH

As mentioned previously, general differences in achievement on international mathematics assessments between Korean male and female students have significantly narrowed. However, the achievement gap between boys and girls at higher performance levels is still wide, and the proportion of females enrolled in special programs for gifted students at certified institutions is less than 30 %. Moreover, according to the Korea National Statistical Office, the percentage of females who enroll in natural science and engineering courses in university stands at about 30 %, and only 6.5 % of female graduates from natural science and engineering majors find jobs in the fields of science and engineering. To rectify this trend, Korea launched the Women into Science and Engineering [WISE] Project, an education program aimed at supporting girls in the upper level score brackets to continuously improve their mathematical capability and remain in those brackets.

Before launching the WISE Project, the researchers set the following assumptions about the mathematical learning propensities of female students based on prior research: girls are relatively weak at geometry and problem-solving but have a strong ability in algebra (Cheung, 1989; Hyde, Fennema, & Lamon, 1990; Hanna, 1989; Kaur, 1990); girls tend to believe mathematics is a male domain and parents share their daughters' beliefs (Bhanot & Jovanovic, 2005; Wentzel, 1998); and girls prefer group activities to individual activities for mathematics learning, and they also prefer intuitive and specific operations to abstract and formal approaches (Campbell & Evans, 1997; Chamdimba, 2003; Gray, 1996; Leedy, LaLonde, & Runk, 2003). In this chapter we intend to examine the appositeness of these assumptions in Korea. We focus on aspects unique to Korean students, especially the evaluation of the potential of female middle school students participating in the special program for the gifted as part of the WISE Project.

The research foci can be summarized as follows:

1. Does the problem-solving ability of the girls participating in this study differ across mathematical content areas?
2. What type of support provided by family members is given for the mathematical learning of the female students participating in this study?
3. What is the achievement and potential of the girls participating in this study?

METHODOLOGY

The research involved detailed observations and in-depth interviews with three female students and their parents to investigate how girls with high achievement scores learn mathematics, the extent of their potential, which factors affect their mathematical learning and attitudes, and how the WISE project affects their mathematical learning and attitudes towards mathematics.

The three students (hereafter referred to as P1, P2, and P3) were selected as participants because of their active involvement in the education programs of the WISE poject and their strong desire to pursue future careers in the fields of science or engineering. The three were in their first year of middle school and were from middle socioeconomic backgrounds. They obtained high scores in their school mathematics examinations and had participated in the WISE project since April 2008.

P1 decided to become a scientist after attending the robot camp, one of the WISE programs. Her father majored in science, while her mother majored in art, and she had a younger brother. Her parents showed low confidence in their mathematics skills and expressed a dislike of mathematics, so she got answers to her mathematics questions from school teachers or instructors at private institutes (hagwon). P1 claimed the repetitive calculations in elementary school made her lose interest in mathematics and she became bored. Her interest was sparked by solving puzzles or atypical problems designed to foster creativity. In her own words, the bi-weekly WISE mathematics program, conducted every other Saturday, helped enhance her mathematical skills by having her attempt problems that improved resourcefulness. As a result, she said she was now having fun studying mathematics.

P2 also wanted to become a scientist and studied mathematics assiduously. She was very good at mathematics in elementary school, but she was frustrated in her disappointing scores in middle school. Both of her parents majored in liberal arts, but her father liked mathematics when he was in school. At this stage, he tried to help P2 solve mathematical problems to the point of asking others for help. Her mother was not good at mathematics and disliked it very much. P2 believed her elder brother did well in mathematics because he resembled their father and her weak performance was due to her resemblance to their mother. She liked geometric problems and atypical problems like those studied in the WISE education programs.

Like P1 and P2, P3 also wanted to become a scientist, but she had very low confidence in mathematics. She found it difficult to display her full ability in mathematics tests because of pressure, and felt great anxiety when presented with unfamiliar problems. Her parents were good at mathematics but were too busy to support her learning. Hence, her older sister who did well in mathematics helped with her studies. She worried that she was the only family member who was not good at math as both her younger brother and older sister did well in math. Even though she enjoyed participating in the WISE project, she was constantly concerned about being excluded from the project if her school scores slipped below her current mark. She believed that this was related to her non-participation in private afterschool education classes, something that her classmates did.

The three students were provided with opportunities to facilitate creativity and improve their problem-solving capabilities through 90-minute classes twice a month. The bi-weekly classes had been running since April, 2008. Also, during the summer vacation, the girls took part in mathematics and science camps offered for intensive learning. At the mathematics camp, the researchers asked the three girls to solve 10 separate tasks and interviewed them based on the results in order to assess their levels of interest and problem-solving ability in various mathematics content areas. In order to examine their patterns of participation in group activities and

their problem-solving characteristics more specifically, the researchers provided, recorded, and analyzed classes in geometry, algebra, and a strategy
game. The interviews with parents were also analyzed and compared with
the views of the students. The three girls were interviewed three times for
this study after WISE project classes. The parent interviews and the student
interviews were conducted at the mathematics camp.

RESULTS

Content Area Versus Problem Type

All three students seemed satisfied with the biweekly WISE project education programs (see Appendix 1). P1 said the types of problems in the programs were interesting because they were new, and she felt her interest and
sense of achievement growing through the tackling of new and complicated problems. P2 felt great satisfaction with problems requiring inference,
but claimed teacher assistance was crucial to solving difficult problems. P3
also said problems addressed in the WISE project were interesting because
they were creative. None of them discussed preference or level of difficulty
based on content area; instead they seemed to draw particular attention to
the fact that the problems were new types of questions.

Because the bi-weekly program did not supply sufficient data to evaluate
preferences or problem-solving capabilities according to content area, the
researchers utilized an additional 10 tasks (see Appendix 2). These consisted of two questions on geometry (QG1, QG2), two on finding regularity
(QP1, QP2), two on algebra (QA1, QA2), two on probability and statistics
(QS1, QS2), and two on thinking logically (QL1, QL2). To validate the impact of subject matter, the researchers chose food, something girls are generally interested in, as the subject matter for QL1 and QL2. In Table 13.5,
the extent of each girl's solutions to the ten tasks is shown.

TABLE 13.5 Details of Problem Solutions

	QG1	QG2	QP1	QP2	QA1	QA2	QS1	QS2	QL1	QL2	Total
P1	2	1	2	0	1	2	0	2	2	2	14
P2	2	2	0	2	2	0	0	2	2	0	12
P3	2	1	2	2	1	0	0	2	2	2	14

Note: 2 = Correct answer, 1 = Incomplete answer but solutions correctly attempted,
0 = No trace of a correct solution
Source: MEST (Ministry of Education, Science and Technology)

As shown in Table 13.5, P1 correctly solved QG1, QP1, QA2, QS2, QL1, and QL2 but had the most difficulty with QP2, while P2 correctly solved QG1, QG2, QP2, QA1, QS2, and QL1 and found QP1 to be the most difficult, which was easily solved by the other two girls. P3 correctly solved QG1, QP1, QP2, QS2, QL1, and QL2 but had great difficulty solving QA2.

All three students easily solved questions QG1, QS2, and QL1. The students showed the greatest interest in QL1 in which the subject matter referred to preference for a certain type of pizza. Interestingly, the reason for interest was not the subject matter but the fact that the problem dealt with logic. In particular, P2 said she enjoyed this type of problem the most, and P1 and P3 expressed confidence and pleasure in solving it as well. All three girls easily answered QG1. For this question, the students viewed a building block from top, front, and side and figured out the number of cubes needed to make the block. As for QL2, although P2 did not answer the question due to time constraints, the other two solved it enthusiastically. P1 even claimed it to be the most interesting of the 10 problems. The subject matter of the question was a chef's distribution of dishes on a table.

Although all three subjects said they had never encountered a problem like QS2 before, they accurately identified the conditions of the problem and correctly estimated the probability of both a single event and product event. Based on their personal experiences, the girls appropriately explained how they inferred the proportion of female mathematics teachers and female mathematics teachers with glasses.

On QS1, a question dealing with uncertainty, all three students gave incorrect answers, even though they all claimed they had experienced solving this type of problem before. It was especially noteworthy that no one found the sample space for the tossing of two dice.

QA2 and QG2 were problems the students felt were difficult or simply disliked (see Appendix 1). QA2 was about finding a number that satisfied given conditions through intuitive insight. P1 was the only student to solve the problem, but she expressed a strong dislike of it. As shown in Figure 13.2, P1 first discovered that the 100th place in this sequence was 1 and then found the correct number through a trial and error strategy.

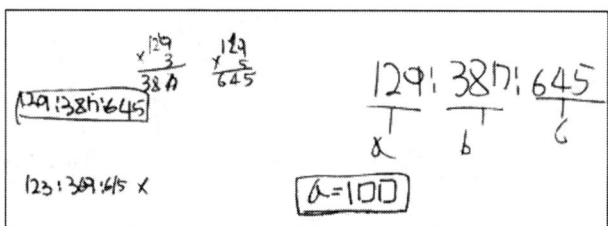

Figure 13.2   Answer to QA2 by P1.

Like QG1, QG2 was about finding the shape of a figure based on views from the top, front, and side. Students found it difficult because they had never been exposed to this type of problem before, and its structure was complicated. Only P2 solved the problem. P1 and P3 did not provide the right answer but considered various situations like those shown in Figures 13.3 and 13.4.

The mean achievement scores of the three students in the various mathematics content areas were: geometry 10, finding regularity 8, algebra 6, probability and statistics 6, and thinking logically 10. Unlike what the research of Hyde, Fennema, and Lamon (1990) and Kaur (1990) showed, the three girls had a relatively high level of problem-solving ability in geometry and logic, while their capabilities for algebra, probability, and statistics were relatively low.

To investigate the students' learning style characteristics when they did individual work and cooperative work, researchers gave the students two hours to solve another three questions and assisted them whenever needed.

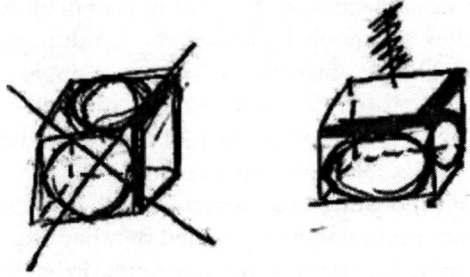

Figure 13.3   Answer to QG2 by P1.

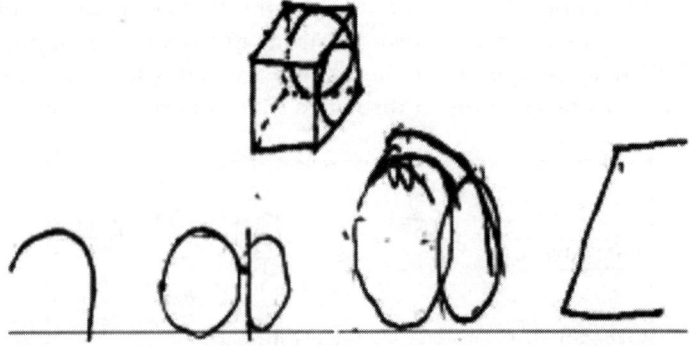

Figure 13.4   Answer to QG2 by P3.

P1 Likes Geometry

P1 was most enthusiastic while solving the problem on splitting a circle (hereinafter referred to as the Circle Problem; see Appendix 2). What was particularly interesting about P1's actions was that she interlinked geometric thinking with an algebraic approach to solve the problem. The following excerpt may indicate how P1 predicted a maximum of 15 regions splitting a circle with 4 lines:

> If 2 lines are drawn, the number of points where the 2 lines meet is one. If the number of lines is 3, there's 2 intersection points in each line, with 4 lines, there's 3 points, so I think if 5 lines are drawn there may be 4 intersection points in each line. So I drew the lines to have 4 intersection points in each line. Answer is 15.

However, P1 did not do well solving the problem that required a repetitive algebraic rule, the Diffy Problem (hereafter referred to as DP Problem; see Appendix 2). As soon as she saw the question, her confidence waned, and she changed her answer several times because she was not convinced it was correct. She looked bored, presented her answer later than others, and was not interested in whether she had the right answer or not.

Furthermore, although P1 displayed interest in the third task about finding winning strategies (hereafter referred to as the Game Problem; see Appendix 2), she had difficulty coming to the right solution. The question was about exploring winning strategies, but she failed to contribute to group activities with her intuitive answers. P1 wrote the reason for her answer on the Race to 20 as follows:

> One who goes second. If I make the other player give an even number, he or she will call the losing number.

It is not clear from the excerpt whether she simply guessed at the answer and ignored its flaws, and so failed to expand it into a more detailed statement. Nevertheless, P1 chose this problem as the most interesting task.

P2 Likes Inference Problems

Among the 10 content area tasks, P2 completed two problems on geometry perfectly, but solved only one of the two problems for each of the remaining areas. She tended to give up on her solutions even though there was no indication that they would lead to failure. For QP1, finding regularity in the course of building a square with matchsticks, she gave up even

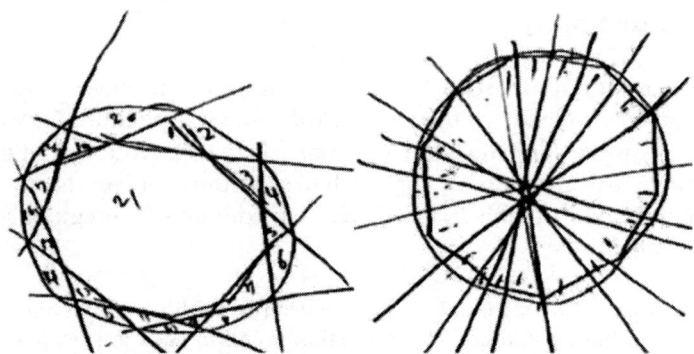

Figure 13.4 How P2 split the circle.

though problems such as QP1 were typically addressed in mathematics textbooks in elementary school. At the interview, P2 expressed a liking of these
types of problems but claimed she simply got frustrated by her failure to
find a pattern for QP1. She also did not come close to solving QL2, to which
the other two students gave perfect answers, because she lacked time.

For the Circle Problem, P2 only utilized repeated trial and error approaches without seeking any generalizations. As illustrated in Figure 13.5,
she spent most of the allotted time splitting the circle regardless of the
given condition that the circle should be split into the maximum number
of regions.

In contrast, P2 actively participated in the process of problem solving for
the DP Problem and the Game Problem. She was confident in her answers
and keen to lead group activities. Moreover, she used interactions with other students to correct her errors and provided feedback when necessary to
make contributions to the problem-solving process. However, like P1, she
found it difficult to grasp the situation and find clues to the solution to the
Game Problem, although she was clearly interested in the question. Of the
three students, she was the most tenacious in solving problems and succeeded in making generalizations. P2 tended to respond enthusiastically to
problems that provided learning opportunities like inference, manipulation, and interaction, rather than to the content area of the problem.

P3 Likes Familiar Types of Problems

P3 correctly solved QP1 and QP2 by finding regularity, and QL1 and
QL2 by use of logic. Like the others, she solved problems involving geometry, QG1 and QS2, easily. When asked about the level of question difficulty,
she responded that familiar questions, ones she had solved before, were

easy, and new types of problems were difficult. Although her score for the 10 tasks was equivalent to P1's and higher than P2's, she did not actively participate in group activities because she considered herself the worst, in terms of ability, among the group. She worried most about her mathematics scores among the three girls and often expressed concerns about not participating in private education.

With respect to the Circle Problem, she alleged that she could not find the maximum number of regions because probability was infinite. Although she was good at finding rules (QP1, QP2) in familiar problems, she lacked confidence encountering unfamiliar problem situations. P3 repetitively drew circles in the same way to find the maximum number of regions and focused on each drawing rather than their relationship or regularity. When she felt she could not complete the problem in the allotted time, she inquired about the time limit. She exerted more effort on the DP Problem than the Circle Problem. She was still reluctant to talk to others about her individual activity results and, in her interview, she said that she was ashamed of her performance. The same was true for the Game Problem. Even though she correctly answered "the one who goes first wins," she had no explanation to back up her answer. Essentially, she was good at playing games with others, but was not interested in exploring a winning strategy at all.

P3 felt uneasy and became passive as soon as she faced a new problem type. She said the Game Problem was interesting, but did not show any interest in finding useful clues that would help in the solution process. During her interview, P3 declared the most important factor in learning mathematics was continuously solving exercises in order to become familiar with mathematics problems, and not the effort exerted or persistence exhibited by an individual to solve one problem.

Necessity Versus Interest

Each of the three students dreamed of becoming a scientist, and they said they studied mathematics because it was a prerequisite for science majors. To excel at mathematics, they deemed it necessary to study hard at school and to solve as many examples as possible at private institutes. They also said that the study of mathematics was excessively burdensome and did not voice any indication of an intrinsic interest in mathematics. The following is an excerpt from the interview with P1 about her interest in mathematics and studying mathematics.

> What I like most is to solve math tasks in the library once that section has been covered and then explain it to my friends once a week in the math class.... I like math when my school teacher explains math laws through humor. For

example, he used a cooking sequence to explain the commutative law. We have to put hot water into the instant cup of noodles. He said to eat instant noodles, we should never swallow the hot water first and then eat the un-cooked noodles. It was funny.

The mothers of the students emphasized the importance of mathematics because good mathematics scores provided wider college options. The mothers of P1 and P2 were concerned about the mathematics scores of their daughters because they themselves had been extremely weak in mathematics. The mother of P3 was good at mathematics in school and worried about the lower level of mathematical skills of her daughter compared to her son who was in grade 2. The three mothers encouraged their daughters to study mathematics intensively, and attributed their sons' successes in mathematics to interest in the subject and natural mathematical intuition. The mother of P2 described her views on the future careers of her son and daughter as follows:

> My first son likes math and science, so I advised him to major in aerospace or mechanical engineering, but he said "No" without hesitation. I also encouraged him to consider biology to resolve world food problems since genetic engineering will one day be a prevalent field. Still, he didn't respond... I haven't discussed P2's dream or given any advice for a future career. She once said she wanted to be a painter because she liked drawing. That was her first choice, but she soon realized the difficulty of making a living from painting and the amount of money it would take to become a painter. Then she decided to become a scientist.

All of the mothers had low expectations of their daughters' abilities to succeed in mathematics or even to enjoy mathematics, and were skeptical of their daughters' future career choices as scientists. In contrast, they were convinced of their sons' innate mathematical talent and interest in mathematics, and so urged their sons to enter the sciences and engineering. P2's father energetically supported his daughter's mathematics studies but did not actively help her ponder her future career path. The fathers of P1 and P3, when young, were not fascinated by mathematics, nor were they attentive to their daughters' mathematical studies or future career paths. On the whole, it was the perspective of the mothers that had the greatest impact on their daughters' attitudes toward mathematics, development of their learning habits, and the direction of their future career paths. Moreover, it was the emphasis the mothers placed on the need for mathematics, more than their daughters' interests in mathematics regardless of their actual mathematical performance in school, that was influencing the girls' attitudes.

It seemed that the three girls approached mathematics in terms of necessity rather than interest from their early years of elementary school. P1 had been solving mathematics exercises at private institutes since elementary school and believed that the best way to study mathematics was to repeatedly complete calculation exercises. P2 believed that she needed to prepare in advance for school classes at a private institute and repetitively solve exercises in school to achieve a high level of mathematics proficiency. P3 worried that she could not participate in a private institute because of her family's inability to afford private education and believed that she must solve as many mathematical exercises as possible to score well in mathematics.

What is common to all three participants was the comment that the WISE project provided intriguing mathematics problems. Because they were new, required inference, and involved group work, the problems were different from those solved in school or at private institutes. Hence, although it seemed that the three girls studied mathematics solely because it was a required subject, they may have been eager to develop an interest in mathematics. On the other hand, they wanted to deal with more interesting and easier problems in the WISE project classes because of fears of solving unfamiliar problems. As mentioned above, the participants occasionally did not identify the clues that would aid in solving complicated and abstract situations which eventually led them to give up. This trend highlighted the need for combining familiar content with unfamiliar content when instructing students.

The father of P2 worked on mathematics questions with his daughter. He was an ethics teacher who went so far as to seek the advice of a mathematics teacher at his place of work to help his daughter with her studies. The reason P2 remained confident when solving new problems, although her mother expressed strong doubts about her daughter's mathematical aptitude, appears to be her father's continuous support. Specifically, the fact that P2's father had a negative view of private institutes and a high interest in mathematics sustained P2 and nurtured the development of a positive attitude towards mathematics learning. P3's father, however, provided little support and showed little concern for her study of mathematics. She was assisted by her elder sister whose view was that girls lagged behind boys in high school in terms of mathematics scores, and that women were not inherently good at mathematics. Worse still, her mother considered her mathematical ability to lag behind that of her younger brother in elementary school, and therefore she provided little support in her daughter's mathematical studies.

The mothers had controlled their children's education and supported their mathematics learning. Mothers believed their daughters were going through the same learning process they once did, so they strived to avoid past mistakes made during their school days. Hence, they focused more on the ways to achieve good mathematics grades or the reasons why mathematics was important than on ways to stimulate interest in mathematics.

Because these mothers assumed their daughters, like themselves, were not interested in mathematics, they stressed top performance in mathematics and the reasons why it was vital in all walks of life. If their daughters scored low on a mathematics test, they became quite impatient when tutoring their children because they fully recognized the importance of learning mathematics. As a result, these female students paid keen attention to mathematics scores and lost confidence in a future connected to mathematics. In other words, excessive pressure appeared to make these female students lose interest in mathematics.

Achievement Versus Potential

As mentioned above, P1, P2, and P3 were female students from small cities and were all studying mathematics because they wanted to become scientists; they were in the upper-level score brackets for their grade levels. For them, mathematics was merely a required subject, and the best way to attain high marks was through continuous practice. Their mothers believed women needed to expend greater effort to be as good in mathematics as men who, they felt, were naturally gifted with respect to mathematical ability, and this innate talent naturally enabled men to enjoy mathematics. It is often said that women prefer writing and painting to mathematics and have a natural talent in those areas. Thus, mothers try to teach their daughters to overcome their innate inferiority in mathematics by advising them to practice solving as many exercises as possible and to exert continuous efforts.

The WISE project education programs supported girls with high mathematics achievement levels by developing problems that helped them improve their mathematical capability, and encouraged the cultivation of positive attitudes toward the subject. After the four-month program, however, it was found that their confidence did not improve as expected. The observations and interviews with the three participants in the study showed that the girls appeared to lose confidence in the course of solving new problems. Their mothers continuously sent overt messages that men were inherently better than women in mathematics, so the three participants attributed their failure to solve problems in the WISE program to their lack of mathematical sense. They understood that they needed to solve many creative mathematics problems, but were frustrated by repeated failure. Moreover, they felt ashamed of their inability to contribute to group activities because they did not identify any clues leading to the solutions of particular problems.

From this study, we concluded that although new and challenging tasks to stimulate girls' potential were provided, the level of uneasiness increased as the number of unsolved tasks increased. This finding illustrated the frustration girls feel in the scores attained at schools and from private institutes. Students in the upper achievement ranks receive private education from their early years in school to promote their performance levels in the higher grade levels. In most cases because private institutes aim too high, students often still get low scores on tests. This result may pose severe blows to girls. In an interview, P2's mother described her daughter's mathematics scores from a private institute in the following manner:

> I sent her to a private institute specializing in math during the winter vacation when she was a 6th grader. She frequently took tests there, but her marks were in the 30s and 40s. So, I called the instructor, because I was very concerned about her performance. He said the tests were more difficult than those she took in school.

The girls became dismayed at the grades they received from the institutes because they regarded the low scores to be a true reflection of their performance levels and did not consider the relatively higher marks they attained in school. The reasons why these upper mathematics achievement bracketed students enrolled in private education appeared to be their obsession with the need to learn school-level content in advance. The typical learning pattern for these students was to learn school content in advance and to review what had already been learned in school at the institutes. This had a negative impact on female students. The advanced learning, aimed at enhancing the potential for educational growth, actually contributed to lower levels of achievement, resulting in the girls losing confidence in mathematics.

The pattern of participation by a student in the WISE education programs was found to be closely related to the interest and support of their parents. In particular, the way parents supported the mathematics learning of their children was considered crucial to boosting their potential. No difference in mathematics performance was found for the two girls in the study receiving the private education and the girl who did not, but a clear difference in confidence materialized for the girl whose parent led the course of learning together with the child.

CONCLUSION

In this study the mathematical capability of three students participating in the WISE project was examined, their learning patterns and perspectives on mathematics were explored, and the support of their parents considered. There were five distinct findings that vividly portray the realities facing Korean female students who achieve high scores in mathematics and their future potential in mathematics.

First, the girls' problem-solving abilities differed according to the types and characteristics of the problems examined and not the content area covered. All three students, including P1 who said she liked geometric problems, found it interesting yet challenging to solve new and open-ended problems. This finding indicates that content area alone may not affect gender differences in mathematics achievement; rather, it is the complex relationships among factors, such as difficulty, problem type, and content area, that may affect gender differences. The WISE project's aim to provide repetitive challenging problems to enhance capabilities may, at times, have resulted in undermining the girls' confidence in mathematics. Therefore, consistent support for girls to improve their confidence in mathematics by practicing many challenging problems is needed.

Second, the three girls believed it was desirable for girls to study mathematics through continuous practice, and that they must do this in order to perform as well as boys who had innate talents in mathematics. The girls were also inclined to perceive that they were not good at mathematics despite their high scores in school mathematics examinations. Their low confidence may have followed from their mothers' underestimation of their daughters' mathematical abilities. This finding supports the claim by Bhanot and Jovanovic (2005) that student confidence in mathematics was more likely to depend on their parents' beliefs about their mathematical ability than on their actual test scores. To perform well in mathematics, they also thought that they needed to enroll in mathematics courses at private institutes to learn school mathematics content in advance and to review knowledge gained in school. This attitude seemed to be affected by their mothers' failures in, or concerns about, their own mathematical studies. The perspectives of mothers about mathematical learning focused on the need for mathematics rather than on an intrinsic interest for the subject. Mothers' views about mathematics appeared to lower the confidence levels and raise the anxiety levels of their daughters. Therefore, female students and their parents, especially mothers, need to be educated about mathematics and mathematical learning.

Third, the girls preferred group activities to individual activities. This pattern was similar to previous research findings that found girls preferred cooperative activities, while boys preferred competitive activities (Chamdimba, 2003; Geist & King, 2008). The girls also showed profound interest in discovering and developing mathematical ideas through interactions in groups, and felt responsible to make more contributions during group activities. This outcome can be enhanced through effectively teaching girls via supportive joint studies in separate programs like WISE.

Fourth, the female participants possessed a strong will to discuss their future career paths and to realize their dreams. However, their families appeared to pay less attention to the career path orientation of their daughters than their sons. In addition, as many studies on gender and parental influence previously reported, the three girls' parents tended to underestimate their daughters' competency in mathematics. Indeed, the parents were not receptive to the notion that girls could learn mathematics and science sufficiently well to continue into careers in the fields of science or engineering. As grade level increased, the girls and their parents had a tendency to believe that women became poorer at mathematics and therefore tended to delay deciding on future career paths. These views highlight the need to identify trends by consistently performing case studies on girls in the WISE program.

Fifth, even if one parent paid attention to the mathematical studies of a daughter and enthusiastically helped with problem solving, this may raise her potential immensely because the daughter might cultivate a strong will and positive attitude towards mathematical studies. P2 was a case in point. Although P2 possessed a lower achievement level than the other two, she was an aggressive learner, quick to rectify errors through participation in group discussions, and led debates in group activities.

The gender gap in mathematics achievement among Korean students remains substantial amongst the highest performing students. According to the findings of this study, low confidence levels and the limited learning opportunities of such methods as continuous practice for girls who were high performers, seemed to be a direct result of erroneous judgments and views about mathematics learning that were held by their mothers. Therefore, further studies are crucial to the development of programs that educate the parents of high achieving girls to identify and foster the potential of their daughters. In addition, educators need to develop intervention programs to encourage female students to learn mathematics not by continuous practice but through solving challenging problems.

REFERENCES

Barnes, M. (1997). Classroom views of gender differences. In B. Doig & J. Lokan (Eds.), *Learning from children: Mathematics from a classroom perspective* (pp. 41–61). Melbourne: ACER.

Becker, J. R. (2001). Single-gender schooling in the public sector in California: Promise and practice. In B. Atweh, H. Forgasz, & B. Nebres (Eds.), *Sociocultural research on mathematics education: An international perspective* (pp. 367–378). Mahwah, NJ: Lawrence Erlbaum Associates.

Bezuk, N. S., Whitehurst-Payne, S., & Aydelotte, J. (2000). Successful collaborations with parents to promote equity in mathematics. In W. G. Secada (Ed.). *Changing the faces of mathematics: Perspectives on multiculturalism and gender equity* (pp. 143–148). Reston, VA: The National Council of Teachers of Mathematics.

Bhanot, R. & Jovanovic, J. (2005). Do parents' academic gender stereotypes influence whether they intrude on their children's homework? *Sex Role, 52*(9), 597–609.

Brew, C. (2003). Mother returning to study mathematics: The development of mathematical authority through evolving relationships with their children (pp. 65–100). In L. Burton (Ed.), *Which way social justice in mathematics education?* Westport, CT: Praeger.

Campbell, K. T. & Evans, C. (1997). Gender issues in the classroom: A comparison of mathematics anxiety. *Education, 117*(3), 332–339.

Chamdimba, P. C. (2003). Gender-related differences in working style during cooperative learning in secondary school mathematics: A Malawian case study (pp.153-167). In L. Burton (Ed.), *Which way social justice in mathematics education?* Westport, CT: Praeger.

Cheung, K. C. (1989). Gender differences in the junior secondary (grade7) mathematics curriculum in Hong Kong. *Educational Studies in Mathematics, 20*(1), 97–103.

Crombie, G., Sinclair, N., Silverthorn, N., Byrne, B. M., DuBois, D. L., & Trinneer, A. (2005). Predictors of young adolescents' math grades and course enrollment intentions: Gender similarities and differences. *Sex Role, 52*(5), 351–368.

Eccles, J., Adler, T. F., Futterman, R., Goff, S. B., Kaczala, C. M., & Midgley, M. C. (1985). Self-perceptions, task perceptions, socializing influences, and the decision to enroll in mathematics. In S. F. Chipman, L. R. Brush, & D. Wilson (Eds.), *Women and mathematics: Balancing the equation* (pp. 95–122). Hillsdale: Lawrence Erlbaum Associates.

Ercikan, K., McCreith, T., & Lapointe, V. (2005). Factors associated with mathematics achievement and participation in advanced mathematics courses: an examination of gender differences form an international perspective. *School Science and Mathematics, 105*(1), 5–17.

Fennema, E. & Sherman, J. (1977). Sex-related differences in mathematics achievement, spatial visualization and affective factors. *American Educational Research Journal 14*(1), 51–71.

Fennema, E. (2000). *Gender and mathematics: What is known and what do I wish was known?* Prepared for the Fifth Annual Forum of the National Institute for Science Education, Detroit. Retrieved September 8, 2008 from http://www.wcer. wisc.edu/archive/nise/News_Activities/Forums/Fennemapaper.htm

Forgasz, H., Leder, G. C., & Thomas, J. (2003). Mathematics participation, achievement, and attitudes: What's new in Australia? In L. Burton (Ed.), *Which way social justice in mathematics education?* (pp. 241–260). Westport, CT: Praeger.

Geist, E. A. & King, M. (2008). Different, not better: Gender differences in mathematics learning and achievement. *Journal of Instructional Psychology, 35*(1), 43–53.

Gray, M. (1996). Gender and mathematics: Mythology and misogyny. In G. Hanna (Ed.), *Towards gender equity in mathematics education: An ICMI study* (pp. 27–38). Dordrecht, The Netherlands: Kluwer Academic Publishers.

Guo, Y. (2005). *Asia's educational edge: Current achievement in Japan, Korea, Taiwan, China, and India.* New York: Lexington Books.

Hanna, G. (1989). Mathematics achievement of girls and boys in grade eight: Results from twenty countries. *Educational Studies in Mathematics, 20*(2), 225–232.

Hanna, G. (2003). Reaching gender equity in mathematics education. *The Educational Forum, 67*(3), 204–214.

Hyde, J. S., Fennema, E., & Lamon, S. J. (1990). Gender differences in mathematics performance: A meta-analysis *Psychological Bulletin, 107*(2), 139–155.

Joffe, L., & Foxman, D. (1986). Attitudes and sex differences- some APU findings. In L. Burton, (Ed.), *Girls into maths can go* (pp. 38–50). London: Holt.

Kaiser, G., & Rogers, P. (1995). Introduction: Equity in mathematics education. In P. Rogers & G. Kaiser (Eds.), *Equity in mathematics education* (pp. 1–10). London: The Falmer Press.

Kaur, B. (1990). Girls and mathematics in Singapore: The case of GCE 'O' level mathematics. In L. Burton (Ed.), *Gender and mathematics: An international perspective* (pp. 98–112). London: Cassell.

Kiamanesh, A. R. (2006). *Gender differences in mathematics achievement among Iranian eighth graders in two consecutive international studies (TIMSS 99 & TIMSS 2003).* Retrieved September 10, 2008 from http://www.iea.nl/fileadmin/user_upload/IRC2006/IEA_Program/TIMSS/Kiamanesh.pdf

Korea National Statistical Office (KNSO). (2008). Retrieved September 25, 2008 from http://www.nso.go.kr/

Leder, G. C. (1990). Gender and classroom practice. In L. Burton (Ed.), *Gender and mathematics: An international perspective* (pp. 1–8). London: Cassell.

Leder, G. C. (1992). Mathematics and gender: Changing perspectives. In D. A. Grouws (Ed.), *Handbook of research on mathematics teaching and learning* (pp. 597–622). New York: Macmillan.

Leedy, M. G., LaLonde, D., & Runk, K. (2003). Gender equity in mathematics: Beliefs of students, parents, and teachers. *School Science and Mathematics, 103*(6), 285–293.

Mendick, H. (2003). Choosing maths/doing gender: A look at why there are more boys than girls in advanced mathematics classes in England. In L. Burton (Ed.), *Which way social justice in mathematics education?* (pp. 169-187). Westport, CT: Praeger.

MEST (Ministry of Education, Science and Technology). Retrieved September 25, 2008 from http://www.mest.go.kr/

Meyer, M. R., & Koehler, M. S. (1990). International influences on gender differences in mathematics. In E. Fennema & G. C. Leder (Eds.), *Mathematics and gender* (pp. 60–95). New York: Teachers College Press.

Mullis, I. V. S., Martin, M. O., Gonzalez, E. J., & Chrostowski, S. J. (2004). *TIMSS 2003 international mathematics report.* Chestnut Hill: International Study Center Lynch School of Education Boston College.

OECD. (2001). *Knowledge and skills for life: First results from PISA 2000.* Retrieved August 12, 2008 from http://www.oecd.org/dataoecd/44/53/33691596.pdf

OECD. (2004). *Learning for tomorrow's world: First results from PISA 2003.* Retrieved August 12, 2008 from http://www.oecd.org/dataoecd/1/60/34002216.pdf

OECD. (2007). PISA 2006: *Science competencies for tomorrow's world.* Retrieved August 12, 2008 from http://www.oecd.org/dataoecd/15/13/39725224.pdf

Shuard, H. (1986). The relative attainment of girls and boys in mathematics in the primary years. In L. Burton (Ed.), *Girls into maths can go* (pp. 23–37). London: Holt.

Stallings, J. (1985). School, classroom, and home influences on women's decisions to enroll in advanced mathematics courses. In S. F. Chipman, L. R. Brush, & D. M. Wilson (Eds.), *Women and mathematics: Balancing the equation* (pp. 199–233). Hillsdale: Lawrence Erlbaum Associates.

Wentzel, K. R. (1998). Parents' aspirations for children's educational attainments: relations to parental beliefs and social address variables. *Merrill-Palmer Quarterly, 44*(1), 20–32.

APPENDIX 1

QA2: Please find integers a, b, c that satisfy all of the following conditions:

1. a, b, c are 3-digit integers
2. the number of each digit of a, b, c are configured as 1, 2, 3, ... 9.
3. the numbers may not be repeated (for example, a = 123, b = 456, and c = 789 is possible, but a = 112, b = 345, and c = 678 or a = 123, b = 456, and c = 678 are not possible)
4. a:b:c = 1:3:5

QG2: Figure 1 shows tape wrapped around part of a cube. I, II, and III are projection drawings of the taped cube in Figure 1 from the top, front, and side.

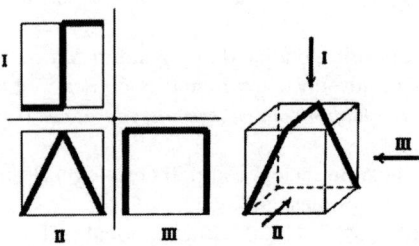

Figure 1

If the view from the top, front, and side are as follows below (Figure 2), describe how the tape is wrapped around the cube in Figure 1 (the length of used tape should be as short as possible).

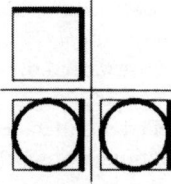

Figure 2

APPENDIX 2

Circle Problem

How many regions you can get splitting a circle with straight lines.
1. Basic Activity
    a. Below is a circle. What is the maximum number of domains you can get if you split the circle below with 3 straight lines?

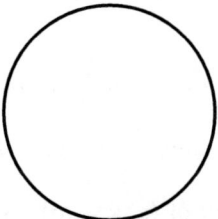

    b. If you draw 5 straight lines in the circle, what is the maximum number of domains you can get?
    c. What is the maximum number of domains you can get if you divide the circle with 10 straight lines?
2. In-Depth Activity
Name the maximum number of domains a circle can be divided into by n number of lines.

Diffy Problem

The term diffy is derived from difference and designed to practice subtraction. However, diffy has many other surprising attributes. Let's take a look at the various conditions of diffy and explain how these conditions are completed.

Definition of Terminology: Diffy foot, diffy head, diffy activity, diffy number

With ordered pairs made of four numbers (a, b, c, d), we call four pairs made by subtracting the adjoining two numbers ($|a-b|$, $|b-c|$, $|c-d|$, $|d-a|$) **'diffy foot'**, or F(a, b, c, d). We call (a, b, c, d) **'diffy head'** of ($|a-b|$, $|b-c|$, $|c-d|$, $|d-a|$), or H($|a-b|$, $|b-c|$, $|c-d|$, $|d-a|$).

In the example above, the diffy foot of (11, 7, 8, 12) is (4, 1, 4, 1), which can be expressed as:

$$F(11, 7, 8, 12) = (4, 1, 4, 1)$$

And diffy head of (4, 1, 4, 1) is (11, 7, 8, 12) and can be expressed as follows:

$$H(4, 1, 4, 1) = (11, 7, 8, 12)$$

The activity for calculating diffy foot is called '**diffy activity**'. When diffy foot becomes $(0, 0, 0, 0)$, diffy ends. The number of diffy activities until diffy foot is $(0, 0, 0, 0)$ is called the '**diffy number**' and is expressed as $N(11, 7, 8, 12)$. In the example, the diffy number is $N(11, 7, 8, 12) = 3$.

1. Basic Activity
   a. Are the diffy foot of (a, b, c, d) and (a+p, b+p, c+p, d+p) always the same?
   b. Are the diffy numbers of (a, b, c, d) and (d, c, b, a) always the same?
   c. Are the diffy foot of (a, b, c, d) and (−a, −b, −c, −d) always the same?
   d. Are the diffy numbers of (a, b, c, d) and (na, nb, nc, nd) always the same?
2. In-Depth Activity
   What is the condition for the diffy number of (a, b, c, d) to be 2?
3. Exploring Activity
   Name conditions for the diffy number of (a, b, c, d) to become 3.

Game problem

1. Race to 20
   Please find winning strategies.
   a. Does the one who goes first have an advantage? Or does the one who goes second have an advantage?
      And what is the reason?
   b. Is there a 'winning' strategy that guarantees victory?
      If so, please explain how you can win in either situation (go first, go second).

CHAPTER 14

GENDER AND HIGH ACHIEVERS IN MATHEMATICS

Who and What Counts?

Gilah C. Leder
La Trobe University, Australia

Helen J. Forgasz
Monash University, Australia

INTRODUCTION

Documentation in the early 1970s of often subtle, yet consistent gender differences in mathematics performance and participation in optional, often advanced, mathematics courses in favor of males served as an effective catalyst for action in many countries. "There is," wrote Lubienski (2008, p. 351), "a history of government education policies and intervention arising in response to data on achievement disparities." Collectively, the initiatives introduced have certainly been effective. Females' participation rates in mathematics and other related fields and careers long considered to be male domains have gradually improved.

International Perspectives on Gender and Mathematics Education, pages 315–339

Over time, a number of robust research findings have emerged. There is considerable overlap in the performance of females and males. Gender differences in mathematics performance are rarely reported before or during the early years of elementary school. From the beginning of secondary schooling and beyond, males often, but not unfailingly, outperform females on tests in mathematics. Whether performance differences are found seems to depend on the format and content of the tests (e.g., Cox, Leder, & Forgasz, 2004), whether high or low cognitive level items predominate (e.g., Hyde, Lindberg, Linn, Ellis, & Williams, 2008), whether standardized tests or classroom-based data are compared (e.g., Kenney-Benson, Pomerantz, Ryan, & Patrick, 2006) and the composition of the group. For example, gender differences in favor of males are frequently found when the sample consists of high achievers (Leder, Forgasz, & Taylor, 2006; McGraw, Lubienski, & Strutchens, 2006; Mullis & Stemler, 2002)—a finding of particular relevance for this chapter with its emphasis on high achievers in mathematics.

When government interventions have occurred, they have often been fuelled by results derived from large-scale databases (Lubienski, 2008). Continued monitoring of such data sets for trends in gender differences in mathematics performance and participation thus often has an impact well beyond the interested research community. Furthermore, when such monitoring occurs by scholars outside the field of mathematics education, "the problem framing and proposed solutions are often inconsistent with scholarship in our field" (Lubienski, 2008, p. 352).

The Australian Context

Thomson and De Bortoli (2008) highlighted the change in disparity between male and female Australian students' performances in PISA 2003 and PISA 2006. They wrote:

> In the PISA 2003 assessment, gender differences in favour of males in mathematics were evident in a large number of countries; however, this was not the case in Australia, and it was noted that "Australia seems to have been able to contain [the] widening of gender disparity with age in mathematics" (Thomson & De Bortoli, 2004, p. 32). However in PISA 2006, there is evidence of a decline in the scores of 15-year-old females and no associated decline in the score for males, resulting in a significant gender difference and one that is higher than the OECD average. *The decline in scores for females appears to have come from the higher end of achievement;* 18% of females achieved at Level 5 or 6 in 2003, compared to 13% in 2006. For males the corresponding proportions were 22% and 20% respectively. (p. 244, emphasis added)

These results, as well as the finding of "a gender difference in mathematics in favour of males that has not existed for many years," Thomson and De Bortoli further argued, indicate "that gender needs to be reconsidered as an issue for Australian education" (p. 244). The report in the more recent PISA testing of a decreasing number of high performing females is particularly provocative.

Our Study

In this chapter we examine data from two large data sources—Victorian (Australia) grade 12 public examination results and the Australian Mathematics Competition [AMC]—to explore recent gender differences among high achievers in mathematics. To supplement this quantitative information, we also include some qualitative data, gathered through questionnaires administered to a small group of extremely high-achieving students, about the experiences, perceptions, and career trajectories of some of the very highest achievers in the AMC. The specific sources of data on which we have drawn in this chapter are described in a later section.

In focusing on the highest achieving students (those in the top 1–2%), our aims were:

1. to determine if patterns of gender difference favoring males that have been reported in earlier literature still persist
2. to trace the personal and career trajectories, over time, of the very highest achieving females, and consider whether they consider their gender *per se* to have been a disadvantage and to have influenced their study and career choices.

PREVIOUS RESEARCH: PROVIDING A CONTEXT

Gender Differences in Mathematics Learning

Gender differences in mathematics learning are continuing to attract attention from researchers and the wider public.

The subject . . . —sex differences in science and mathematics—is controversial. Although a topic of discussion for years, this issue was brought to public attention again in 2005 by the much-quoted comments of Lawrence Summers, then president of Harvard University, suggesting that at least part of the reason for the dearth of women at the top in math and science professions is a lack of intrinsic aptitude. Any claim that genetic differences between the sexes are somewhat responsible for the underperformance of women in some valued field is going to provoke debate, and this was no exception. Both academic and public debate ensued. (Barnett, 2007, p. i)

Prompted, in part, by Summers' remarks and the resulting controversy, Halpern et al. (2007) embarked on an extensive review of relevant research. In this search, they explained, they relied extensively "on findings that have been replicated and on the relative transparency of the scientific method" (p. 2). Like many other researchers before them, these authors stressed that a substantial overlap in the scores of males as a group and females as a group, was invariably reported in the research surveyed for their review. Several of their conclusions, undoubtedly familiar to those interested in gender differences in mathematics learning, are particularly relevant as the focus in this chapter is on high achieving students. "Sex differences in science and math achievement and ability," they reported, "are smaller for the mid-range of the abilities distribution than they are for those with the highest levels of achievement and ability" (Halpern et al., 2007, p. 1). Indeed, the authors further argued that it was particularly the relatively large gender difference in performance "at the highest end of the distribution" on high-stakes mathematics tests that has attracted the most consistent media coverage. Reasons for both the overlap and differences in performance of males and females were found to be multifaceted: "There is no single factor by itself that has been shown to determine sex differences in science and math. Early experience, biological constraints, educational policy, and cultural context each have effects" (Halpern et al., 2007, p. 41).

Mathematics High Achievers,[1] Affect, and Gender Differences in Performance

In a study with sixth-grade students, Preckel, Goetz, Pekrun, and Kleine (2008) found that gender differences in attitudes toward mathematics were stronger in their group of gifted students than in their comparison group of average-ability students. Reflecting on these findings, the authors speculated:

> [I]t might be that gifted students are more aware of their aptitudes, and of social expectations how to use these aptitudes, such that gender-role stereotypes and the gender-linked development of academic motivation become more pronounced in this group of students.... [I]t may be that gifted girls, more so than average-ability girls, tend to develop and use their abilities only if type of ability and gender-linked social expectations are congruent to each other, which may contribute to explaining why there is such a dramatic loss of female talent in the domains of mathematics, the inorganic sciences, and engineering. (Preckel et al., 2000, p. 155)

The gender break up in the samples of Finnish students in Nokelainen, Tirri, & Merenti-Valimaki's (2002) study of the self-attributions of mathematically gifted students illustrates, again, the dominance of males amongst the highest achievers. Of 77 former mathematics Olympians (adults, median age 37) only 9 (12%) were female, and of 52 high school future Olympians (median age 17) who had taken part in national competitions in mathematics, 9 (17%) were female.

Mathematics High Achievers, Persistence in Advanced Courses, and Career Paths

Putative gender differences in the development of exceptionally talented mathematics students have also attracted sustained research attention. In their impressive, longitudinal study of talented teenagers, Csikszentmihalyi, Rathunde, and Whalen (1993) speculated that "many promising young mathematicians—and especially young women—become disengaged from math not because they are unable to keep up with the cognitive challenges but because they cannot bear the supercharged atmosphere of the math clubs" (p. 108). Once disengaged from mathematics in high school, they further noted, "he or she is also likely to drop out of the *domain* altogether" (p. 108). More recently, through an intervention program, Morrow and Schowengerdt (2008) identified factors with the potential to maintain the interest of mathematically talented high school girls in advanced mathematical studies: the development of a mathematical voice and of a broader view of advanced mathematics, being positive about taking risks, and being mathematically challenged in a supportive environment.

Data gathered as part of the Study of Mathematically Precocious Youth [SMPY], founded by Julian Stanley in 1971, has also served as a unique source for monitoring the longer term development of high school students who obtain an extremely high score on specific mathematics tests and possible gender differences in the pathways chosen beyond school (see, e.g., Lubinski & Benbow, 2006; Lubinski, Benbow, Webb, & Bleske-Rechek, 2006). Although a detailed account of these findings is beyond the scope of this chapter, the following summary reflects their thrust:

> [I]n the SMPY cohorts, although more mathematically precocious males than females entered math-science careers, this does not necessarily imply a loss of talent because the women secured similar proportions of advanced degrees and high-level careers in areas more correspondent with the multidimensionality of their ability-preference pattern (e.g., administration, law, medicine, and the social sciences). By their mid-30s, the men and women appeared to be happy with their life choices and viewed themselves as equally successful. (p. 316)

The continuing lower representation of females among those entering careers in mathematics and related areas—science, technology, engineering, and mathematics (STEM)—continues to attract research attention. Hyde et al. (2008) examined data from some seven million school students in 10 American states on tests assessing cognitive performance and concluded: "Gender differences in math performance, even among high scorers, are insufficient to explain lopsided gender patterns in participation in some STEM [science, technology, engineering, and mathematics] fields" (p. 495). Drawing on data from various mathematics competitions, Andreescu, Gallian, Kane, and Mertz (2008) pointed to the influence of socio-cultural and environmental factors on students', and particularly females', willingness to continue studying mathematics beyond the high school level, and thus to have the option of pursuing careers in mathematics and related fields. More generally, Lacampagne, Campbell, Herzig, Damarin, and Vogt (2007) noted in their review of issues affecting gender equity in mathematics, that "particularly missing in the research are reasons for women's lack of persistence in continuing in mathematics from the undergraduate to the graduate program" (p. 250).

Mathematics and Gender-Role Stereotyping

One of the variables identified very early as a contributing factor to females' under-representation in mathematics and science-related careers was "sex-role congruity"—that is, whether girls believed that these careers fitted with their own images of what was appropriate for females (see Leder, 1992, for a summary of a range of contributing factors). The stereotyping of mathematics as a male-dominated domain has been inferred from the images of mathematicians drawn by children. Picker and Berry (2000) gathered drawings of "mathematicians at work" from 476 grade 7 (12–13-year-olds) students from the U.S., UK, Finland, Sweden, and Romania. Male images predominated, and the mathematicians were portrayed as authoritarian, obsessed with numbers/mathematics, lacking friends, and possessing supernatural powers. They were often illustrated as somewhat disheveled (e.g., messy hair and unkempt beards), bespectacled, unfashionably clothed, and with pens in their pockets. About a quarter of the U.S. students drew or made some reference to Einstein-like characters; just over 20% of the "mathematicians" were represented as teachers. Picker and Berry (2000) concluded that "[t]he dominant image of a mathematician that emerged from this study is...that of a white, middle aged, balding, or wild-haired man. This points to both a gender and a racial gap in pupils' images of mathematicians" (p. 89). More recently, Piatek-Jimenez (2008) examined the beliefs of five undergraduate female mathematics students and found they shared beliefs of mathematicians as extremely intelligent, obsessed with mathematics, and socially inept. It was claimed that "the images of mathematicians held by these women appear to align with many of the beliefs Picker and Berry...found amongst middle school students."

In Summary

That gender differences in mathematics learning persist, particularly among high-achieving students, is confirmed by the research findings reviewed in this section. Despite some improvements in females' participation and performance in mathematics over time, it seems that they are still less likely than males to persevere with advanced mathematics or enter mathematically based careers. In the remainder of this chapter we draw on recent Australian data to explore these issues further.

OUR DATA SETS

Throughout this chapter, our focus is on high-achieving students. For our first two data sets—sources 1 and 2—we have turned to performance data from the end-of-secondary-school assessment program in Victoria, Australia.

Our other information is drawn from data gathered from a long-running, national mathematics competition with a sustained high penetration in Australian schools, the Australian Mathematics Competition [AMC]. In the first of these explorations, source 3, the focus is again on the mathematics performance of students in their final year of secondary school. For source 4, we move beyond comparisons of performance statistics and report predominantly qualitative data. These were gathered from the highest achievers in the AMC some time after the competition success and after leaving school to gain some insights into their subsequent career directions and personal lives. Our emphasis is particularly on the females among the former AMC medalists. Issues covered include their career choice and factors influencing that choice, their motivations, and self-descriptions of aspects of their lives.

Source 1

The Victorian (Australia) Certificate of Education (VCE) is a two-year program of study. At grade 11, students' achievements are assessed in their schools and their results recorded as having satisfactorily or not satisfactorily completed the requirements. At grade 12 (final year of schooling) the second year of VCE studies involves a high-stakes public examination program, and the results are used for university selection and for pathways into other post-school options.

Data on enrollments in the VCE and in each of the subjects offered in grade 12 are publicly available on the Victorian Curriculum Assessment Authority (VCAA) website (VCAA, 2009). At the grade 12 level, there are three mathematics subjects offered: Specialist Mathematics (considered the most challenging), Further Mathematics (considered the least challenging), and Mathematical Methods, the subject serving as the key pre-requisite mathematics subject for entry into mathematics and science-related university courses.

For some time, students have been required to use graphics calculators in the examinations for Mathematical Methods. For three years, 2002–2004, a trial of an alternative parallel subject, Mathematical Methods (CAS), was conducted in which students used the more sophisticated computer algebra systems (CAS) calculators instead of graphics calculators. In the year 2004, the final year of the trial, a small number of schools participated; however, the number of schools was greater than in the previous two years.

From 2005 to 2007, the Mathematical Methods (CAS) subject ran parallel to the Mathematical Methods subject (identical curricula with respect to content), and schools had the choice in which of the two subjects to enroll students. At the time of writing, it was expected that Mathematical Methods (CAS) would be the only option available from 2010 onwards— that is, the use of the graphics calculator was being phased out. The number of schools, and of students, enrolled in Mathematical Methods (CAS) has increased steadily since the introduction of this parallel-running, alternative subject.

Enrollment data for 2004 and 2007, by gender, for the grade 12 VCE cohort and for the two parallel mathematics subjects, Mathematical Methods and Mathematical Methods (CAS), are shown in Table 14.1.

As can be seen in Table 14.1, more females than males were enrolled in the VCE in 2004 and 2007, with a slight increase in the proportion of females over the time period. However, there were more males than females enrolled in Mathematical Methods and Mathematical Methods (CAS) in both years, with the proportions of males increasing in both subjects over the time period.

Each of the two subjects had three assessable components in 2004 and 2007: a school-based assessment (mainly test-based), and two examinations—the first included multiple choice and short answer questions, and the other involved extended solutions. Due to a change in policy, in 2007 the first examination was technology free; in the second examination, students could use their calculators. The results for each of the three assessable tasks are publicly available from the VCAA website. The results are

TABLE 14.1   Enrollments in VCE, Mathematical Methods, and Mathematical Methods (CAS) by Gender, 2004 and 2007

Enrollments	2004			
	Male	Female	Total	M:F
Grade 12 VCE	23543	26432	49975	.89[a]
Mathematical Methods	9769	8216	17985	1.19
Mathematical Methods (CAS)	247	151	398	1.64
Mathematical Methods (% of grade 12 VCE cohort)	41.49	31.08	35.99	1.33
Mathematical Methods (CAS) (% of grade 12 VCE cohort)	1.05	.57	.80	1.84
	2007			
Grade 12 VCE	22588	26252	48840	.86
Mathematical Methods	7806	6389	14195	1.22
Mathematical Methods (CAS)	988	587	1575	1.68
Mathematical Methods (% of grade 12 VCE cohort)	34.56	24.34	29.06	1.42
Mathematical Methods (CAS) (% of grade 12 VCE cohort)	4.37	2.24	3.23	1.96

[a] M:F ratios < 1 mean more females than males; M:F ratios > 1 mean more males than females

reported as grades: A+, A, B+, B, C+, C, D+, D, E+, E and UG (ungraded). The frequencies and percentages of males and females achieving the two top grades, A+ and A, and the male-to-female ratios are shown for 2004 and 2007 in Table 14.2.

The data in Table 14.2 reveal that, in the vast majority of cases, higher proportions of males than females achieved the two top grades in both subjects in both years. It was clear, however, that the dominance of males was even greater in the Mathematical Methods (CAS) subject. Also evident was that for Mathematical Methods (CAS) from 2004 to 2007 the M:F ratios increased for the A+ grade for each assessment task; in examination 2 in which calculators were used in both years, the M:F ratio also increased for the A grade. It should be noted that the same patterns were evident in the data for 2002 and 2003 and for the grade B+ as well (see Forgasz & Griffith, 2006).

The data allowing the comparison in achievement on the two parallel running Mathematical Methods courses suggest that at the very highest levels of achievement the introduction of the more sophisticated technology, the CAS calculators, as part of the assessment regime seems to have increased the gender gap in performance in favor of males. This particular technology-driven curriculum initiative, it seems, warrants careful monitoring to ensure that females are not disadvantaged.

TABLE 14.2 Mathematical Methods and Mathematical Methods (CAS) Results at Top Two Achievement Levels (A+ and A), by Gender, 2004 and 2007

| | Mathematical Methods | | | | | Mathematical Methods (CAS) | | | | |
| | Male | | Female | | | Male | | Female | | |
	N	%[a]	N	%	M:F[b]	N	%	N	%	M:F
2004										
Grade	Task 1 (school-assessed)									
A+	1747	18	1261	15	1.20[c]	42	17	20	13	**1.31**[d]
A	2003	21	1939	24	0.88	57	23	28	19	**1.21**
Grade	Examination 1									
A+	1102	12	793	10	1.20	35	14	16	11	**1.27**
A	1306	14	1030	13	1.08	40	17	19	13	**1.31**
Grade	Examination 2									
A+	1013	11	593	7	1.57	33	14	12	8	**1.75**
A	1074	11	874	11	1.00	34	14	20	14	1.00
2007										
Grade	Task 1 (school-assessed)									
A+	1288	17	988	16	1.06	191	19	73	13	**1.46**
A	1655	21	1430	23	.91	206	21	121	21	**1.00**
Grade	Examination 1—technology free									
A+	876	12	598	10	1.20	137	14	47	8	**1.75**
A	947	13	793	13	1.00	141	15	70	12	**1.25**
Grade	Examination 2—using calculators									
A+	838	11	522	8	1.38	160	17	47	8	**2.13**
A	895	12	679	11	1.09	148	15	72	13	**1.15**

[a] Within gender cohort percentages. Gender cohort sizes found in Table 14.1
[b] Male to female ratio (M:F) = percentage Males : percentage Females
[c] Shaded ratios: when M:F > 1 i.e., greater % males than females
[d] Bolded M:F ratio: higher M:F ratio for the two subjects

Source 2

Another way in which VCE results are reported is by *study score*—an overall score for each subject. Study scores are the product of a process of statistical moderation in which scores are standardized across all subjects such that the maximum score attainable is 50, the mean is 30, and the standard deviation is approximately 7. Thus 50 is considered a perfect score for a subject. Each year, in a large lift-out section of the daily press, the names of students, and the schools they attended, are published for those achieving scores of 50 down to 40 for each VCE subject. The results are organized by subject. Hence the data available for the various mathematics subjects offered can be examined, as well as for all other subjects.

It should be noted that for each VCE subject, scores between 46 and 50 represent 2% or less of the cohort for each subject. Since enrollment numbers for the Mathematical Methods (CAS) subject were relatively small compared to the other subjects, very few students obtained scores of 50 down to 46. Hence for the analysis of the 2007 VCE results reported here, the data for Mathematical Methods CAS and Mathematical Methods (as parallel options) were combined. The enrollment data, by gender, for all grade 12 VCE students, and those taking Specialist Mathematics, the combined Mathematical Methods courses, and Further Mathematics are shown in Table 14.3.

As noted earlier, there were more females (53.8% of the cohort) than males undertaking the VCE at the grade 12 level in 2007. However, it is also clear that there were more males (55.7% of the cohort) than females enrolled in the combination of the two Mathematical Methods subjects; to be equitably represented, 53.8% of the combined Mathematical Methods subject cohorts should have been female.

Based on the data from a lift-out section of the Victorian daily newspaper, *The Age* (December 19, 2007), a breakdown, by gender, of those obtaining the highest study scores (50–46) for each of the three VCE mathematics subjects—Specialist Mathematics, Further Mathematics and the combination of Mathematical Methods and Mathematical Methods (CAS)—is shown in Table 14.4. The gender of students was determined from the given names listed. For some, it was impossible to identify whether it pertained to a male or a female and was thus designated "gender unknown."

The data in Table 14.4 clearly reveal much higher proportions of males than females achieving each score from 50 down to 46 in each of the grade 12 VCE mathematics subjects. That means that the highest achieving 2% in each subject is clearly male dominated. In fact, when the data shown in Ta-

TABLE 14.3   Grade 12 VCE and Mathematics Subject Enrollments by Gender (and Cohort Percentages), 2007

Category of enrollment	All	Male		Female	
		N	%	*N*	%
Grade 12 VCE students	48840	22588	46.2%	26252	53.8%
Specialist Mathematics	4804	3012	62.7%	1792	37.3%
Mathematical Methods & Mathematics Methods (CAS)	15427	8600	55.7%	6827	44.3%
Further Mathematics	24787	11623	46.9%	13164	53.1%

TABLE 14.4 Top-Scorers in the Various Grade 12 VCE Mathematics Subjects by Study Score, and Gender

Study score	All	Male	Female	Unknown
Specialist Mathematics				
50	14	**12 (86%)**[a]	1 (7%)	1 (7%)
49	5	**5 (100%)**	—	—
48	12	6 (50%)	6 (50%)	—
47	13	**11 (85%)**	2 (15%)	—
46	21	**15 (71%)**	6 (29%)	—
Mathematical Methods (combined)				
50	36	**24 (67%)**	9 (25%)	3 (8%)
49	29	**24 (83%)**	5 (17%)	—
48	27	**18 (67%)**	6 (22%)	3 (11%)
47	41	**29 (71%)**	11 (27%)	1 (2%)
46	66	**38 (58%)**	19 (29%)	9 (14%)
Further Mathematics				
50	60	43 (72%)	13 (22%)	4 (7%)
49	36	**19 (53%)**	14 (39%)	3 (8%)
48	48	**30 (63%)**	17 (35%)	1 (2%)
47	60	**37 (62%)**	22 (37%)	1 (2%)
46	108	58 (54%)	48 (44%)	2 (2%)

[a] Bolded = gender group with higher representation (based on data published in *The Age*, December 19, 2007)

ble 14.4 are compared to the enrollment data in Table 14.3, it is again quite clear that the proportions of males achieving each of the scores of 50 down to 46 is higher than their enrollment proportions in each of the mathematics subjects. It should also be noted that male dominance is most marked in the most challenging mathematics subject, Specialist Mathematics.

To put these findings into perspective, the male-to-female ratio of the 98 students achieving a study score of 50 in English, a subject traditionally considered to be a female domain and one with the highest grade 12 enrollment numbers (N = 42762 of which 19814 or 46.3% were male and 22948 or 53.7% were female) was approximately 1:1; the relative dominance of males in mathematics was not replicated by females in English.

In summary, earlier findings of male dominance in the very highest achievement levels in mathematics (e.g., Halpern et al., 2007) have again been demonstrated in the analyses presented here for the top 2% of achievers in the Victorian grade 12 VCE results from 2007. Not only was this true for the most demanding subject, Specialist Mathematics, but for all three subjects offered at this level.

Source 3: The Australian Mathematics Competition: Background and Performance Data

The Australian Mathematics Competition (AMC) has been held annually since 1978. The first competition attracted some 60,000 secondary school students from 700 schools across Australia. More recently, in each year over 20% of the Australian secondary school population has entered the competition. In addition, students from close to 40 other countries now also participate in the AMC. Reference to this latter group is, however, beyond the scope of this chapter.

The AMC is open to students of all standards and each entrant receives some form of acknowledgement of success in keeping with the AMC's over-riding aim of rewarding outstanding performance, as well as giving "average" students a sense of achievement (Australian Mathematics Trust [AMT], 2007). Each year a small number of students, about 1 for every 10,000 students entered, receives a medal. Details of the other categories of awards are shown in Table 14.5.

The AMC papers are composed by a specially convened committee, drawn from teachers and university academics. Three separate Competition papers are set for students in grades 7 and 8, in grades 9 and 10, and in grades 11 and 12, respectively. The papers contain a balance of questions in arithmetic, algebra, and geometry with which students should have classroom experience, as well as problems of a general nature that may require translation of an issue set in a real-life context to one able to be solved as a mathematical problem through a synthesis of mathematical skills. The questions are graded, with the early questions accessible to students of all standards, while problems placed later in the papers are challenging to the most elite students.

TABLE 14.5   Achievement and Participation Data for Grade 12 Participants in the Grade 11–12 AMC Paper, 2004

	Number of students within merit category			
AMC awards	M	F	All	M:F
Prize (top 0.3%)	57	7	64	8.14
HD[a] (top 2%)	268	87	355	3.08
D (top 15%)	1906	884	2790	2.16
C (top 50%)	4068	2951	7019	1.38
Participation	4850	5012	9862	.97
Totals	11149	8941	20090	1.25

[a] HD = High Distinction; D = Distinction; C = Credit

The data in Table 14.5 indicate the number of Australian grade 12 students who entered the competition and sat the grade 11–12 competition paper in 2004 and their levels of success in the competition.

From Table 14.5, it can be seen that gender differences in performance are most acute at the highest levels of achievement: prize, HD, and D. When the difference in the numbers of males and females entering the competition at grade 12 is taken into account, the M:F ratios shown in the last column of Table 5 diminish slightly, but certainly do not disappear.[2]

As mentioned earlier, each year a medal is awarded to a small number of the most outstanding students. "These are awarded on the judgment of the committee to students who are outstanding within their region (Australian State or Territory or other country), within their year group and internationally" (AMT, 2007, n.p.). There are few females among the medalists (Leder, Forgasz, & Taylor, 2006) even though roughly equal numbers of females and males enter the AMC, especially in the earlier secondary school years. For example, in 2004, 22 medals were awarded to students from schools in Australia;[3] five of these students—all boys—were in grade 12. In 2008, 26 medals were awarded to students from Australian schools; three of these medalists—two boys and one girl—were in grade 12 (AMT, 2008).

In summary, the 2004 AMC data for grade 12 students—representative of patterns discernable for other years—reveal clearly how males dominate at the very highest levels of achievement in this competition.

In the next section we shift the focus beyond the mathematics performance of AMC medalists and delve into aspects of the lives of this mathematically elite group. Of particular interest were their self-descriptions, their motivations, their career choices, and the factors influencing those choices.

Source 4: Australian Mathematics Competition Medalists: Beyond Performance Data

The long-running AMC, with its high penetration into Australian schools, has provided not only a unique record of student performance in mathematics but, each year, also celebrates the successes of a small number of exceptionally high achievers. Between 1978 and 2006, based on performance in the AMC, 690 medals were awarded to students attending a secondary school within Australia. The actual number of recipients was in fact less, 420 students, since it has not been unusual for a particular student to win a medal in more than one year. With the support of the AMC's organizers, previous medalists were alerted, by letter sent to the last available address, to the URL of an online survey specifically aimed at former medalists, and were encouraged to complete it. Of the letters sent out, 52 were returned as undeliverable. In addition, any completed survey in which the items relating to an AMC medal were not answered was discarded.[4] The total of 90 useable surveys, 80 from males and 10 from females, represents a response rate of at least 25%, described as acceptable by, for example, Hamilton (2003) and McBurney and White (2004). The sample was considered representative as respondents included medalists from 1979 to 2005, mirrored the ratio of single and multiple medalists found in the larger sample, and reflected a male dominance among the AMC medalists consistent with the gender ratio shown for prize winners in Table 14.5.

THE MEDALISTS' SURVEY

The survey covered five broad areas: background; school and university; career/vocation (actual or intended); work habits; and some general issues about self. Some items were open-ended: for example, "What did winning a medal mean to you?"; "What are your favorite leisure pursuits?" Others were in 5-point Likert response format to which students responded SA (strongly agree) to SD (strongly disagree). For example, "Once I undertake a task, I persist;" "If at all possible, I'd rather work alone than with others to complete a task." A second survey, which enabled more detailed explorations of issues already opened up, was sent to a small subgroup of the respondents.

In what follows we focus on selected findings pertaining particularly to the female medalists. After the presentation of some contextual background information, with data from male respondents given for comparison, voice is given to two of the female medalists, both of whom chose a career for which they perceived mathematics to be critical yet did not describe themselves as mathematicians. Both of these women completed the initial and follow-up surveys.

Male and Female Medalists: Group Data

As mentioned above, 90 useable surveys were received from medalists: 80 from males and 10 from females. Four of the latter were born outside Australia. For the males the figure was 23%, virtually identical to the proportion of 22.2% given in the census figures for Australia's population as a whole (Australian Bureau of Statistics, 2008).

Two (20%) of the female medalists listed mathematics as a favorite subject at school compared with 60% of males who did so. English was listed by three of the females and by 10% of the males.

None of the females described themselves as a mathematician. Four were medical practitioners; four were working in, or towards, mathematically related fields (medical scientist, meteorologist, statistician/physicist, and software engineer); while the other two worked in fields that did not involve mathematics. In contrast, some 15% of the males described themselves as mathematicians. Many others worked in mathematics-related professions, including actuary, engineer, hedge fund trader, software developer, and computer scientist.

Although the majority of the female respondents indicated that a male (uncle, father, lecturer, mathematics teacher) had influenced their choice of career, the following excerpt may be indicative of the factors influencing the more constrained career choices (compared to the males) of the female medalists:[5]

> Unfortunately, Australia does not value academic achievement, and although there has been talk of a maths/science teacher crisis for decades (one of the year 8 maths teachers at my high school was actually primarily a metalwork teacher) little has been done to promote maths and maths teaching as a career option. More promotion is needed for maths generally, as leading to a wide range of careers. When I was at school it was assumed that a uni maths degree would lead only to a teaching career or academia. (The careers counselor was totally useless.) Other maths graduates I know have had similar experiences.

There was much overlap in the motivations and working habits of the female and male medalists. They thrived when doing difficult, challenging, and highly skilled work. Once started, they persisted with a task. Their motivation and task commitment was high. Many liked competitive work, as well as cooperative work. They strongly rejected the suggestion that they might sometimes "work at less than my best so others won't resent me." They were divided whether they worried "that my success may cause others to dislike me," though the males disagreed more strongly than the females with this statement.

None of the AMC medalists, neither females nor males, mentioned any negative aspects of winning a medal. Many indicated that winning a medal gave them great satisfaction and pride in having their mathematics achievement recognized. That teachers and schools, as well as the medalists themselves, reacted variously to the recognition of high mathematical achievement is captured evocatively by the following two excerpts, both from female medalists:

> The AMT sent some extra challenging problems, but it wasn't really followed up. I did do some of them. If my school had given me any encouragement or some time off the incredibly boring school maths classes to do them, I would probably have done a lot more. So actual benefits—negligible.... My best experience of maths education was the one year I spent in Japan as an 8-year-old

in 1976, which gave me a huge head start, followed by two enlightened maths teachers in 4th + 5th grade primary school back in Australia who allowed me to work through every text book they had at my own pace. I finished grade 10 maths half-way through 5th grade. Unfortunately, none of my subsequent teachers were as inspired. In 6th grade they couldn't find anything for me to do. High School was much worse—I had to start on year 7 maths like everyone else, and was bored to tears for the rest of my school days. I like a challenge, but there wasn't one. (Leonie,[6] who became a freelance orchestral musician)

I think (it) got me an invitation to participate in the Tournament of the Towns—which in turn meant regular exposure to (a) more challenging mathematics, and (b) other extremely talented students. I gained a great deal from this program. At school, I got somewhat embarrassed by the fuss and teacher pride, but on the other hand my teachers were happy to let me do other things in class once I finished class work. (Barbara, who is intending to become a statistician or physicist)

Many of the males, on the other hand, indicated that winning a medal opened mathematically exciting doors:

Selection into the Mathematical Olympiad training programme, with many flow-on benefits, including: learn much more mathematics and at a higher level, meet like-minded people many of whom are now good friends, encouragement to continue with mathematics. (Martin, now doing statistical research at a highly prestigious university outside Australia)

Summary

There was much overlap in the group responses of the male and female medalists—for example, in the collective responses to the questionnaire items used to probe motivations and working habits. However, for other items, small but consistent differences emerged. Both males and females indicated that mathematics had been their favorite subject at school, many had studied mathematics at university, and several had entered mathematically dominated occupations. On each of these items, proportionately more males than females indicated that this was the case. No female medalist described herself as "a mathematician," but almost one in six of the male medalists did. None of the respondents considered having been a medalist to have been a disadvantage, although more males than females pointed to specific advantages in and beyond school gained as a result.

In the next section these group data are supplemented with more detailed sketches of two of the female medalists. These brief accounts afford glimpses into some of the factors which intentionally or opportunistically played a role in shaping their paths into adulthood.

TWO FEMALE MEDALISTS HAVE THEIR SAY

As medal recipients, Kim and Amanda are two of a stringently selected group of females whose outstanding performance in the Australian Mathematics Competition provided a rare and distinctive dimension to their school days. In the short biographies that follow, we touch on their work preferences, careers, and more personal matters.

Kim

Kim, who was a medalist in grade 12, is now in her early twenties and describes herself as a software engineer. Most of her family members work in either scientific or engineering fields. She credits Mr. X, "The teacher who ran the after-school maths club in Hobart," as the person who really fostered her interest in mathematics, a subject that to her is "all about interesting problem solving." When asked if there were any particular factors that influenced her decision not to become a mathematician, she replied that she "discovered in Year 12 that I enjoyed computer science even more than mathematics." She continued:

> I knew that I wanted to do scientific research of some sort since early high school. The only difficulty was deciding on a field, since I was interested in too many areas of science—mathematics, physics, chemistry, biology, eventually also computer science.... In my undergraduate degree, I tried to choose majors that would cover as many fields as possible, so I could put off the decision. Eventually I realised that I liked computer science best of all, and did my honours in computer science.

And is mathematics important in her current work?

> Yes, definitely. My work involves programming software for ship motion modelling, which includes making accurate predictions of waves, tides, currents, etc. at a port. A strong mathematical background makes it much easier for me to understand the equations used for modelling, as well as to pick up mistakes—both in the original equations, and in my implementation of those equations.

To date, she has had no gaps in her career. The most important factors which have influenced, and will continue to influence, her career choice are the requirement that her work uses her intellect, makes the best use of her talents, allows her to combine career and family, and leaves time for other things in her life. For her to work optimally, the physical location is less important than the need for it to be "without interruptions or distractions." When asked to reflect further on working creatively she noted:

> I find that going for a walk is the most useful thing I can do when I get stuck on a really difficult problem, since that breaks my mind out of the unproductive loop it is stuck in, and lets me get a fresh perspective on the problem. I've had too many occasions when I finished an exam, only to have the answer to a question I couldn't solve at the time pop into my mind during my walk home.

> Though in my case, the longest period of time with no visible progress has been about four days, probably because I've never worked on a project where weeks or months without visible progress would have been acceptable!

The need to work to a firm deadline, without the luxury of a long period of gestation, is a comment made by many of the medalists, both female and male.

Kim describes herself as "intelligent, reserved, logical, geeky, random, a long-term planner." Her favorite leisure pursuits include reading, computer games, board games, and role-playing games. She is keenly aware that her high achievements have had an element of "trail blazing" but views this positively.

> One advantage [of having been a member of the opposite sex] would be not having to put up with awards I've won being presented with a comment such as "this is the first time a female student has won this award two years in a row," as though my being female was more important than the achievement itself. But on the other hand, if I were not female, I wouldn't have the opportunity to puncture the egos of those people who believe women can't be good programmers or good mathematicians.... Somehow, any people who genuinely are smarter than me are also smart enough not to have such opinions. I've also been told by a couple of friends that when they get annoying comments about women being bad at maths/science/computers, they just remind the person of my existence, which shuts them up quite effectively. So it is a good thing that I exist, and that I am female. I'm not sure whether being an example in this way can be called an advantage, but it is definitely a good thing.

In Summary

Kim clearly considered her mathematical knowledge to be invaluable in her daily work. Her leisure occupations reflect her interest in computers. Though keenly aware that her high mathematics achievement was viewed as unusual and special by some of her peers, she felt supported by friends who prized and valued her achievements.

Amanda

Amanda, who won a medal in grade 8, is in her mid-twenties. She is a post-graduate student in medical science, close to completing her doctoral studies. She considers developing her own research projects to be a major challenge in the immediate future. Winning a medal not only gave her "the satisfaction of both knowing that I could do something in a perfectionist way, and that there was recognition for it," but also the opportunity to attend a mathematics summer school organized for outstanding mathematics students. Science was her favorite subject at school, and she now has "a small regret that I didn't spend more time on mathematics." She fondly remembers two "great" teachers at high school who particularly fostered her interest in mathematics. When asked if there were any factors that influenced her decision not to become a mathematician, she replied "Probably doing maths at high school and feeling like I had to compete with people who were not only very good at it, but also tremendously enthusiastic and energetic about pursuing it." She further reflected that

> It's probably true that mathematical problems are solved in a subconscious leap after periods of little progress, but I prefer the biological research that I am doing now, as even while there are still problems that require musing over and that intellectual leap to solve, there are still concrete experiments that can be done in the mean time to try to close the gap.

Nevertheless, mathematics is important for Amanda's research, which "involves applying statistical methods to gene expression data." Presenting at an international conference has been a career highlight to date. "Requires use of my intellect," "involves helping others," and "makes best use of my talents" were the most important factors that influenced her choice of her career. She works best "in an environment with other people who are very focused on working as well." Being a female, she indicated, has made "no major differences yet, but career gap wise, it would be easier to be male." Amanda's hobbies of diving, snorkeling, and reading are indicative of the interests she has beyond her studies.

In Summary

Like Kim, Amanda, too, considered mathematics to be important in her current work. Despite recognizing the benefits gained from winning a medal, mathematics was never her favorite subject. Her reference to the enthusiastic and energetic competition seemingly embraced by many of her mathematically high achieving peers at school, but rejected by her, is evocative of Csikszentmihalyi et al.'s (1993) observation that "many promising young mathematicians—and especially young women—become disengaged from math not because they are unable to keep up with the cognitive challenges but because they cannot bear the supercharged atmosphere of the math clubs" (p. 108). The greater appeal of science was reflected in Amanda's choice of career, but neither mathematics nor science interests were readily apparent from her leisure pursuits. Although indicating that, to date, being a female had not proved to be a disadvantage, she speculated that longer term that might change: "career gap wise, it would be easier to be male."

SUMMARY REFLECTIONS

Given the inevitably small sample, it is difficult to generalize about the medalists as adults. Out of the 90 respondents, only one talked of personal difficulties. On the other hand, a number talked about the challenge of moving from one fixed-term contract to another. Some indicated that their careers were not necessarily the outcome of carefully structured, long-term plans: "I didn't make a decision 'not to become a mathematician'; I made a series of decisions to study other things and work in other fields." None gave a negative response to "what winning a medal" brought them. Many talked of the importance, specifically or in general terms, of mathematics to their jobs or daily lives. All gave multiple responses to the question about their leisure pursuits and many, presumably interested in what happened to their fellow medalists, asked to be given some feedback about the findings of the survey. Only one was without work—that is, was neither still studying nor currently employed. Career decisions were sometimes made in, and through, circumstances beyond the control of the individual. "I believe," one wrote, "that a model of career progression which contains 'gaps' as opposed to 'normal' periods of work is outdated." Nevertheless, the comments made by Lubinski et al. (2006) about the SMPY cohorts they followed, also apply to the sample of AMC medalists:

> although more mathematically precocious males than females entered math-science careers, this does not necessarily imply a loss of talent because the women secured similar proportions of advanced degrees and high-level careers in areas more correspondent with the multidimensionality of their ability-preference pattern (e.g., administration, law, medicine, and the social sciences. (p. 316)

That being a female inevitably plays a part in shaping that career path is sug-gestively highlighted in another of the female medalist's (Renee's) musing:

> Clearly, Australia (like other Western nations) continues to sustain disparities in men's and women's achievements in the workplace, public life and the economy (amongst other things). However . . . I do not believe it is possible to

separate my being female from other aspects of my identity and life. If I were male I simply would not be the same person.

FINAL WORDS

From the VCE data presented in this chapter, it is very clear that, as in the past, there is a significant gender difference favoring males in patterns of enrollment, particularly in the most demanding (optional) mathematics courses offered at the grade 12 level. A similar, persistent pattern of male dominance at the very highest levels of achievement was also identified. With respect to the mandating of hand-held calculators within the mathematics curriculum and for use in examinations, it appears that females' underperformance relative to males' at the very highest levels of achievement may be exacerbated as the level of sophistication of the calculators increases. The PISA results from 2006 (Thomson & De Bortoli, 2008) appear consistent with a decline in overall performance for Australian students and a widening gender gap indicating that girls have fallen even further behind. Although speculative, the increased emphasis on the use of sophisticated hand-held technologies for mathematics learning may be contributing to this. Further research is called for.

Despite the consistent gender differences favoring males in participation and in mathematics achievement at the highest levels, it is also clear from the AMC data presented that females are capable of high achievement in mathematics and of building constructively on that achievement. Relatively little is known about the lives beyond school of the AMC medalists, and particularly the females among them. It is not easy to determine whether being recognized as an exceptionally good mathematics student made a difference longer term to their lives and career paths. Many of the respondents, both females and males, talked of the boost to self-esteem and delight in having high achievement recognized. For some, winning a medal influenced the mathematics curriculum made available during their time at school and the choice of subjects at university—particularly if that choice was reinforced by good teaching and inspiring teachers. The role of teachers and schools in recognizing the mathematical talent of the highest achieving students and intervening with appropriate curricular options for these exceptional students appears to make a difference in their longer term persistence in pursuing mathematical studies and embarking on related careers. This seems especially pertinent for females; this inference is consistent with the reported findings from a planned intervention beyond school reported by Morrow & Schowengerdt (2008).

In conclusion, the data presented in this chapter clearly signal that gender remains a factor differentiating the outcomes of mathematical studies at the very highest levels of achievement, at least in the Australian context. For equity and social justice to prevail, with the expectation that all students have the opportunity to meet their potential and contribute to their societies, continued close monitoring of large-scale achievement data is critical. Charting trends over time, while simultaneously recognizing that factors other than gender alone are implicated, is crucial. Societal changes as well as the specific and localized contexts in which patterns and trends are followed must also be considered. We also urge continued work on smaller, more focused research studies to tease out and unpack the underlying reasons and factors contributing to any patterns and trends noted that point to gender inequities.

NOTES

1. As noted earlier, we have defined "high achievers" as those in the top 1–2%. In the review of previous work, the terminology used by the authors has been adopted. For a detailed discussion of definitions of "gifted" see, for example, Heller and Schofield (2000).
2. The M:F ratio of 1.25 shown in Table 14.5 for participation in 2004 in the grade 12 section of the AMC is comparable to the M:F ratios of 1.19 and 1.22 shown in Table 14.4 for entries in Mathematical Methods in 2004 and 2007, respectively.
3. In keeping with the focus of this chapter, no further reference is made to medalists from other countries.
4. Although only medalists were specifically directed to the survey site by letter, others who perused the AMT Website also had access to the survey and might have explored it out of curiosity.
5. The original Australian-English spelling is retained when the medalists' own words are reproduced.
6. All names are pseudonyms.

REFERENCES

Andreescu, T., Galligan, J. A., Kane, J. M., & Mertz, J. E. (2008). Cross-cultural analysis of students with exceptional talent in mathematical problem solving. *Notices of AMS, 55*(10), 1248–1260.

Australian Bureau of Statistics. (2008). *Census of population and housing.* Retrieved September 26, 2008. from http://www.abs.gov.au/websitedbs/d3310114.nsf/Home/census.

Australian Mathematics Trust (AMT). (2007). *Events. Medals.* Retrieved February 20, 2007, from http://www.amt.canberra.edu.au/amcfact.html

Australian Mathematics Trust(AMT). (2008). *Events.* Retrieved September 18, 2008, from http://www.amt.canberra.edu.au/amc2008.html

Barnett, S. M. (2007). Complex questions rarely have simple answers. *Psychological Science in the Public Interest, 8*(1), i–ii.

Cox, P. J., Leder, G. C., & Forgasz, H. J. (2004). The Victorian Certificate of Education—Mathematics, science and gender. *Australian Journal of Education, 48*(1), 27–46.

Csikszentmihalyi, M., Rathunde, K., & Whalen, S. (1993). *Talented teenagers.* Cambridge: University of Cambridge Press.

Forgasz, H., & Griffith, S. (2006). CAS calculators: Gender issues and teachers' expectations. *Australian Senior Mathematics Journal, 20*(2), 18–29.

Halpern, D. F., Benbow, C., Geary, D., Gur, D., Hyde, J., & Gernsbacher, M. A. (2007). The science of sex-differences in science and mathematics. *Psychological Science in the Public Interest, 8*(1), 1–52.

Hamilton, M. B. (2003). *Online survey response rates and times.* Retrieved February 24, 2008 from http://www.supersurvey.com/papers/supersurvey_white_paper_response_rates.pdf

Heller, K. A., & Schofield, N. J. (2000). International trends and topics of research on giftedness and talent. In K. A. Heller, F. J. Mönks, R. J. Sternberg, & R. F. Subotnik (Eds.), *International handbook of giftedness and talent* (2nd ed.) (pp. 123–137). Oxford: Elsevier Science Ltd.

Hyde, J. S., Lindberg, S. M., Linn, M. C., Ellis, A. B., & Williams, C. C. (2008). Gender similarities characterize math performance. *Science, 321,* 494–495.

Kenney-Benson, G. A., Pomerantz, E. M., Ryan, A. M., & Patrick, H. (2006) Sex differences in math performance: The role of children's approach to schoolwork, *Developmental Psychology, 42*(1), 11–26.

Lacampagne, C. B., Campbell, P. B., Herzig, A. H., Damarin, S., & Vogt, C. M. (2007). Gender equity in mathematics. In S. S. Klein (Ed.), *Handbook for achieving gender equity through education (2nd ed.)* (pp. 235-252). London: Taylor and Francis.

Leder, G. C. (1992). Mathematics and gender: Changing perspectives. In D. A. Grouws (Ed.), *Handbook of research in mathematics teaching and learning* (pp. 597–622). New York: Macmillan.

Leder, G. C., Forgasz, H. J., & Taylor, P. J. (2006). Mathematics, gender, and large scale data: New directions or more of the same? In J. Novotná, H. Moraová, M. Krátká, N. Stehliková (Eds.) *Mathematics in the centre. Proceedings of the 30th conference of the International Group for the Psychology of Mathematics Education Vol 4* (pp. 33–40). Prague, Charles University: PME.

Lubinski, D., & Benbow, C. P. (2006). Study of mathematically precocious youth after 35 years. Uncovering antecedents for the development of math-science expertise. *Perspectives on Psychological Science, 1*(4), 316–345.

Lubinski, D., Benbow, C. P., Webb, R. M., & Bleske-Recheck, A. (2006). Tracking exceptional human capital over two decades. *Psychological Science, 17*(3), 194–199.

Lubienski, S. T. (2008). On "gap gazing" in mathematics education: The need for gaps analyses. Research commentary. *Journal for Research in Mathematics Education, 39*(4), 350–356.

McBurney, D. H., & White, T. L. (2004). *Research methods.* Belmont CA: Wadsworth.

McGraw, R., Lubienski, S.T., & Strutchens, M.E. (2006). A closer look at gender in NAEP mathematics achievement and affect data: Intersections with achievement, race/ethnicity, and socioeconomic status. *Journal for Research in Mathematics Education, 37*(2), 129–150.

Morrow, C., & Schowengerdt, I. (2008). Stepping beyond high school mathematics: A case study of high school women. *ZDM. The International Journal on Mathematics Education, 40*(4), 693–708.

Mullis, I. V. S., & Stemler, S. E. (2002). Analyzing gender differences for high-achieving students on TIMSS. In D. F. Robitaille & A. E. Beaton (Eds.) *Secondary analysis of the TIMSS data* (pp. 277–290). Dordrecht, The Netherlands: Kluwer.

Nokelainen, P, Tirri, K., & Merenti-Valimaki, H-L. (2002). Self-attributions of mathematically gifted. In R. Craven, H. W. Marsh, & K. B. Simpson (Eds.), *Self-concept research: Driving international research agendas. Collected papers of the second biennial self-concept enhancement and learning facilitation (SELF) research centre international conference.* Sydney: University of Western Sydney, SELF Research Centre. Retrieved September 15, 2008, from http://web.archive.org/web/20031123181849/http://self.uws.edu.au/Conferences/2002_CD_Nokelainen1,_Tirri_&_Merenti-Valimaki.pdf

Piatek-Jimenez, K. (2008). Images of mathematicians: A new perspective on the shortage of women in mathematical careers. *ZDM. The International Journal on Mathematics Education, 40*(4), 633–646.

Picker, S., & Berry, J. (2000). Investigating pupils' images of mathematicians. *Educational Studies in Mathematics, 43*(1), 65–94.

Preckel, F., Goetz, T., Pekrun, R., & Kleine, M. (2008). Gender differences in gifted and average-ability students. Comparing girls' and boys' achievement, self-concept, interest, and motivation in mathematics. *Gifted Child Quarterly, 52*(2), 146–159.

Thomson, S, & De Bortoli, L. (2008). *Exploring scientific literacy: How Australia measures up. The PISA 2006 survey of students' scientific, reading and mathematical literacy skills.* Hawthorn, Australia: ACER Press.

Victorian Curriculum and Assessment Authority [VCAA]. (2009). *Senior secondary certificate statistical information.* Retrieved April 6, 2009, from http://www.vcaa.vic.edu.au/vce/statistics/subjectstats.html

CHAPTER 15

WHAT ARE HIGH ACHIEVING YOUNG WOMEN'S PERCEPTIONS OF MATHEMATICS OVER TIME?

Amanda Lambertus and Susan Bracken
North Carolina State University

Sarah Berenson
University of North Carolina–Greensboro

I really liked Multivariable Calculus, because I like seeing things in 3-D and being
able to rotate them around graphs and I thought it was really fascinating
and really liked my professor, too.
—Shelly, Pre-Med, Junior

INTRODUCTION

This chapter addresses how high-achieving young women perceive the role of mathematics in relation to their education and future career choices. We triangulated a total of 19 interviews collected from nine young women

International Perspectives on Gender and Mathematics Education, pages 341–362
Copyright © 2010 by Information Age Publishing

across time. Interviews were conducted during the young women's middle school, high school, and/or college years. Finally, we discuss the implications of these findings to the *Principles and Standards for School Mathematics* (National Council for Teachers of Mathematics [NCTM], 2000).

The research presented here is a direct result of a 10-year study following girls, ages 11–15, who attended a mathematics and science summer program. In the first few years of the study, our purpose was to encourage these young women to continue on the advanced mathematics track, meaning that they would be in a position to take calculus no later than their final (senior) year in high school. Many of the young women in our study did continue on to complete calculus while enrolled in high school. However, out of more than 250 young women, only two planned to or chose to major in mathematics as undergraduates. This lack of interest in college-level mathematics does not seem to be an isolated event for the young women in this study. In recent decades, more efforts have been put in place to retain women in the study of higher level mathematics: "while helpful, so far [they] have only delayed the stage when young women tend to leave the track. They used to drop out in high school, as soon as they were given some choice of electives. Now they wait until college" (Burton, 1995, p. 121).

HISTORY OF GIRLS ON TRACK

Girls on Track was originally designed to be a three-year multi-institutional intervention program for middle school girls. The research team developed a two-week summer program and year-round activities to increase motivation and interest in learning mathematics. Through continuous grant support from the National Science Foundation, the program was allowed to extend into a 10-year longitudinal study. The time has allowed researchers to follow and track the middle school participants through high school, college, and in some cases into graduate school. Because the program covers a large span of time, a historical perspective of Girls on Track will provide a context for understanding the scope of the larger, longitudinal study. Some of the findings that were previously reported are also presented in this section.

In 1999, the Center for Research in Mathematics and Science Education at North Carolina State University began a longitudinal study of middle school girls who were selected to take Algebra 1 in seventh or eighth grade. In North Carolina, Algebra I is usually taken in the ninth grade (the first year of high school). Further, it is taken by those students planning to take more advanced mathematics at the high school level, possibly completing Advanced Placement Calculus or beyond (Dulaney, 1996). There were 40 high-achieving girls in the first cohort attending a two-week summer camp. Those young women are now enrolled in their final year of college or graduated in spring 2008. In subsequent years, cohort sizes enrolled in the two-week summer school were 78 girls in 2000, 84 girls in 2001, 40 girls in 2002, and 42 girls in 2003 (n=284). The summer program focused on solving community problems such as population growth, traffic, and trash disposal in an urban county. These types of problems were specifically designed to give the girls experiences with mathematics in realistic contexts. On the final day of the camp, the girls used technology and mathematics to frame their problems and present their findings to their parents. In the first three years of the project, 45 girls were interviewed to investigate how they and their parents perceived their school achievement.

Howe and Berenson (2001) reported that there were four characteristics that emerged from these interviews. Every girl reported strong support from her family in terms of academics. Typical is the answer Brittany gave to the question about what her parents would do if she got a bad grade in mathematics:

> They wouldn't get mad at me, they would ask me why I got the bad grade and then help me to improve. I don't think they would get mad at me, I think they would help me. They would probably hire a tutor.

A second factor was that these girls all wanted to understand what they were doing in mathematics. One said she wanted to know "if I'm really learning and not just memorizing." The desire to understand, and their opinions of their teachers, came out in response to the question "How was math for you this year?" The girls were not asked directly about their teachers but all gave an opinion that, in almost every case, depended on whether the teacher helped them understand the topic or concept being studied. Assertiveness was the third characteristic to emerge from these 45 interviews. These are not the "good girls" who are passive in class, remaining quietly in the background while the teacher asks the boys the hard questions. Their desire to understand, as well as their desire to get a good grade, prompted them to ask questions in class and to seek help offered by teachers before and after class. One girl said, "If I don't get it, I don't get it and I'm just going to raise my hand." When they got lower grades than they wanted, they went to the teacher and asked why they got the low grade and what they could do to bring it up. Finally, these girls believed in hard work. What they all did to bring up a grade or maintain an A, was to work hard or harder.

Four years later, in 2004, 39 telephone interviews were conducted with high school juniors and seniors (young women in their final two years of high school) who were in cohorts 1 and 2 as middle school students. Berenson, Vouk, Michael, Greenspon, and Person (2004) reported results from one aspect of the interviews. They believed that the girls showed generally positive attitudes towards high school mathematics (see Table 15.1). As a re-

TABLE 15.1   Attitudes Towards Mathematics in High School: Cohorts 1 and 2 (N = 39)

Attitude	Number
Expressed confidence in mathematical ability	21
Had a great mathematics teacher	8
Enjoyed the challenge in mathematics	7
Liked mathematics	5
Used "fun" to describe mathematics	4
Felt nervous about next mathematics Class	5
Thought mathematics was difficult/hard	3
Didn't like mathematics because of the teacher	2

sult of grounded theory analysis (Glaser, 1998), eight themes arose that described the girls' attitudes towards mathematics. Five of these themes show a common like and enjoyment of mathematics. The three final themes did not display the same positive attitude toward mathematics as the first three. It seems reasonable that girls who did not enjoy mathematics or believed it was difficult would possibly be nervous about their next mathematics course.

Using data obtained from school records, it was reported that among high school girls in the longitudinal study, 119 or 84% were on in a position to be prepared to take calculus by their final year of high school. The 20 girls who took Algebra 1 in the seventh grade in 1999–2000 were all on track to take calculus by their junior year in high school.

In addition to the previous 39 phone interviews, another round of individual interviews was held in 2004 with 34 high school students, and these data formed the basis for two research papers that were presented at a conference in Oxford, UK in 2005 and later as book chapters (Howe, Berenson, & Vouk, 2007; Berenson, Williams, Michael & Vouk, 2007). During these interviews, data were collected concerning what these high school girls valued in terms of career attributes. An overwhelming theme that emerged from the high school girls' conversations was that time was a very important factor in making career decisions. As part of the millennium generation, these girls valued quality of life issues as early as high school:

> I guess I would look for a job that didn't involve that much work. Obviously every job involves work but not a job that required me to work six or seven days a week. I really think that even if you love your job, it shouldn't be more than 8 or 9 hours a day of work and you should have weekends and free time and you know a life separate from work. So something that can accommodate that and a good salary.

Many of these high-achieving young women also valued flexibility of working hours and being one's own boss.

In 2005, we interviewed by telephone 86 of the girls involved in the longitudinal study, aged 16–20, who were juniors and seniors in high school, or young women just beginning their college career or in their second year of college. Two papers reported at the International Group for the Psychology of Mathematics Education in 2006 indicated the mathematics pathways that these high-achieving girls took in high school and the relationship between their mathematics course selections and their career choices (Berenson, Michael, & Vouk, 2006). Approximately half of these 86 girls were planning to or had committed to major in science, technology, engineering, or mathematics (STEM) and were thinking of pursuing STEM careers. Sixty percent had taken Calculus 1 or beyond, and only four of the young women were undecided as to their career choices (Berenson, Michael, & Vouk, 2006). We recognize that through high school and the early years of college these career choices are subject to change.

Collapsing career categories, into STEM careers and non-STEM careers, we conducted a chi square test $[\chi^2_{.05 \ .1} = 3.897]$ to compare frequencies of mathematics courses taken and young women choosing STEM or non-STEM careers. We compared those girls taking Algebra II through calculus to those taking advanced calculus courses in high school. Those girls taking advanced calculus courses are more likely to choose STEM careers rather than non-STEM careers. Girls taking lower level mathematics (Algebra II through calculus) are equally likely to choose STEM as non-STEM careers (Table 15.2). Interestingly, one of the 86 interviewed intended to major in mathematics and one in teaching mathematics (Berenson, Michael, & Vouk, 2006).

When asked to list strengths they would bring to the workplace, they described themselves as hardworking, determined, organized leaders who were good with people. Disappointingly, only one out of 86 girls described herself as "smart" and considered this an attribute she would bring to the workplace.

Along with career choice driving mathematics course selection, the decision to continue taking mathematics beyond the Algebra 2 level might be determined by previous mathematics success. A quantitative study conducted by Wilson, Mojica, Slaten, and Berenson (2006) found that decisions

TABLE 15.2 Mathematics Courses Taken in High School and Choice of Careers

Mathematics taken	STEM careers	Other
Algebra 2—Calculus 1	23	29
Advanced Placement Calculus, and beyond	20	10

among girls in this longitudinal study to take mathematics beyond Algebra 2 correlated with measurements of mathematics success that included standardized tests, course selection, and course grades. This quantitative study used high school transcript data, Preliminary Scholastic Aptitude Test (PSAT), and Scholastic Aptitude Test (SAT) scores collected throughout the study on each of the girls. The transcript data included courses taken, year courses were taken, grade received in the courses, as well as End-of-Course test grades. The PSAT is a practice test for the SAT commonly taken by students in their third (junior) year of high school; the SAT is one of the tests used for college admission.

After eight years, the longitudinal study revealed that high-achieving girls were confident, hard working, and assertive, and they had strong parental support for their academic and career choices. Most set high expectations of themselves academically. Half of these girls enjoyed mathematics from middle grades through high school, indicating that they enjoyed being challenged. Approximately half of the sample of 86 girls were planning to prepare for STEM careers, with a majority choosing the medical field. These high-achieving girls were not interested in engineering (with the exception of bio-medical engineering), physics, or computing careers. They also showed little interest in pursuing mathematics degrees and careers. These findings support those of the U.S. Department of Education (2000) in which 50% of women in their final (senior) year of high school stated that they would not enroll in any more mathematics courses. The girls in this study from cohorts 1–4 are now in college, have graduated from college, or are attending graduate school.

TRIANGULATION ACROSS TIME—RESEARCH METHODS

This study focused on nine girls with whom there were multiple interviews throughout the history of the Girls on Track program. The first interviews were conducted during the camp when the participants were in middle school. The second round of interviews occurred in 2004 when the girls were in high school, and a third set of interviews was held in late 2006 and early 2007 when the girls from the first two cohorts were in college.

As described previously, based on previous data conducted in this longitudinal study, these young women were found to be high achievers, many taking advanced mathematics courses from the middle grades into high school. All interview participants who fitted these criteria were contacted by e-mail and/or by phone. A total of 19 interviews over a 10-year span with nine young women were analyzed. The young women had been enrolled in private or public high schools and universities. Using pseudonyms for the students, their career intents and at what stage the interviews took place are summarized in Table 15.3.

All interviews conducted were semi-structured and conducted in person. During the semi-structured interviews, a series of structured questions was asked to each interviewee, with the researchers also using open-ended questions to probe more deeply (Gall, Gall, & Borg, 2003). Since the interviews were conducted over the span of the 10 years, different interview protocols were used to interview participants at each educational stage: one protocol each for middle school, high school girls, and college women. Despite using different protocols, all of them explored participants' experiences, interests, and activities. All interviews were audio-taped, and field notes were taken by interviewers during the interviews. Field notes taken by the researcher included interview content and interesting points that arose and needed to be followed up on later in the interview. The interviews lasted from 30 to 90 minutes.

The interviews were transcribed verbatim and analyzed using NVivo qualitative software (Weitzman, 2003). Much of the initial analysis of the

TABLE 15.3 Participants, Education Level, Educational Institution, and Career Intentions

Pseudonym	Stage of Interviews	Major/Career Interest	Middle/High School, University Types
Ray	High School College	Business/Italian Psychology, Medicine, Law, or Teaching	Public High School Public University
Meg	Middle School College	Investment banking	Public Middle School Private University
Katya	High School College	International Business	Private High School Public University
Zena	Middle School High School College	Biology/ Pre-Medicine	Public Middle School Public High School Private University
Wendy	High School College	Business	Public High School Public University
Kay	Middle School College	Accounting	Public Middle School Private University
Yasmin	Middle School College	Anthropology/Pre-Medicine	Public Middle School Public University
Ginger	High School Freshman and Senior Years	Television or Foreign Affairs	Public High School
Amber	Middle School High School	Science industry or Patent Law	Public High School Public University

TABLE 15.4   Codes for Describing Girls' Beliefs and Perceptions of Mathematics

Codes	Explanation
Thinking, problem solving, like to be challenged	Girls like to be challenged, perceive that mathematics increases their thinking skills, also some talked about the problem solving skills they developed in higher mathematics courses
Educational tool	Advance study in other areas such as chemistry or physics, business degrees, as well as in future mathematics classes.
Relation to STEM career pursuit	Mathematics will help them in their pursuit of STEM-related careers.
Confidence in abilities, enjoyment, and satisfaction	The girls are good at mathematics, and therefore, it helps them build confidence in their academic abilities. They also express that they have natural ability and genuinely enjoy mathematics.
Connections to realistic contexts	The girls talk about liking classes that are connected to other areas. Prefer applied mathematics classes that are relevant for their majors and career interests.

interview transcripts took place in regular team meetings over the course of four months. The researchers discussed codes and definitions of these codes. They also coded several interviews together. For the first and second level coding processes, serial tagging, analyzing each transcript one at a time, and parallel tagging, reading and comparing each response to the same question, were employed (Baptiste, 2001). Initially, the data were labeled, tagged, and coded based on emerging themes in the data (Baptiste, 2001). The research team used 21 interviews from 10 girls conducted over a span of several years to create a series of codes about the girls' beliefs and perceptions of mathematics (see Table 15.4).

To contribute to the validity, reliability, and veracity of the study, strategies of verification included using incremental evidence (Morse, Swanson, & Kuzel, 2001), triangulation between researchers (Merriam, 2002), and adequately engaging in data collection so that the data become saturated (Merriam, 2002). In order to address issues relating to reliability, multiple coders were used, coding was checked and refined at both the first and second levels of analysis, and inter-rater reliability was established (Morse et al., 2001).

DISCUSSION OF FINDINGS

Throughout this longitudinal study, these high achieving young women continued to be successful in advanced mathematics courses. Beginning in middle school, they were successful in Algebra 1 in seventh or eighth grades, moving through the advanced mathematics sequence to take calculus in high school. Some of the young women in this study reported that they tested out of many of their college mathematics courses. College or university students in the United States take placement tests for mathematics. These tests help to place the students in appropriate mathematics courses so they are not repeating material. Another way in which students can earn college credit is to take Advanced Placement courses in high school and pass exams with high enough scores (universities and colleges determine what is an high enough score) for achieving college credit for such courses.

Most described enjoying their challenging mathematics courses, and while none of the young women was majoring in mathematics, they spoke positively about their mathematics experiences. The young women's enjoyment of mathematics complements Brush's (1980) findings. Brush found that students' reasons for liking mathematics were related to being able to understand the material and finding mathematics fun, challenging, and interesting.

Looking for links between their enjoyment and success in mathematics, we asked for their perceptions of the connections between mathematics and their chosen careers. The following themes emerged from the analysis of these data (discussed in Table 15.4):

- Mathematics as a tool for thinking and problem solving
- Mathematics as an educational tool
- Mathematics as a tool for supporting their STEM career pursuits
- Mathematics as a tool for building confidence and enjoyment in the subject
- Mathematics as a tool for connecting to realistic contexts

These themes are not discrete. A response to the interview questions could have fallen into one or more themes.

Mathematics as a Tool for Thinking and Problem Solving

Some of the young women expressed the notion that their study of mathematics had enhanced their ability to think, analyze, solve problems, and think abstractly about ideas. The young women's notions on how mathematics helps them in this regard were more pronounced at the college level, after they had taken several mathematics courses throughout their educational experiences. However, there were two young women who said that they believed mathematics was a tool while in middle school.

Meg said that she enjoyed mathematics and felt that mathematics was about "solving patterns." Another participant, Yasmin, thought positively about mathematics because there were multiple approaches for solving problems. If she did a problem two different ways and found two different answers, she would try it a third time to find out what she did wrong and go back and fix the problem.

During the second set of high school interviews, there was only one of the nine girls who said that mathematics could be a tool for thinking and problem solving. As a high school senior taking calculus, Ginger said that she liked the challenge and felt that it pushed her to be her best academically. The calculus class was valuable to her because of "the knowledge that [she] take[s]" from it, and how she was going to use that knowledge in the future.

The college interviews provided a more in-depth perspective on how the young women in the study viewed mathematics as a tool for problem solving and thinking. They discussed this view from a variety of perspectives, including how they felt it would relate to their careers and how they felt about their skills with relation to their course work. For example, Meg's description stemmed from how she saw mathematics relating to her chosen career (investment banking):

> I think a lot of it is just being able to problem solve correctly. How to read a problem and think through the kind of steps that you would need to take and solve. So, I think a lot of the math classes we're taking helped to develop my analytical and quantitative and critical thinking skills.

Meg was referring to mathematics classes that she had taken in high school and college. These classes included Multivariable Calculus and Statistics. Although these courses were not specifically related to her major, Economics, she found that they helped develop the thinking skills she needed for her major.

Another young woman, Kay, did not have to take any mathematics courses in college. She stated that she took mathematics-based courses in accounting and finance. She felt that these mathematics-based courses and her high school mathematics courses would help her prepare for problem solving within the technology of her field because:

> Um, so when it's with accounting it's all numbers-based and being very organized and that kind of thing and the [software] programs we used were all just a way to draw information together.

She will still need to know the mathematics and the meaning behind all the numbers to make sense of the software program's output.

To these young women, their study of advanced mathematics in high school and college provided them with intellectual tools to be used in their chosen careers. They recognized that their success in the communities in which they had chosen to practice required high levels of problem-solving and critical thinking skills. They readily acknowledged that their study of advanced mathematics in high school and college had contributed to their habits of mind in relation to thinking.

Mathematics as an Educational Tool

At the middle school level, the young women discussed how the mathematics they were learning at the time would help them with their mathematics later. They discussed strategies for learning mathematics. Yasmin talked about working on problems that she had solved incorrectly and trying to solve them using different techniques. This is similar to the example given above in "Mathematics as a tool for thinking and problem solving" in that Yasmin was trying to find different methods for solving problems she had trouble with in the first place. So while she did not directly speak to mathematics as helping her develop her problem-solving skills, she used mathematics to improve her resources and techniques for solving problems. Yasmin, however, believed that the technique of reworking problems would help her understand the mathematics better, and therefore receive better grades in class. Another young woman, Amber, talked about the after-school help she sought from her teacher. The after-school practice would help her understand the mathematics better than just listening during class time.

The interview of the young women at the high school level provided glimpses of how the girls viewed mathematics as an educational tool. Ginger (high school freshman) believed that the honors mathematics track (meaning that she would take Calculus in her senior year) was appropriate for her. She stated that it was "always going to help me out a lot whether I decided to do math related [careers] or not, it's going to help me. So I definitely want to stay above in math." Later, when Ginger was a senior, she said that the reason for taking advanced mathematics classes was because she was interested in mathematics, and felt that she excelled in that area, but also, that she would not need to take mathematics in college and she could study other areas in which she had an interest.

A third young women, Katya, felt that Algebra I and II were more basic courses and helped her understand mathematics classes she took later as well as the Physics class. However, she expressed the belief that calculus would help her in the future with college science and mathematics courses. Even at the high school level, the students were thinking about mathematics in terms of aiding their understanding in more advanced mathematics courses and science courses. Most of these young women will test out of their mathematics requirements at the college level.

At the college level, these young women gave additional meaning to the perception of mathematics as a tool, in this case, for advanced study in the physical sciences and in pursuing business degrees. The thoughts expressed here showed an immediate use for mathematical tools in contributing to their continued academic success in college and graduate school. Ray, a business and Italian major, stated that her statistics and mathematics courses helped her in preparing to further her education:

> Well, I realized that a lot of business has to do with financing and stuff like that. So, a lot of statistics and math goes into that, and accounting, if I want to do anything like that. And later on, if I want to pursue getting an MBA, a lot of math is tied into that.

Ray was not the only young women to comment that her high school mathematics courses helped her prepare for mathematics-based courses related to her major or career interest. Katya, also a business major, stated:

> I definitely feel that taking the math that I have, I feel comfortable in many of the, many of the required math courses that are forthcoming, but, I don't necessarily see a heavy involvement in math in what I plan on doing. But, I definitely realize the benefits of having those math courses, because from now on if there is a math requirement or math-related requirement, like Statistics, I feel more comfortable going into it and don't feel as scared as I feel some of my peers are.

Similar to both Ray and Katya, Wendy felt that her business major was supplemented by the mathematics courses she had taken in high school. She used her mathematical knowledge and skills in non mathematics-based classes as well as those that were mathematics-based.

> Well especially with business, it [mathematics] really has applied to all of my Econ [Economics] and accounting classes. I remember having to go back to, like, the basic Algebra and like, 'oh, what rules do I have to do,' because it's always, like, solving for variables, like, 'how much gas is this company using,' so definitely in my business related courses, I've used a lot of math.

Some of these young women were required to take courses related to research or quantitative studies. Instead of choosing mathematics courses to fill these requirements, the young women chose courses that would enhance their majors. Statistics courses were common choices. Zena (biology major) stated that she was going to choose a second statistics class to fill this requirement instead of a computer science or mathematics course. The statistics course she chose was specifically designed for a biology major. She felt this class would be the most relevant to her goals.

The young women acknowledge the hierarchical nature of mathematical understanding and the applicability of these ideas to other difficult disciplines such as physics and chemistry. This was true for those high-achieving young women who had not yet chosen a specific career as well.

Mathematics as a Tool for STEM Careers

Six of the young women spoke of the usefulness of mathematics to their chosen careers, and these were women in their final two years of college. When the young women were younger (in middle or high school), they did not have a clear image of how mathematics or science would be helpful in their careers. This may be a result of the fact that they were still exploring their career options and were not sure what their goals were. As the girls entered college those career goals became more focused as they selected majors. However, there was still a great deal of decision making concerning careers. Yasmin (pre-medicine major), for example, felt that "math is an integral part of medicine and why you get a certain dose that you get or why, I don't know, why medicines are the way they are." However, she did not know how to relate specific higher-level classes such as Differential Equations to medicine. One experience that did help her make connections between calculus and medicine was her specialized calculus class for biology majors. She said:

> ...we learned all of the concepts, all the problems and the examples we used were like...we're trying to calculate how often a synapse fires or calculating the optimal existence point of a species and stuff like that. So, it was—I thought a lot more useful, a lot more applicable just because I could see examples in how it would be, I mean, of any worth. It wasn't a cone draining sand or two trains start at the same point and go in opposite directions and gets wherever first.

Wendy, a business major, felt that mathematics was an integral part of her macroeconomic and other economic courses. This feeling of using a specialized type of mathematics in business courses was echoed by other women pursuing business, finance, or accounting degrees. One young woman, Kay, was hesitant to connect her accounting and finance courses to mathematics, saying that they were "math-based, but those are business kind of courses." She went on to express that business and accounting required a lot of basic mathematics and being able to work with and understand numbers, but she did not feel that her higher-level mathematics courses (calculus) related to being an accountant. These types of statements raise the question: How do these young women define mathematics?

For the young women in this study, there was only a tentative notion that the mathematical ideas they had learned would be directly useful in their chosen careers. It was unusual for the young women in this sample to experience mathematical courses at the college level that were tailored for certain careers, such as the case of Yasmin. This is surprising since so many of these young women attended large universities where the number of students committed to a particular major could support specialized calculus courses for different areas of study. For example, in universities attended by these young women there are calculus courses for business and social science majors and calculus for life and management sciences. The course descriptions for these courses are different than descriptions for courses listed as Calculus I. Since we do not know exactly what type of mathematics courses these young women completed, it is possible that they did not take advantage of such opportunities in their universities.

Mathematics as a Tool to Build Confidence and Satisfaction

In sixteen of the interviews, the young women expressed beliefs related to the notion that mathematics is a tool to build one's confidence and provide satisfaction. The young women expressed the notions of being confident in their abilities from middle school through college. These young women were high achievers in their high school mathematics courses; this was demonstrated in their perceived confidence in feeling more prepared to take mathematics in courses over their peers. However, excelling at mathematics not only built confidence in mathematics but in other subjects as well. Their confidence in mathematics did not seem to waiver with their experiences with different teachers. However, this confidence and high achievement in mathematics did not seem to encourage the young women to pursue mathematics as a college major.

Three of the young women discussed mathematics as a tool for building confidence in middle school. However, for these young women, that confidence was associated with a strong notion that the girls had "worked hard" for their grades, and that they were proud of their abilities. Kay said that mathematics had been "pretty fun and not too hard" when asked about her experiences in middle school. During middle school, one girl, Amber, took a mathematics exam to discover her abilities and knowledge, and the results of the test showed that she had the same knowledge base as an third year mathematics student in high school. As a result of this test, Amber knew that she was good at mathematics. These two young women, along with Meg and Yasmin, stated that they knew they could earn A's and B's in mathematics classes and would be disappointed with anything else. The girls also attributed earning lower grades to their own lack of knowledge or understanding of the material, an undedicated work ethnic, and being distracted during class. They did not place responsibility of their earning lower grades on the teacher.

At the high school level, two young women, Ray and Ginger, described their mathematics ability as being a confidence builder and adding to their academic satisfaction. Because they felt that they had natural ability for mathematics, the young women enjoyed their mathematics classes.

By the time the young women entered college, most of them felt that they were "good at math," and their transcripts showed that they had been able to excel at mathematics during high school. Several of the girls took two semesters of Advanced Placement Calculus, along with Advanced Placement Statistics; one took a Calculus III, a third semester of calculus offered through one of the local universities, and one completed Differential Equations at the college level when enrolled in high school.

Some of the young women, Ray, Meg, Wendy, and Kay, discussed the level of their mathematics-based finance and business classes as using basic mathematics and that the mathematical concepts used were not challenging. The few girls who talked about their statistics courses during college had similar feelings. For example, when discussing her college statistics course, Wendy described her statistics class as being, "the T-test and the Z-test and it wasn't too difficult, it was pretty simple."

Other young women spent more time describing why mathematics made them feel confident. Katya's final response during her college interview revealed that she benefited from all of her advanced mathematics courses:

> But, I definitely, like, despite my not choosing a career that was intense in math and science, I definitely feel that going into several math classes in high school, several upper level classes in high school, has just helped me be confident, not just in, like, later math classes, like math classes that I have to take now, but also just in my coursework in general, because it's, it's sort of, it's almost that quantifiable competency that I feel that I'm capable of that sort of helps me say, "okay, I did this so obviously I'm capable of other things, too." ... I definitely realize the benefits of having those math courses, because from now on if there is a math requirement or math-related requirement, like Statistics, I feel more comfortable going into it and don't feel as scared as I feel some of my peers are.

Meg described how she felt her personal attributes influenced her course choice with regard to mathematics.

A lot of it is influenced by what I like to do. I mean, I like classes that are very analytical in number—in nature, very quantitative in nature. And by that I mean I like classes where there's a lot of like calculator number crunching and that kind of thing.... I like working through problem sets rather than— and I like working through problem sets and taking exams rather than reading books and writing papers.

It is evident that for these high-achieving young women there was a sense of satisfaction that came with their success in advanced mathematics courses. For some, that satisfaction was translated into positive feelings towards mathematics. Most of the young women in this study were interested in careers that help or work with people and "make a difference" (Etzkowitz, Kemelgor, & Uzzi, 2000).

Mathematics as a Tool for Connecting to Realistic Contexts

During the middle school years, the young women were specific in the things that they liked and disliked about mathematics and science. They gave explicit examples of the types of classroom activities they enjoyed and how those activities fitted into a broader scheme of the world. The young women expressed an interest in knowing how mathematics was applicable to other areas, such as medicine, chemistry, or operations research, as well as making connections to their own lives.

For some, the Girls on Track summer camp was the first time that they had experienced real-world applications of mathematics, including the community problems discussed earlier. This may have opened their eyes to the possible uses of mathematics that had not previously crossed their minds. Amber, for example, said:

I like how they [camp counselors] take math into a different setting and not take it from sitting in a classroom and not being able to do anything and taking sports algebra and having to figure out things with math and have fun at the same time.... It is lots of fun to put math into a different setting.

In ten of the nineteen interviews, views were expressed about this same enjoyment, of seeing mathematics in a different setting than the classroom. Yasmin echoed Amber's sentiments and went further to say that it was helpful to understand the different ways mathematics could be used in the real world. During middle school, the girls wanted to know more than just the mathematical concepts but also how those concepts were applied in a variety of situations.

Two girls in the high school interviews specifically talked about mathematics and their enjoyment of learning mathematics, especially when it was connected to applications. Ginger expresses a desire to use the knowledge she had learned. Being able to use the knowledge was what she saw as valuable. Ray said that she found "hands-on experiments" made classes interesting, and she enjoyed it when the mathematical concepts were related back to everyday uses.

The college interviews revealed the most about how the young women felt mathematics should connect to real life and not simply be learned in isolation. The college interviews showed that the young women chose to take mathematics-based courses related to their fields of interest, or they preferred to take statistics courses over all other types of classes. They seemed to feel that these courses would meet their requirements for graduation and help them in the pursuit of their chosen careers. However, this desire to relate their course work with their interests was intertwined with other codes such as *mathematics as an educational tool.* For example, a college student may have been required to take calculus, but then stated how refreshing it was that the course was related to biology (in the case of a calculus course for biology majors). This would mean that they viewed the calculus class as an educational tool since it was required, and that they liked the application aspect of it being related to biology, and possibly their future careers.

WHAT WE HAVE LEARNED ABOUT THESE HIGH-ACHIEVING YOUNG WOMEN

These young women saw the role of mathematics as instrumental to their participation in their chosen communities of practice (Wenger, 1998). Few elected to join mathematical communities, relying upon the requirements of their respective communities to prescribe their mathematical studies in college. Mathematical knowledge was viewed as a tool for success in other courses such as physics, chemistry, and finance. The young women in college acknowledged the value of mathematics in developing habits of mind that improved their problem-solving and reasoning abilities. This belief that they studied mathematics to improve their knowledge aligns with Schoenfeld's (1989) findings, that college students study mathematics for intrinsic, rather than extrinsic, reasons. Some acknowledged that the study of advanced mathematics in high school improved their self-confidence through their successes in difficult courses. Overall, these young women who enjoyed and were successful in mathematics viewed mathematics as a tool for success in their chosen career communities.

As we examined the data, we begin to understand the value that these high-achieving young women placed on their mathematical experiences; the overriding theme was that mathematics was viewed as a tool. However, the tool had different uses depending on the perceived needs of the individual. Of the 19 interviews analyzed, mathematics as an educational tool was discussed in 14 of them. Six of the young women interviewed described this use of mathematics in the two or three interviews held with them at the different educational levels (middle school, high school, and college). However, the theme that mathematics was a tool for the pursuit of a STEM career only occurred in the interviews when they were in college. This may be because the college-age young women in this study were starting to think seriously about their possible careers, narrowing choices, and planning the courses they would need to complete at the college level.

RELATING OUR STUDY TO THE PRINCIPLES AND STANDARDS OF SCHOOL MATHEMATICS

The young women in this study used their mathematical knowledge and decisions concerning high school course taking to "get ahead" in college. They completed basic mathematics requirements prior to college enrollment and were able to choose mathematics courses or mathematics-based courses in college that were related to their areas of interest. These young women would have started school in the mid 1990s, meaning that they would have received much of their mathematics education under the influence of the National Council for Mathematics Teachers' (NCTM) Principles and Standards for Mathematics (NCTM, 2000). In North Carolina, the standard course of study for mathematics is based on the "philosophy of teaching and learning mathematics that is consistent with the current research, exemplary practices, and national standards" (Department of Public Instruction, 1999, p. 1). The Principles and Standards emphasize understanding of mathematics as well as skill attainment across all grade levels and all areas (Weiss, 2001). In addition, the Principles and Standards were written as guidelines to revise the curriculum and teaching practices in the 20th century, based on increased (research-based) knowledge on how students learn mathematics.

There are six principles NCTM (2000) describes on which teachers should base classroom decisions: Equity, Curriculum, Teaching, Learning, Assessment, and Technology. We feel that two of these principles are particularly relevant to this study, the equity principle and the learning principle. In addition, there are also 10 standards outlined by NCTM: Number and Operation, Algebra, Geometry, Measurement, Data Analysis & Probability, Problem Solving, Reasoning and Proof, Communication, Connections, and

Representation. Of these ten, we would like to look more closely at two of them as well: Problem Solving and Connections. We would like to discuss these four areas from the perspective of these young women and perceptions of mathematics.

The learning principle states that "students must learn mathematics with understanding, actively building new knowledge from experience and prior knowledge" (NCTM, 2000, p. 20). The young women stated that learning mathematics and understanding the mathematical concepts were important to them. They also said that they were able to use their mathematical knowledge in a variety of situations, and this had enabled them to build confidence in their abilities, find satisfaction in learning mathematics, and enjoy the learning of mathematics.

The equity principle states that "excellence in mathematics education requires equity—high expectations and strong support for all students" (NCTM, 2000, p. 12). The young women in this study expressed enjoyment in the opportunity to study mathematics at high school. They often exceeded the high school mathematics requirements and ventured into the world of college mathematics as high school students. Most attended magnet schools, schools whose curriculum is enhanced through innovative approaches or themes. Within these schools, academic study was highly valued. They young women then proceeded to large, research-intensive universities. They had participated in mathematics- and science-related extracurricular activities, such as the Girls on Track camp or Governor's School, a state-wide leadership summer program for high school juniors. The availability of courses and support from the school, as was received by these young women, was likely to have contributed to them feeling they could choose the mathematics courses appropriate for them at the college level. They were not daunted by the prospects of taking college mathematics.

The problem solving standard has several aspects, including to "build new mathematical knowledge through problem solving" (NCTM, 2000, p. 52). Our young women were able to use their mathematical knowledge and skills as a tool for thinking and problem solving. Several of the young women discussed the idea that mathematics had made them not only better quantitative thinkers, but better thinkers in general. The young women were also able to use their understanding of mathematics to build new knowledge in related areas—for example, business or physics. They were extending their problem-solving abilities to other areas.

The connections standard states that "instructional programs from prekindergarten through grade 12 should enable all students to recognize and use connections among mathematical ideas, . . . [and] recognize and apply mathematics in context outside of mathematics" (NCTM, 2000, p. 64). Our young women demonstrated that they were able to value the connections mathematics makes to life, their career choices, and to other subject areas.

They wanted to make those connections and valued teachers and courses that enabled them to do so. They also took courses designed to make connections between mathematics and other subjects, such as the calculus class for biology majors.

One intention of the Standards was to have students leave the study of mathematics liking and appreciating their experiences. These young women saw the value that mathematics added to their scholarship and education. The Standards' expectations for students were for them to be able to use their mathematical knowledge, skills, and understanding for decision making and to effectively participate in society (Weiss, 2001). The young women in this study have successfully demonstrated that they used their mathematical knowledge to aid in their course decisions at the college level and in their ability to successfully navigate those courses.

ACKNOWLEDGEMENTS

Funded, in part, by grants from the National Science Foundation [#09813902, #0204222, and #0624584]. The views expressed here are not necessarily those of the National Science Foundation.

REFERENCES

Baptiste, I. (2001, September). Qualitative Data Analysis: Common Phases, Strategic Differences [42 paragraphs]. *Forum Qualitative Sozialforschung / Forum: Qualitative Social Research* [On-line Journal], *2*(3). Retrieved November 16, 2007 from http://www.qualitative-research.net/fqs-texte/3-01/3-01baptiste-e.htm

Berenson, S., Michael, J., & Vouk, M. (2006, July). *The relationship between high school mathematics and career choices among high achieving young women.* In J. Novotná, H. Moraová, M. Krátká, & N. Stehliková (Eds.), *Mathematics at the centre. Proceedings of the 30th Conference of the International Group for the Psychology of Mathematics Education* (pp. 1–220). Prague, Czech Republic: Charles University.

Berenson, S., Vouk, M., Michael, J., Greenspon, P., & Person, A. (2004, April). *Women and information technology: A seven year longitudinal study of young women from middle grades into college.* Paper presented at the Annual Meeting of AERA. San Diego, CA.

Berenson, S., Williams, L., Michael, J., & Vouk, M. (2007). Examining time as a factor in young women's information technology career decisions. In C. Burger, E. Creamer, & P. Meszaros (Eds.), *Reconfiguring the firewall: Recruiting women to information technology across cultures and continents* (pp. 65–76). Wellesley, MA: AK Peters, Ltd.

Brush, L. (1980). *Encouraging girls in mathematics.* Cambridge, MA: Abt Books.

Burton, N. (1995). Trends in mathematics achievement for young men and women. In I. M. Carl (Ed.), *Seventy-five years of progress: Prospects for school mathematics* (pp. 115–127). Reston, Va.: National Council of Teachers of Mathematics.

Department of Public Instruction (1999). *North Carolina standard course of study: Mathematics.* Retrieved June 17, 2009 from http://www.ncpublicschools.org/curriculum/mathematics/

Dulaney, C. (1996). Should students take Algebra I in middle school? *Evaluation and Research Report No. 96E.08.* Retrieved June 17, 2009 from http://www.wcpss.net/evaluation research/reports/earlier_years/9608students_algebra_midsc.pdf

Etzkowitz, H., Kemelgor, C., & Uzzi, B. (2000). *Athena unbound: The advancement of women in science and technology.* Cambridge, MA: Cambridge University Press.

Gall, M. D., Gall, J. P., & Borg, W. R. (2003). *Educational research: An introduction* (7th ed.). Boston: Allyn and Bacon.

Glaser, B. (1998) *Doing grounded theory: Issues and discussions.* Mill Valley, CA: Sociology Press.

Howe A., Berenson, S, & Vouk, M. (2007). Changing the high school culture to promote interest in IT careers among high achieving girls. In C. Burger, E. Creamer, & P. Meszaros (Eds.), *Reconfiguring the firewall: Recruiting women to information technology across cultures and continents* (pp. 51–64). Wellesley, MA: AK Peters, Ltd.

Howe, A. C. & Berenson, S. B (2001). Characteristics of high achieving middle school girls towards mathematics. Unpublished manuscript. Raleigh, NC: Center for Research in Mathematics and Science Education.

Merriam, S. B. (2002). Assessing and evaluating qualitative research. In S. B. Merriam (Ed.), *Qualitative research in practice: Examples for discussion and analysis* (pp. 18–33). San Francisco: Jossey-Bass.

Morse, J. M., Swanson, J. M., & Kuzel., A. J. (Eds.). (2001). *The nature of qualitative evidence.* Thousand Oaks, CA: Sage.

National Council of Teachers of Mathematics. (2000) *Principles and standards for school mathematics.* Reston, VA: NCTM.

Schoenfeld, A., (1989). Explorations of students' mathematical beliefs and behavior. *Journal for Research in Mathematics Education, 20*(4), 338–355.

U.S. Department of Education (2000). *Trends in educational equity for girls and women* (NCES No 2000-030). Washington DC: Office of Educational Research and Improvement. Retrieved May 3, 2009 from http://nces.ed.gov/pubs2000/2000030.pdf

Weiss, I. (2001). *Investigating the influence of standards: A framework for research in mathematics, science, and technology education.* Washington, DC: National Research Council.

Weitzman, E. A. (2003). Software and qualitative research. In N. K. Denzin & Y. S. Lincoln (Eds.), *Collecting and interpreting qualitative materials* (2nd ed., pp. 310–339). Thousand Oaks, CA: Sage.

Wenger, E. (1998). *Communities of practice: Learning, meaning, and identity.* New York, NY: Cambridge University Press.

Wilson, P.H., Mojica, G.M, Slaten, K.M., & Berenson, S. (2006, July). High school course pathways of high achieving girls. In J. Novotná, H. Moraová, M. Krátká,

& N. Stehliková (Eds.), *Mathematics at the centre. Proceedings of the 30th Conference of the International Group for the Psychology of Mathematics Education* (pp. 1–430). Prague, Czech Republic: Charles University.

SECTION IV

TERTIARY STUDENTS

CHAPTER 16

THE INFLUENCE OF HIGH SCHOOL AND UNIVERSITY EXPERIENCES ON WOMEN'S PURSUIT OF UNDERGRADUATE MATHEMATICS DEGREES IN CANADA

Jennifer Hall
University of Ottawa, Canada

INTRODUCTION

This chapter discusses a study conducted in Ontario, Canada, which explored the high school and university mathematics experiences of women currently enrolled in upper years of undergraduate mathematics degree programs. The findings highlight the importance that the women placed on supportive social relationships, their preference for applied mathematics, and how they confront feeling "othered" in a variety of ways. Data collection occurred via individual semi-structured interviews with six women,

International Perspectives on Gender and Mathematics Education, pages 365–390
Copyright © 2010 by Information Age Publishing
365

and these interviews focused on the women's families, peers, personal characteristics, and experiences with the formal education system. I begin by providing statistical data, theoretical perspectives, and related literature on gender issues in mathematics, particularly at the post-secondary level. I next describe the context of this study, from the broader provincial education system to the university at which the study occurred. The remainder of the chapter is allocated to discussions of the methodology, findings, and conclusions of this study.

In Canada, women continue to be underrepresented in mathematics participation at the post-secondary level. Although approximately 60% of undergraduate students are female (Statistics Canada, 2008a), the proportion of women in mathematics is substantially lower. From the 1992–1993 academic year to the 2004–2005 school year, the proportion of women in mathematics and the related fields of computer and information science actually decreased from 30% to 25% in terms of enrollments and from 33% to 27% in terms of graduations (Statistics Canada, 2008b, 2008c), the latter of which is shown in Figure 16.1.

These statistics are particularly striking when compared to other fields that were previously male-dominated, such as medicine, law, and business; these fields now have equitable participation and graduation rates in many countries (Dworkin, 2001; Freeman, 2004). Furthermore, when making comparisons with the neighboring country of the United States, the proportion of women participating in undergraduate mathematics degree programs in Canada seems to be even more of an anomaly: Women received 45% of the bachelor's degrees in mathematics awarded in the

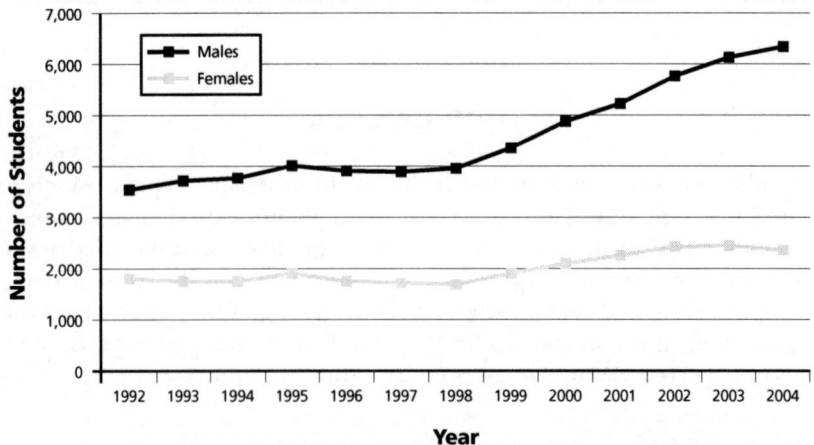

Figure 16.1 Number of bachelor's degrees granted in mathematics, computer and information science, by sex, by year, in Canada.

2005–2006 academic year in the United States (National Center for Education Statistics, 2008).

These statistics raise questions about exactly who the women in undergraduate mathematics degree programs in Canada are and what led to their enrollment and perseverance in this field that is still so male-dominated. This study addresses these questions by exploring the supports and challenges these women faced during their high school and university mathematics experiences.

Situating the Problem

Over time, researchers have taken various theoretical perspectives to explain gender differences in mathematics, such as differences in participation, achievement, and affective variables. Reports on the history of gender issues in mathematics (Hanna, 2003; Leder, 1992) note that most early research attributed the inferior performance and underrepresentation of women in mathematics and science to innate biological factors. However, a seminal piece of research (Fennema & Sherman, 1977) challenged these biological explanations, which shifted future studies to focus on how society and culture influence achievement in and attitudes toward mathematics. Recent research has examined the manner in which mathematics is taught, and explored notions of teaching in gender-connected ways. Although some current researchers still align themselves with biological explanations for gender differences in mathematics and others contend the "gender problem" is solved, I believe there is still much work to be done to ensure equity in this domain, particularly at the post-secondary level.

Gender issues are connected to social and cultural factors, which is a concept embedded in both my research question and the manner in which I conducted my research. I align myself with researchers (e.g., Boaler, 1997; Morrow & Morrow, 1995) who believe that teaching practices need to change in order to ensure more meaningful, successful, and positive mathematics experiences for both females and males. I also share the opinion of feminist researchers (e.g., Becker, 1995; Belenky, Clinchy, Goldberger, & Tarule, 1986; Gilligan, 1982; Whitten & Burciaga, 2001) who believe that instructional practices should include a greater focus on conceptual understanding and connecting mathematics to women's lives rather than a focus on the memorization and rule-following behaviors that are too often associated with mathematics teaching and learning. Furthermore, similar to some feminist researchers (e.g., Boaler, 1997; Burton, 1995; Whitten & Burciaga, 2001), I believe that mathematics classrooms that are supportive, inquiry-based learning communities that value a variety of approaches to doing mathematics allow female students' voices to be heard.

In terms of factors that may contribute to a lack of women selecting mathematics for university study, several studies have found that females at the high school level often have negative attitudes toward mathematics (Fennema & Sherman, 1977; Forgasz & Leder, 1996; Lupart, Cannon, & Telfer, 2004; McGraw, Lubienski, & Strutchens, 2006; Weinburgh, 1995) and a lack of confidence in their abilities (Damarin, 1993; Fennema & Sherman, 1977; Richardson & Suinn, 1972). Recent research from the United States has also suggested that females are self-selecting away from mathematics due to an interest in fields that relate to organic materials (Lubinski & Benbow, 2006). Furthermore, societal stereotypes, such as the notion that mathematics is a field that lacks human interaction, may turn female students away (Morgan, Isaac, & Sansone, 2001; Morrow & Morrow, 1995). Conversely, enrollment in mathematics-related fields is linked to having supportive family members and teachers, passion for and skill in the subject area, and specialized educational opportunities (Gill, 2000).

Several factors relate to perseverance in mathematics-related fields of study at the university level. Faculty support and quality teaching are linked to perseverance in these fields (Gavin, 1996; Seymour, 1995; Zeldin & Pajares, 2000), whereas a negative educational experience in these regards is linked to attrition and feelings of dissatisfaction (Herzig, 2004; Sax, 1994; Seymour, 1995). Students may also leave mathematics-related fields due to a loss of interest in the subject area or a feeling of being overwhelmed by the academic demands (Seymour, 1995), whereas other studies link attrition and dissatisfaction to women's feelings of not fitting in (Damarin, 2000; Herzig, 2004; Rodd & Bartholomew, 2006). Alternately, having a large cohort of female peers in the program is a support for perseverance (Frazier-Kouassi et al., 1992; Jackson, 1992; Mastekaasa & Smeby, 2008; Sharpe, 1992). Studies have also noted the positive impact of supportive relationships with family members and peers (Blair, 1991; Gavin, 1996; Hall, 2006; Zeldin & Pajares, 2000). Finally, certain personal characteristics, such as competitive attitudes, determination, and serious academic attitudes, have been linked to women's perseverance in mathematics-related fields (Gavin, 1996; Hall, 2006; Rodd & Bartholomew, 2006; Seymour, 1995; Zhao, Carini, & Kuh, 2005).

Few studies related to women's enrollment and perseverance in mathematics and related fields at the undergraduate level have been conducted in Canada. Of the existing Canadian studies on this particular issue, very few specifically focus solely on mathematics and use qualitative methods to obtain a more detailed understanding of women's experiences. Therefore, my study contributes to the field by providing in-depth portraits of women's experiences in mathematics at the high school and university level and highlighting the voices of women in a male-dominated field of study in Canada.

Context of the Study

Before describing the details of the study, I provide context by offering some general information about education in Canada and specific details about the university at which the study occurred.

In Canada, education falls under the purview of individual provinces and territories. Each province or territory sets its own elementary and secondary school curriculum, although there are many similarities across the country. Generally, students are enrolled in elementary and secondary school from the age of four (Junior Kindergarten) to the age of 18 (Grade 12). In Ontario, the province in which this study occurred, students are required to take mathematics in each year of elementary school (Junior Kindergarten to Grade 8). At the high school level (Grades 9 to 12), students are required to take three mathematics courses.

Ontario has 19 publicly funded universities that offer a variety of degree programs at the undergraduate and graduate level. The study was conducted at a bilingual university in Ontario that offers bachelor's, master's, and doctoral degree programs in English and French, Canada's two official languages. Approximately one third of students at the university are in French-language programs. In 2007, the university had approximately 35,000 students registered, of which approximately 31,000 were undergraduate students.

In the mathematics department, undergraduate courses are offered in both English and French. However, at the graduate degree level, courses are only offered in English since they are offered jointly with an English-language university. Graduate students are permitted to do their research and write their theses in French. The department offers specializations, majors, and minors in mathematics and in statistics. All participants in this study had a specialization or major in mathematics. The required mathematics courses for these programs relate to various applied and theoretical topics, such as calculus, algebra, analysis, logic, set theory, ring theory, and statistics.

METHODOLOGY

This study addressed the question: "In what ways do women who were educated in Canada and who are nearing completion of undergraduate mathematics degrees feel they have been supported and challenged in their high school and university mathematics experiences?" To investigate this question, semi-structured individual interviews were conducted with women in upper years of undergraduate mathematics degree programs. To be eligible to participate, women were required to be in the third or fourth year of their studies and to be educated in Canada for all levels (elementary school, high school, and university). The latter requirement was included to ensure the participants had similar experiences in terms of their formal education.

The participants were recruited from the university's mathematics department through mass emails and classroom presentations. Five women who fulfilled all of the study's requirements responded to the call for participants. A sixth woman, a master's student in mathematics, also volunteered to participate and was included in the study as she fulfilled all other participation requirements besides the year of study requirement and was still immersed in university mathematics culture.

The interviews were conducted between March and May of 2008 on the university campus, and had a mean duration of approximately 80 minutes. Each participant took part in a single-session, semi-structured individual interview that was audio-taped to allow for subsequent transcription. The interview protocol included eight main questions (with related follow-up questions) that focused on four key dimensions that had been shown in previous studies to be linked to women's choice of and perseverance in mathematics as an undergraduate field of study: family (e.g., Blair, 1991; Gavin, 1996; Gill, 2000); peers (e.g., Frazier-Kouassi et al., 1992; Hall, 2006; Jackson, 1992); personal characteristics (e.g., Rodd & Bartholomew, 2006; Seymour, 1995; Zhao et al., 2005); and the formal education system (e.g., Gill, 2000; Herzig, 2004; Sax, 1994). By examining the combination of these four dimensions in my research, I felt I would develop a more complete understanding of the women's lives and mathematics experiences.

Participants

All participants were between the ages of 20 and 22 at the time of the interviews, and five of the six participants were in their third or fourth year of undergraduate Mathematics degree studies. In Table 16.1, the participants' university programs of study and career plans are briefly outlined.

Feminist researchers (DeVault, 1990; Lather, 1988) have found that female participants tend to open up to female researchers easily, and I found that I was indeed able to quickly build rapport with my participants. I also have some other similarities to my participants: I hold an Honors Bachelor of Science degree in Applied Mathematics and Physiology, and I was educated in Ontario for all levels of my education. I am also a young woman very close in age to my participants. As a woman with a similar educational background to my participants, I believe the participants opened up to me more so than they would have with an "outsider," be it a male or non-mathematician.

TABLE 16.1   Description of Participants' University Programs and Career Plans

Name	Year	Program	Language	Career plans
Allison	4th	Mathematics/English	French	High school teaching
Brooke	4th	Mathematics/Biology	English	Intelligence/Forensics
Chantal	3rd	Mathematics	French	Statistics
Dana	3rd	Mathematics/Economics	French	Actuarial science
Elise	2nd year—master's degree	Mathematics (Mathematics undergraduate degree)	French	Statistics
Felicity	4th	Mathematics	French	University teaching/Business

DATA ANALYSIS

Data analysis began with verbatim transcription of the interviews (174 single-spaced pages of data in total). Participants were then asked to perform a member check on their choice of the full transcript or a summary. Member checks were included in this study not only in order to "solicit informants' views of the credibility of the findings and interpretations" (Creswell, 1998, p. 202), but also as a way to increase participants' involvement in the research process and construction of data (Lather, 1988). Transcription began after each individual interview, which meant there was an overlap of time during which some interviews were being transcribed while others were taking place. However, analysis did not begin until all transcripts were completed.

The transcripts were initially analyzed by using color-coded highlighting (corresponding to each dimension of the study) and making margin notes. I first created a "portrait" of each participant across the four dimensions of the study (Family, Peers, Personal Characteristics, and the Formal Education System), using her words as much as possible to provide a sense of her individual voice, and then explored the data for all six participants across the four dimensions. However, for the purpose of this chapter, the data for all participants are provided together for each dimension.

For each dimension, I first found subtopics that were common to all participants and then created tables that summarized the findings for the subtopics within each dimension. For instance, within the family dimension, subtopics included parents' education, parents' careers, and parents'

	Allison	Brooke	Chantal	Dana	Elise	Felicity
High school mathematics experiences						
Transition to university						
...						

Figure 16.2 Example of partial table summarizing trends in sub-topics in the formal education system dimension.

feelings about mathematics. I then created tables that summarized the findings for the subtopics within each dimension and examined the data within each subtopic across the six participants (See Figure 16.2 for an example of a partial table).

Finally, I conducted a further level of analysis as I noticed three broad themes that related to several dimensions. These themes, which will be explored later in the chapter, related to the importance of supportive relationships, the participants' preference for applied mathematics, and issues surrounding "othering."

Findings

In this section, I first discuss the findings related to each to the four dimensions of the study: family, peers, personal characteristics, and the formal education system. Then, I describe the findings of three themes that encompass multiple dimensions.

By Dimension

Family

Family members positively impacted the participants' mathematics experiences in many ways. Participants described themselves as having close relationships with their families, and felt they had been well supported by their families in both their mathematics experiences and lives more broadly.

The participants generally felt that their fathers had more positive attitudes and were more competent in mathematics than their mothers. For instance, Dana described her father as someone who is "really good in math" and has "always liked math," whereas she stated that "I don't think it's my mom's favorite thing, for sure. She doesn't mind it, but (pause) it's not something she's passionate about." Several participants fondly recalled working on mathematics homework with their fathers. Felicity described working on mathematics homework with her father and brother as follows:

"It was fun. You can just picture us at the kitchen table, trying to figure something out." All reported that their parents were generally very supportive of their degree choices, valued education, and wanted them to have successful careers. The participants also felt that their parents did not hold gender-stereotypical views of mathematics or any other fields. For instance, with regard to her choice of university field of study, Chantal stated that her parents "never put any restrictions, so it's pretty loose." All reported that their parents expected them to go to university, although no participants reported that their parents explicitly suggested they study mathematics.

Parental education and careers were quite varied and did not seem to play a direct role in the participants' mathematics experiences; in fact, only one parent has a mathematics-related job. Five of the twelve parents finished their formal education at the high school level, and only two parents have education above the bachelor's degree level. Although the parents were generally not highly educated, they did place significant value on education and were very supportive of their children's educational endeavors. Notably, three participants' parents received the majority of their education in another country (Morocco, Lebanon, and Haiti), and are first-generation immigrants to Canada. Perhaps outside circumstances, such as issues surrounding immigration, prohibited some of the parents from obtaining higher education. Possible implications of this finding will be explored in detail later in the chapter.

Siblings also positively impacted the participants' mathematics experiences, particularly when the participant had an older sibling. Two of the four participants with older siblings have older siblings who hold degrees and have careers in mathematics-related fields. Nearly all of the participants mentioned that they worked with their siblings at homework time. This was not only a bonding time, but it also allowed siblings to assist one another with mathematics. For example, Dana stated that "when I go home, I definitely help him [her younger brother] with his homework. Every single time I've gone back home, I'll sit down with him."

Peers

The participants' close friends were generally not interested in mathematics. Most participants found that they did not fit in with the other women in mathematics, whom they viewed as very withdrawn and serious. The participants described the men in mathematics and the women in other subject areas very differently from the women in mathematics.

The participants' close friends at both the high school and university level were very diverse in their academic interests, although most were not interested in mathematics. None of the participants' close friends from high school studied mathematics at university. Generally, the participants felt that their close friends in high school did not impact their feelings towards mathematics, and they bonded on a level beyond academic interests. At university, a few participants worked with their peers from their mathematics classes on assignments, but only Elise and Felicity reported being close friends with a group of students from their mathematics classes. Most of the participants' close friends were from outside their academic majors, and even from outside the university.

Most participants felt they were very different from the women in their mathematics classes. The females in mathematics were depicted as being very serious, shy, not very social, driven, and overly focused on mathematics. For instance, Brooke described the women in mathematics as follows:

> The girls in my math class seem very quiet. . . . I've tried to strike up a conversation with some of them, but they've just – they're all kind of very in on themselves. . . . I don't mind asking them for notes or whatever, but a lot of the girls in my math classes seem to be a little more shy. Especially—I'm taking a second year math class right now and those girls especially seem to be especially in on themselves and just like (pause) it's me against everybody else.

Brooke, Dana, and Felicity all stated that they felt that their female peers had "no life" outside mathematics, whereas these participants explicitly described how they valued socializing and being involved in many activities outside school. It was more important to these participants to be well-rounded than to solely focus on their academics and receive top marks.

Interestingly, participants who spoke about the males in mathematics did not find them to be similar to the females in mathematics whatsoever. For example, Brooke described the males in mathematics as "normal college guys, but a little geekier." She felt that they were friendlier and more helpful than the women in mathematics. Elise reported that she finds the men in mathematics much more relaxed and not as focused on being "perfect" academically as the women in mathematics. Overall, the participants provided a very different perception of men and women in mathematics.

The participants also found great differences between the women in mathematics and the women in other subject areas. For instance, Allison described women who were studying English as doing "more judging [than the women in mathematics]" and she stated, jokingly, "I don't know if it's just in the math gene that takes out the cattiness." Dana found the women in economics to be "a lot more relaxed" than the women in mathematics, although "random (pause) all over" in terms of intelligence, whereas she described the women in mathematics as "smarter than the average person in class." Generally, women in other subject areas were seen as being more outgoing than women in mathematics, although they were sometimes linked to other negative traits not ascribed to the women in mathematics.

Personal Characteristics

Common characteristics that supported the women in their mathematics experiences included being very stubborn and determined. For instance, Brooke believes she is more motivated to do something if someone tells her she cannot do it, as she likes to prove others wrong. All expressed a long-standing love for mathematics, and highlighted their aptitude for the subject area. Allison, Elise, and Felicity attributed part of their success in mathematics to being very well-organized. Finally, several participants recognized that their love of the challenge of mathematics problems was instrumental to their success.

The characteristics that hindered the women in their mathematics experiences tended to be very typical of university students in general. For instance, the participants tended to procrastinate, spend too much time socializing, and not be sufficiently wide awake in morning classes to understand the material. The only mathematics-specific challenge reported was Elise's tendency to focus on small mathematical details rather than the "big picture."

Feelings regarding self-confidence in mathematics were an issue for Chantal and Elise, who doubted their abilities at times. Chantal, however, felt that she was able to step back from the situation that had caused the negative thoughts (e.g., a poor mark on a test) and give herself a fresh start. Elise's issues surrounding a lack of self-confidence appeared more substantial, as she felt anxious, felt she needed to perform at all times, and was afraid to ask questions for fear that she may look "dumb or stupid." She described her feelings as such:

> I want to find the answer on my own and I don't need nobody's help....I'm scared of asking for help because sometimes people look at you and they're like, "What? You don't know that? You should know that!" Or, I'm just afraid of that....So, sometimes I don't say anything when I should say something....I always want to understand everything right away. And when I don't, I start to feel like, "Oh my God! Maybe I shouldn't be in math. Maybe I shouldn't be here. I don't understand this and everybody seems to understand."

Notably, Elise received a final grade of A+ in every mathematics course she took, but she stated that she did not even realize she was such an excellent student until professors started to notice her and make comments.

Formal Education System

The participants reported many similarities in their educational experiences. At the high school level, all but Brooke were involved in special mathematics programs. Several participants reported issues with the transition to university, and four made changes to their university programs of study. All the participants noted gender imbalances in their university mathematics classes, particularly with regard to professors. The participants shared common views of what constituted a "good" or "bad" mathematics teacher or professor.

Five participants participated in some sort of enriched work in mathematics at their public high schools, be it in-class or in an extracurricular activity. Chantal and Elise were enrolled in a special program that involved more enriched and varied coursework and a focus on community service. Allison, Dana, and Felicity participated in national or provincial mathematics contests, and Felicity was enrolled in enriched mathematics classes.

Several participants noted that they had a difficult time with the transition to university, citing such issues as being intimidated by the large class sizes, not adjusting well to early morning classes, and not realizing how much work was required. These participants all were able to adjust to university life by their second year of studies. Most recalled having first-year classes with hundreds of students. Their classes now range in size from fewer than 10 to approximately 50 students. All participants prefer smaller classes as they feel more of a sense of connection with the professor and other students.

Several participants made changes to their university program of study; only Allison and Chantal are still enrolled in the same degree program in which they began (mathematics and English; mathematics, respectively). Elise and Felicity began in a more broad-based science and mathematics program and dropped sciences from their schedule, resulting in strictly a mathematics degree program. Both Brooke and Dana began in other fields. Brooke added mathematics as a second major to her initial biology degree and Dana completely changed programs, from biopharmaceutical studies to a double specialization in mathematics and economics. Notably, two of the three participants who are enrolled in a solely mathematics degree program are returning to university once they complete their current degree to obtain a second undergraduate degree in another field (statistics major with an economics minor—Chantal; psychology major—Elise). Of the undergraduate participants in the study, only Felicity plans to do a master's degree in mathematics. The changes to these women's academic programs have been substantial, and in some cases, have affected their career plans.

Although no participants noted that they had issues with the gender imbalance in mathematics, they all stated that it existed, both in terms of mathematics students and mathematics professors. For example, Allison stated that her mathematics classes have always been male-dominated over the four years of her degree, and she has only had two female mathematics professors. However, she stated that she had not experienced any problems with this situation:

> I've never felt it was an issue. (pause) I've never had any of them [the male students] hit on me or be, you know, extremely superior because they're male or

anything.... They've all been really great and we're just students.... There's never been any, you know, "Coddle them because they're girls" or you know, "Treat them differently because..."

In the interviews, I asked the participants to estimate the proportion of female students in their mathematics classes, and their estimations ranged from 20% to 45%. Elise found the proportion of women decreased from approximately 40% in her undergraduate classes to 20% in her graduate-level classes. A few participants stated that they did not have any female mathematics professors whatsoever during the course of their degrees, and the greatest number of female mathematics professors experienced by a participant was two. Clearly, a gender imbalance in mathematics, both in terms of students and professors, was evident in these women's experiences at this university.

The participants all recalled having excellent mathematics teachers and professors, but none mentioned that these educators acted particularly as mentors. Generally, the participants described good mathematics educators as being very passionate about the subject area, able to communicate the material well, organized, personable, and supportive of students. For example, Dana described her favorite mathematics professor as follows:

> You can tell that he actually enjoys what he's doing, that type of thing. He explains very, very clearly. He never, ever just assumes that you know something.... He'll make jokes once in a while. He'll talk about his daughter, his son. He just makes it more personal, I think.

Conversely, poor mathematics educators were seen as those who made students feel stupid by frequently making such comments as "This is easy," and "You should understand this." Poor educators were also viewed as disorganized and unable to communicate the mathematical material to students. At the university level, some participants mentioned that they would skip classes and avoid enrolling in classes with these poor mathematics professors. Participants' views of a quality mathematics educator involved mostly interpersonal aspects, such as communication skills and friendly personalities.

Across Dimensions

In this section, I describe three cross-dimensional themes that emerged during my data analysis. When analyzing the data across participants for each of the four dimensions (family, peers, personal characteristics, and the formal education system), I noticed these themes, which encompass multiple dimensions of the study.

Supportive Relationships

The participants were shown to value caring, supportive relationships, as was evidenced by their closeness to their families, descriptions of "good" mathematics educators, and separation from "non-caring" characteristics they attributed to the other women in mathematics.

The importance that the participants place on relationships is seen most clearly with respect to the participants' families. The participants all described themselves as very close to their families, who they felt had been very supportive of their educational endeavors. Family support has been shown in several studies (e.g., Blair, 1991; Gavin, 1996; Gill, 2000) to positively affect women's enrollment and perseverance in mathematics-related fields of study.

Outside their families, the participants valued mathematics educators who were supportive of students and who were friendly and personable. Similar findings about the importance that women place on such caring characteristics of mathematics professors have also been found in previous studies (Abri, 2006; Gavin, 1996; Herzig, 2004; Sax, 1994; Seymour, 1995; Zeldin & Pajares, 2000). For instance, Seymour (1995) found that the students in her study liked professors who got to know them "as people" and who really cared about them. My participants felt that good mathematics educators also needed to show passion about the subject area and communicate well, characteristics that both relate to an educator's ability to bond with students.

The participants' descriptions of the other women in mathematics were also telling of the value they placed on relationships. Several participants explicitly stated that they believed it was more important to be social and have a well-rounded life rather than strictly focus on mathematics, even if that hindered their academic achievement. This relates to findings of studies (Morgan et al., 2001; Morrow & Morrow, 1995) that report that mathematics is viewed as a solitary field with little interpersonal involvement. Feminist and educational researcher Nel Noddings (1998) stated that "We must explore the unpleasant possibility that many girls do not *want* to be part of the math crowd because its members seem socially inept or aloof" (p. 18, emphasis in original). Even though the women in my study were enrolled in mathematics degree programs, they distanced themselves from the "math crowd"—the "other" women in mathematics. The participants all valued personal relationships with their peers, but tended to obtain their friendships from outside the mathematics department, as they generally did not find they could relate to their peer group in the mathematics department on a social, personal level.

Applied Mathematics

All participants clearly articulated a preference for mathematics that was more applied and "real-world" than theoretical in nature. This was further evidenced by their university degree choices and career choices. Strong language such as "hate" was used in regard to such topics as proofs, working in an infinite number of dimensions, and theory in general. In the existing literature regarding gender issues in mathematics education, women's preferences for certain types of mathematics are not frequently discussed; mathematics is often viewed as a single, homogeneous field. However, in my study, participants expressed strong disdain for being required to take theoretical mathematics courses in their undergraduate degree programs. They wanted to learn mathematics that they felt was practical and real-world applicable. My findings are related to those of a recent study of female engineers and engineering students (Abri, 2006), which found that women had a desire for relevancy in the curriculum and real-world applicability. Other earlier research (Belenky et al., 1986; Gilligan, 1982) highlighted women's preferences for connectedness in their learning. Related, my participants showed their broad range of interests in their degree program choices: three participants are enrolled in double major degree programs and two of the three participants who are in a strictly mathematics degree program plan to obtain a second undergraduate degree in another field upon finishing their current degree.

Participants also favored careers that were connected to applications of mathematics rather than pure mathematics. Only Felicity is considering becoming a mathematics professor, but she is also interested in working in a business-related field. The careers planned by the other five participants are in the fields of actuarial science, intelligence/forensics, statistics (two participants), and high school teaching (mathematics and English). These careers use mathematics, but also require knowledge from other disciplines. Several participants voiced the value that they placed on using mathematics in their careers, but not focusing solely on it. Some studies (Lubinski & Benbow, 2006; Morrow & Morrow, 1995) have shown that females tend to prefer to work with people and organic matter. This relates to the findings of my research, as the participants' career choices all involve working with people.

Being 'Othered'

"Othering" is defined by Weis (1995) as "that process which serves to mark and name those thought to be different from oneself" (p. 18). The participants in my study found themselves to be othered in several different ways, such as being mathematics majors, being female, and for some participants, being members of non-dominant cultural and language groups. Furthermore, even within the small cohort of women in mathematics, some participants felt othered.

The participants in this study were placed in a situation of being othered by the very nature of the subject they chose for university study. In Canada, very few undergraduate students of either sex enroll in and graduate from mathematics degree programs. In the 2005–2006 academic year, only 3.51% of undergraduate enrollments and 4.35% of undergraduate graduations at Canadian universities were in mathematics and the related fields of computer science and information science (Statistics Canada, 2008b, 2008c). Of the undergraduate students in mathematics and related fields, women constituted only 25.3% of enrollments and 26.4% of graduations (Statistics Canada, 2008b, 2008c). Put another way, female undergraduate students in mathematics and related fields constitute approximately one percent of all undergraduate students (0.9% of enrollments and 1.1% of graduations). Thus, the participants are members of a very small proportion of the undergraduate students in general, the cohort of undergraduate mathematics students. Within this small cohort, women only account for approximately one quarter of the students, making them members of an extremely small group. Although no participants explicitly stated that they were bothered by the gender imbalance in their classes, they all noted that it did exist. This raises the question of whether the participants were so accustomed to classes with more males in them that they accepted the situation as "the way things are," or if there is another explanation.

Related to the gender imbalance in mathematics, issues surrounding societal notions of femininity were discussed by many participants. They tended to position themselves on the ends of the femininity–masculinity spectrum, with some participants identifying themselves as tomboys and some participants expressing preferences for "feminine" activities and characteristics. For instance, Allison positioned herself in contrast to the "Barbie girls" at her high school who very much encapsulated stereotypical notions of femininity in terms of appearance, and she noted that these girls tended to either tease her or ignore her. Conversely, both Dana and Elise noted the importance of looking good and feeling feminine. When discussing cultures that are heavily dominated by one sex, studies have found a variety of coping strategies and outcomes for the non-dominant group. Kanter (1977) found that members of non-dominant groups tend to make themselves less visible. This relates to the women in my study who separated themselves from stereotypical feminine culture; whether these were conscious or unconscious choices and actions, rejecting dominant notions of femininity may have made it easier for these women to fit into the male-dominated culture of mathematics. Conversely, Walkerdine (1989) found that women in male-dominated fields may assert their femininity as "a defense against the frightening possibility of stepping over the gender divide" (p. 276). Similarly, Riviere (1986) found that female academics felt the need to act in stereotypically feminine ways in order to be reassured that they are, indeed, women. It is unknown whether either of these purposes was behind the actions and preferences of the women in my study who acted in a stereotypically feminine manner, but they could be contributing factors, even at a subconscious level.

Several participants separated themselves from the other women in their mathematics degree program, describing many ways that they were not like "them." By positioning the other women in mathematics as an "other," the participants distanced themselves from some of the negative characteristics that are stereotypically attributed to mathematicians, such as lacking social skills and a singular focus on mathematics (Damarin, 2000; Picker & Berry, 2000, 2001). These participants prided themselves on being well-rounded individuals with diverse interests, social skills, and outgoing personalities. Dana described how she "would rather do a lot of things intensively (pause) maybe do a little bit not as good at school, but live every day." Most participants considered the other students in mathematics to strictly be acquaintances and academic colleagues. The participants with a second academic major tended to feel that they did not fit in with those female students either, for a variety of reasons. The participants' close friends tended to be from outside their academic majors, and often, from outside the university setting altogether.

Although this study did not attempt to explore the impact of cultural differences on women's mathematics experiences, cultural issues did arise in the interviews as three participants had parents who immigrated to Canada. Thus, the home country's culture had a large influence on the participants, particularly because the participants are all very close to their families. Cultural values have been shown to differ widely across countries, and these values can have significant impacts on both current citizens and those who have emigrated. For example, cross-national differences in the types of education and careers considered appropriate for women and the value placed on mathematics have been found in many studies (e.g., Baker & Jones, 1993; Hanson, Schaub, & Baker, 1996; Huntsinger, Jose, Larson, Krieg, & Shaligram, 2000; van Langen & Dekkers, 2005). Chantal's parents immigrated to Canada from Haiti, which she described as having a matriarchal culture where "anybody does anything." Chantal felt that her parents' views were influenced by Haitian culture, and this impacted how they raised their children. Dana and Felicity also made references to their parents' attitudes being strongly impacted by the views of the home country's culture (Morocco and Lebanon, respectively). Thus, even though these three participants were educated in Canada for all levels, they had a very different cultural experience at home from their peers whose families had lived in Canada for many generations. This is another factor that could contribute to a feeling of difference.

Another way that some participants were othered was through issues surrounding non-dominant language. Five of the six participants were in French-language mathematics degree programs, and they all noted that their upper-level mathematics classes were only offered in English. Furthermore, French-language mathematics textbooks were rarely available, even in French-language classes. Even for the participants who had been bilingual since a very young age, taking mathematics classes in English was a challenge due to the specialized terminology and differences in notation. The participants expressed frustration at the lack of French-language materials and courses at an officially bilingual university. In the city where the study was conducted, nearly 60% of residents are unilingual English speakers (Statistics Canada, 2008d). Even in French-language degree programs at this officially bilingual university, English-language culture pervaded the participants' experiences, and provided another way in which the participants were othered. However, this issue may not apply to most other universities in Canada, as the majority of universities are strictly English-language or strictly French-language institutions.

SUMMARY OF FINDINGS

The findings from the study, both within and across dimensions, are summarized via a Venn diagram in Figure 16.3. Those items that were supports for participants are placed in the left circle, whereas the items that were challenges for participants are placed in the right circle. The items in the overlapping section were supports in some situations or for some

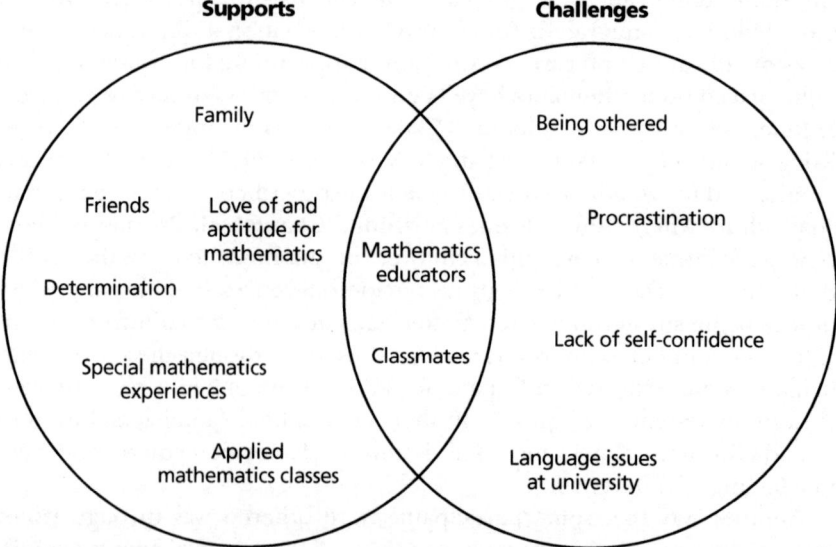

Figure 16.3 Summary of supports and challenges faced by the participants.

participants, whereas they were challenges in other situations or for other participants.

Personal characteristics that supported the women in their mathematics experiences were quite varied. Characteristics that were common to all participants were a great love of and aptitude for mathematics. Some studies (e.g., Gill, 2000; Hall, 2006) have noted the importance of being passionate about and highly skilled in the subject area in order to pursue and persevere in it, and this was clearly evident for the participants in my study. Other characteristics linked to perseverance are determination, competitiveness, and having serious academic attitudes (Gavin, 1996; Rodd & Bartholomew, 2006; Zhao et al., 2005). My participants are all very determined individuals, and two participants described themselves as competitive.

The participants' mathematics skills and interests were further developed through their involvement in either in-class or extracurricular enriched mathematics activities at the high school level. These findings correspond to Gill's (2000) study, which found that participation in after-school programs at the high school level was linked to enrollment in mathematics-related fields of study at university. The participants felt that they had been supported in their mathematics classes by those teachers and professors who were caring, personable, and organized, and who communicated well. Such findings have also been noted in a number of studies that examined perseverance in mathematics-related fields (Abri, 2006; Gavin, 1996; Herzig, 2004; Sax, 1994; Seymour, 1995; Zeldin & Pajares, 2000). As previously described, the participants particularly enjoyed university mathematics classes that were applied in nature, as they could see a real-world connection in what they were learning.

Challenges

The majority of participants felt othered as they did not feel that they fit in with the other women in their mathematics classes. Thus, they were challenged socially as they did not have a peer group in their classes with whom to socialize and bond. The participants' friends were generally from outside their mathematics classes. More broadly, the participants were othered based on their membership in certain groups: mathematics majors, females, and non-dominant language and cultural groups.

As with the personal characteristics that supported the participants, the personal characteristics that challenged the participants also were quite varied. One common problem for the participants was a tendency to procrastinate or to spend too much time socializing. Undoubtedly, this finding is not specific to mathematics; many university students likely have these same tendencies. For two participants, a lack of self-confidence in their mathematics abilities was detrimental, which was a somewhat surprising finding. Although lacking self-confidence in mathematics is frequently cited in the literature regarding females in mathematics more generally (e.g., Damarin, 1993; Fennema & Sherman, 1977; Richardson & Suinn, 1972), studies regarding females in mathematics degree programs tend to show these women as having strong beliefs in their abilities, high levels of self-confidence, and coping strategies (Seymour, 1995; Zeldin & Pajares, 2000).

The participants were challenged in their mathematics experiences by poor mathematics professors—those who belittled the students, were disorganized, and communicated poorly. The participants in French-language programs were challenged by a lack of mathematics classes offered in French, as well as a lack of French resources, even in French-language mathematics classes.

CONCLUSIONS

Implications for Education

A topic frequently explored by mathematics education research is the setting in which mathematics teaching and learning occurs, and how that impacts students' experiences with and attitudes toward the subject area. The educational setting favored by my participants was shown by their preferences for certain styles of mathematics teaching and certain types of mathematics. Specifically, the participants placed great value on having supportive, caring relationships with mathematics educators. This aligns with studies (e.g., Gavin, 1996; Gill, 2000; Seymour, 1995) that have shown how supportive mathematics teachers and professors can strongly impact a woman's decision to enroll and persevere in mathematics at the post-secondary level. Further, the participants in my study all articulated a preference for applied mathematics, and several highlighted the importance of being able to see the real-world use of the mathematics they were learning. This aligns with studies (e.g., Boaler, 1997; Whitten & Burciaga, 2001) that found that instructional practices need to include a greater focus on conceptual understanding and connecting mathematics to women's lives.

More generally, feminist researchers (e.g., Morrow & Morrow, 1995) advocate for cooperative grouping, problem solving, variety in assessments, and the use of real-world contexts in order to make mathematics more meaningful to women. The women in my study expressed a great love for problem solving and a clear preference for mathematics that seemed relevant and real-world applicable. My findings suggest that using the pedagogical approaches suggested by feminist researchers, particularly real-world contexts, may help to interest and maintain women in mathematics.

Directions for Future Research

As I complete this research, several issues still remain, and new questions have been raised. These issues and questions include the possibility of a unique nature of the population studied, issues regarding Canadian culture and multiculturalism, and questions regarding the related populations of men in upper years of undergraduate mathematics degree programs and women who left these programs.

First, due to the type of data collection in this study (volunteer one-on-one, face-to-face interviews), I wonder whether the participants were a unique subset of the women in undergraduate mathematics degree programs. Presumably, shy women would not be prone to volunteer for such a study as it involved speaking to a stranger about their experiences. Also, if the "other" women in mathematics were as mathematics-focused as the participants described, they would likely not want to take time away from their studies to participate in research, particularly near the end of the academic year. Thus, I wonder whether I interviewed women who were indeed a more outgoing, well-rounded type of student, so their view of the "other" women in mathematics as being quiet, shy, and mathematics-focused was accurate, or whether their perception of the other women in mathematics was merely that: a perception. The participants in the study could have very well been talking about each other, and there would be no way to know. It would be interesting to see if the same perceptions of the "other" women in mathematics would be found in a study where all the women in an undergraduate mathematics degree program were interviewed, rather than only volunteers.

The second issue raised by the study relates to culture and ethnicity. Canada has become increasingly multicultural and has a very heterogeneous population in terms of language, culture, and ethnic background. The 2006 Census found that 19.8% of Canadians are foreign-born; the population of foreign-born Canadians increased by 13.6% since the 2001 Census, which is approximately four times higher than the growth rate for the Canadian-born population (Statistics Canada, 2007). Furthermore, 20.1% of Canadians are allophones (those with a mother tongue that is neither English nor French), according to 2006 Census data (Statistics Canada, 2007). Related, more than 200 ethnic origins were reported on the 2006 Census, and 16.2% of Canada's population is comprised of individuals who are visible minorities (Statistics Canada, 2008e). Given these statistics, the number of women in the study whose parents are first-generation Canadians is higher than would be expected. With three of the six women in the study in this situation, further questions were raised regarding who the women are in mathematics in Canada. It is possible that a higher proportion of women from immigrant families (as opposed to multi-generational families in Canada) may elect to study mathematics due to cultural influences from their parents' home countries. As aforementioned, several studies have shown that there are large cultural differences cross-nationally with respect to the value placed on mathematics and the types of education and careers that are considered appropriate for women (e.g., Baker & Jones, 1993; Hanson et al., 1996; Huntsinger et al., 2000; van Langen & Dekkers, 2005). Since the participants' parents played such a large role in their lives, the culture of the parents' home country would presumably have a significant impact. In the future, there may be changes in the number and cultural composition of students who enroll in mathematics at the university level, as Canada becomes an increasingly culturally and ethnically diverse country.

Another issue arose in this study related to the women in mathematics who did not qualify for the study due to its third requirement that the participants must have received all their education (elementary, secondary, and post-secondary) in Canada. When I stated this requirement during my in-class recruitment presentations, I noticed that nearly all the Asian women put their hands down, indicating that they were not eligible to participate. I also had other women contact me about participating in the study who were disqualified by this requirement. Thus, I wondered who all the non-Canadian-educated women in mathematics are—recent immigrants themselves, or perhaps international students who are attending Canadian universities.

The cultural issues raised by this study bring forth further questions: If the majority of women who are studying mathematics at the university level in Canada are those with immigrant parents, those who are immigrants themselves, or international students, what is happening to the women from multi-generational Canadian families who were raised and educated in Canada? Are they selecting away from mathematics at the university level in a greater than statistically expected proportion? Essentially, who are the women who are studying mathematics at the university level in Canada? To begin to understand the complexities regarding the multicultural mosaic that is Canada and how that affects university mathematics enrollment and perseverance of women, future studies may want to explore larger populations, such as all women in mathematics programs in one or several Canadian universities, to better understand precisely the nature of the composition of the female students in mathematics degree programs at the university level, and how cultural issues play a role in these women's choices of mathematics as an academic major.

To fully understand issues surrounding perseverance in mathematics degree programs at the undergraduate degree level, future research might want to examine two other related groups: men, to see if they face similar issues to women in mathematics degree programs, and the counterpoint to these women—women who began in mathematics degree programs but dropped out or changed fields. By strictly examining the unique group of women involved in the study, it is difficult to establish that the findings would not also be similar for the two aforementioned groups; however, I elected to study my unique group of participants as I feel they are understudied in the research literature, particularly in terms of hearing their individual stories. By providing many perspectives on the story, a comprehensive summary of factors that contribute to success and perseverance in mathematics can emerge.

REFERENCES

Abri, J. H. (2006). *Women in engineering: Stories of attrition and retention.* Unpublished doctoral dissertation, Colorado State University, Fort Collins, CO.

Baker, D. P., & Jones, D. P. (1993). Creating gender equality: Cross-national gender stratification and mathematical performance. *Sociology of Education, 66*(2), 91–103.

Becker, J. R. (1995). Women's ways of knowing in mathematics. In P. Rogers & G. Kaiser (Eds.), *Equity in mathematics education: Influences of feminism and culture* (pp. 163–174). London: Falmer Press.

Belenky, M. F., Clinchy, B. M., Goldberger, N. R., & Tarule, J. M. (1986). *Women's ways of knowing: The development of self, voice, and mind.* New York: Basic Books.

Blair, V. (1991). *Differentiating factors of university women persisting and withdrawing from mathematics.* Unpublished master's dissertation, University of Calgary, Calgary, Canada.

Boaler, J. (1997). Reclaiming school mathematics: The girls fight back. *Gender and Education, 9*(3), 285–305.

Burton, L. (1995). Moving towards a feminist epistemology of mathematics. In P. Rogers & G. Kaiser (Eds.), *Equity in mathematics education: Influences of feminism and culture* (pp. 209–226). London: Falmer Press.

Creswell, J. W. (1998). *Qualitative inquiry and research design: Choosing among five traditions.* Thousand Oaks, CA: Sage Publications.

Damarin, S. K. (1993). Equity, experience, and abstraction: Old issues, new considerations. In W. G. Secada (Ed.), *Changing the faces of mathematics perspectives on multiculturalism and gender equity* (pp. 75–83). Reston, VA: National Council of Teachers of Mathematics.

Damarin, S. K. (2000). The mathematically able as a marked category. *Gender and Education, 12*(1), 69–85.

DeVault, M. L. (1990). Talking and listening from women's standpoint: Feminist strategies for interviewing and analysis. *Social Problems, 37*(1), 96–116.

Dworkin, R. W. (2001). Why doctors are down. *Commentary, 3,* 43–47.

Fennema, E., & Sherman, J. (1977). Sex-related differences in mathematics achievement, spatial visualization, and affective factors. *American Educational Research Journal, 14*(1), 51–71.

Forgasz, H. J., & Leder, G. C. (1996). Mathematics and English: Stereotyped domains? *Focus on Learning Problems in Mathematics, 18*(1, 2, 3), 129–137.

Frazier-Kouassi, S., Malanchuk, O., Shure, P., Burkam, D., Gurin, P., Hollenshead, C., Lewis, D. J., Soellner-Youce, P., Neal, H., & Davis, C. (1992). *Women in mathematics and physics: Inhibitors and enhancers.* Ann Arbor, MI: Center for the Education of Women, the University of Michigan.

Freeman, J. (2004). Cultural influences on gifted gender achievement. *High Ability Studies, 15*(1), 7–23.

Gavin, M. K. (1996). The development of math talent: Influences on students at a women's college. *Journal of Secondary Gifted Education, 7*(4), 1–10.

Gill, K. (2000). *Young women's decision to pursue non-traditional science: Intrapersonal, interpersonal and contextual influences.* Unpublished master's dissertation, University of Ottawa, Ottawa, Canada.

Gilligan, C. (1982). *In a different voice.* Cambridge, MA: Harvard University Press.

Hall, J. (2006). *Post-secondary perseverance factors for women with non-traditional undergraduate degrees.* EDU 7190, University of Ottawa, Ottawa, Canada.

Hanna, G. (2003). Reaching gender equity in mathematics education. *The Educational Forum, 67*(3), 204–214.

Hanson, S. L., Schaub, M., & Baker, D. P. (1996). Gender stratification in the science pipeline: A comparative analysis of seven countries. *Gender and Society, 10*(3), 271–290.

Herzig, A. H. (2004). 'Slaughtering this beautiful math': Graduate women choosing and leaving mathematics. *Gender and Education, 16*(3), 379–395.

Huntsinger, C. S., Jose, P. E., Larson, S. L., Krieg, D. B., & Shaligram, S. (2000). Mathematics, vocabulary, and reading development in Chinese American and European American children over the primary school years. *Journal of Educational Psychology, 92*(4), 745–760.

Jackson, A. (1992). Top producers of women mathematics doctorates. *Notices of the American Mathematical Society, 38*(7), 715–720.

Kanter, R. M. (1977). *Men and women of the corporation.* New York: Basic Books.

Lather, P. (1988). Feminist perspectives on empowering research methodologies. *Women's Studies International Forum, 11*(6), 569–581.

Leder, G. (1992). Mathematics and gender: Changing perspectives. In D. A. Grouws (Ed.), *Handbook of research on mathematics teaching and learning* (pp. 597–622). Reston, VA: National Council of Teachers of Mathematics.

Lubinski, D., & Benbow, C. P. (2006). Study of mathematically precocious youth after 35 years: Uncovering antecedents for the development of math-science expertise. *Perspectives on Psychological Science, 1*(4), 316–345.

Lupart, J. L., Cannon, E., & Telfer, J. A. (2004). Gender differences in adolescent academic achievement, interests, values and life-role expectations. *High Ability Studies, 15*(1), 25–42.

Mastekaasa, A., & Smeby, J-C. (2008). Educational choice and persistence in male- and female-dominated fields. *Higher Education, 55*(2), 189–202.

McGraw, R., Lubienski, S. T., & Strutchens, M. E. (2006). A closer look at gender in NAEP mathematics achievement and affect data: Intersections with achievement, race/ethnicity, and socioeconomic status. *Journal for Research in Mathematics Education, 37*(2), 129–150.

Morgan, C., Isaac, J. D., & Sansone, C. (2001). The role of interest in understanding the career choices of female and male college students. *Sex Roles, 44*(5/6), 295–320.

Morrow, C., & Morrow, J. (1995). Connecting women with mathematics. In P. Rogers & G. Kaiser (Eds.), *Equity in mathematics education: Influences of feminism and culture* (pp. 13–26). London: Falmer Press.

National Center for Education Statistics. (2008). *Table 265. Bachelor's, master's, and doctor's degrees conferred by degree-granting institutions, by sex of student and field of study: 2005–06*. Retrieved September 10, 2008 from http://nces.ed.gov/programs/digest/d07/tables/dt07_265.asp?referrer=list

Noddings, N. (1998). Perspectives from feminist philosophy. *Educational Researcher, 27*(5), 17–18.

Picker, S. H., & Berry, J. S. (2000). Investigating pupils' images of mathematicians. *Educational Studies in Mathematics, 43*(1), 65–94.

Picker, S. H., & Berry, J. S. (2001). Your students' images of mathematicians and mathematics. *Mathematics Teaching in the Middle School, 7*(4), 202–208.

Richardson, F. C., & Suinn, R. M. (1972). The mathematics anxiety rating scale: Psychometric data. *Journal of Counseling Psychology, 19,* 551–554.

Riviere, J. (1986). Womanliness as masquerade. In V. Burgin, J. Donald, & C. Kaplan (Eds.), *Formations of fantasy* (pp. 35–44). London: Methuen.

Rodd, M., & Bartholomew, H. (2006). Invisible and special: Young women's experiences as undergraduate mathematics students. *Gender and Education, 18*(1), 35–50.

Sax, L. J. (1994). Mathematical self-concept: How college reinforces the gender gap. *Research in Higher Education, 35*(2), 141–166.

Seymour, E. (1995). Guest comment: Why undergraduates leave the sciences. *American Journal of Physics, 63*(3), 199–202.

Sharpe, N. R. (1992). Pipeline from small colleges. *Notices of the American Mathematical Society, 39,* 3.

Statistics Canada. (2007, December 4). 2006 Census: Immigration, citizenship, language, mobility and migration. *The Daily.* Retrieved August 29, 2008 from http://www.statcan.gc.ca/Daily/English/071204/d071204a.htm

Statistics Canada. (2008a, February 7). University degrees, diplomas, and certificates awarded. *The Daily.* Retrieved July 22, 2008 from http://www.statcan.gc.ca/Daily/English/080207/d080207c.htm

Statistics Canada. (2008b). *Table 477-0013—University enrollments, by registration status, program level, Classification of Instructional Programs, Primary Grouping (CIP_PG) and sex, annual (number),* CANSIM (database), Using E-STAT (distributor). Retrieved January 11, 2008 from http://estat.statcan.gc.ca/

Statistics Canada. (2008c). *Table 477-0014—University degrees, diplomas and certificates granted, by program level, Classification of Instructional Programs, Primary Grouping (CIP_PG) and sex, annual (number),* CANSIM (database), Using E-STAT (distributor). Retrieved January 11, 2008 from http://estat.statcan.gc.ca/

Statistics Canada. (2008d). *Ottawa, Ontario* (table). *2006 Community Profiles.* 2006 Census. Statistics Canada Catalogue no. 92-591-XWE. Retrieved August 24, 2008 from http://www12.statcan.gc.ca/english/census06/data/profiles/community/Details/Page.cfm?Lang=E&Geo1=CSD&Code1=3506008&Geo2

=PR&Code2=35&Data=Count&SearchText=Ottawa&SearchType=Begins&Se
archPR=01&B1=All&Custom=

Statistics Canada. (2008e, April 2). 2006 Census: Ethnic origin, visible minorities, place of work and mode of transportation. *The Daily*. Retrieved August 29, 2008 from http://www.statcan.gc.ca/Daily/English/080402/d080402a.htm

van Langen, A., & Dekkers, H. (2005). Cross-national differences in participating in tertiary science, technology, engineering and mathematics education. *Comparative Education, 41*(3), 329–350.

Walkerdine, V. (1989). Femininity as performance. *Oxford Review of Education, 15*(3), 267–297.

Weinburgh, M. (1995). Gender differences in student attitudes toward science: A meta-analysis of the literature from 1970 to 1991. *Journal of Research in Science Teaching, 32*(4), 387–398.

Weis, L. (1995). Identity formation and the processes of 'othering': Unraveling sexual threads. *Educational Foundations, 9*(1), 17–33.

Whitten, B. L., & Burciaga, J. R. (2001). First steps toward a more feminist, multicultural physics. In E. L. MacNabb, M. J. Cherry, S. L. Popham, & R. P. Prys (Eds.), *Transforming the disciplines: A women's studies primer* (pp. 169–176). Binghamton, NY: The Hawthorn Press.

Zeldin, A. L., & Pajares, F. (2000). Against the odds: Self-efficacy beliefs of women in mathematical, scientific, and technological careers. *American Educational Research Journal, 37*(1), 215–246.

Zhao, C.-M., Carini, R. M., & Kuh, G. D. (2005). Searching for the peach blossom Shangri-La: Student engagement of men and women SMET majors. *The Review of Higher Education, 28*(4), 503–525.

CHAPTER 17

TRY AND CATCH THE WIND

Women Who Do Doctorates at a Mature Stage in Their Lives

Ansie Harding
University of Pretoria, South Africa

Leigh Wood and Michelle Muchatuta
Macquarie University, Australia

**With Barbara Edwards, Lucia Falzon, Sibba Gudlaugsdottir,
Belinda Huntley, Jillian Knowles, Barbara Miller-Reilly,
and Tobia Steyn**

INTRODUCTION

The connection between gender and mathematics has long been on the agenda of many researchers. In fact, during the years between 1970 and 1990, there were more research studies published concerned with gender and mathematics than in any other area (Koehler, 1990). However, a phenomenon that has not been investigated extensively is the trend that shows women choosing to take their doctorates in mathematics later on in life, more so than their male counterparts.

International Perspectives on Gender and Mathematics Education, pages 391–419
Copyright © 2010 by Information Age Publishing
391

The focus of this chapter will be on case studies of seven women from a variety of countries who relate their experiences, both personally and academically, to investigate this trend and give personal perspectives of its impact on their lives. Their stories are inspirational and give focus to the importance of inclusive policies that provide scholarships and opportunities for everyone to achieve their potential.

The phenomenon of women completing a doctorate in mathematics or mathematics education at a mature stage in life is substantiated by the data represented in Figure 17.1, taken from a report issued by the National Science Foundation on United States Doctorates in the 20th Century (Thurgood, Golladay, & Hill, 2006). The data show that the age of female PhDs at graduation over the period 1960–1999 has been consistently higher than the median age of male PhDs.

Also, in a report issued by the National Science Foundation in the U.S. in 1998 (National Opinion Research Center, 1999) it is reported that in traditionally male dominated environments such as Engineering and the Physical Sciences only 6.5% and 7% respectively of doctorate recipients are over the age of 41 compared to 62.9% in the traditionally female dominated field of Education.

The acronym "PhD" is used here interchangeably with the word "doctorate," and the phrases "mature age" and/or "later in life" are defined as referring to an individual's age and/or experiences after the age of 40.

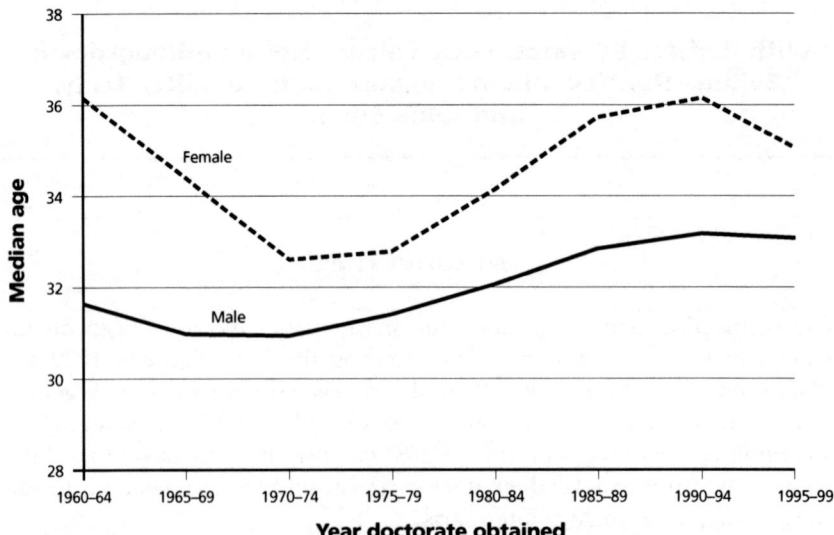

Figure 17.1   Median ages of male and female doctoral graduates for the period 1960–1999 [Modified from: Thurgood, Golladay, & Hill, 2006, p. 23].

LITERATURE REVIEW

The lack of a doctorate means that women are clustered in the lower ranks of academia—doing most of the teaching and making huge contributions but not receiving the material or status awards that come with a research profile. Table 17.1 illustrates the situation at only one university, Macquarie University in Sydney, Australia. The pattern is similar across Australia and probably in other countries.

Some researchers believe that the gender gap in mathematics performance, a well-researched issue, is due solely to genetics (Stanley & Benbow, 1980), but this view has been criticized. A popular reason given for the existence of the gap is the sociological and psychological interplay of cultural forces, expectations, and self-confidence levels that affect women's ability and desire to learn mathematics (Reyes & Stanic, 1988). That is, women have a more negative perception of mathematics than their male counterparts, this being nestled within gender-stereotyped ideas about why they cannot succeed in mathematics (Carr & Jessup, 1997).

Almost as if to counteract this gender gap, we find a surge of women who do doctorates at a mature stage in life. Possible reasons for this phenomenon can be extracted from the literature and will be disputed or supported by the narrations in this study. The question is twofold—on the one hand, why woman do not pursue doctoral studies after a first degree and, on the other, why they then return to pursue doctoral studies at a mature stage of their lives.

There are reasons offered in the literature why many women decide not to pursue doctoral studies early in life. First, men are more confident about learning mathematics than women (Becher, Henkel, & Kogan, 1994; Forgasz, Leder, & Vale, 2000) and women require significantly more confidence than men in order to embark on a PhD program (Mura, 1987). A lack of confidence in their own ability may have led women to lose interest in mathematics earlier in life. A higher level of confidence is then de-

TABLE 17.1 Mathematics and Statistics Departments (Combined)—Macquarie University 2008

Academic level	Male	Doctorate	Female	Doctorate
Professor	5	5		
Associate Professor	3	3	1	1
Senior Lecturer	4	4	2	2
Lecturer	2	1	9	4
Associate Lecturer			8	0

veloped through experiences such as successfully raising a child. This also may explain the move to doctorates in mathematics education rather than mathematics.

There are factors other than confidence, such as lack of financial support, children, and spousal influence (Mura, 1987), that affect women's decisions about studying for a PhD immediately after a first degree. Some researchers have also explored the apparent differential treatment of male and female students by teachers as being a reason for females dropping out of the mathematical pipeline (Hyde & Jaffee, 1998). Female doctoral candidates often feel that they do not fit into the male-dominated worlds of their disciplines (Etzkowitz, Kemelgor, Neuschatz & Uzzi, 2000; Herzig, 2002), and these feelings act as a barrier to female participation in mathematics doctorates earlier in life. A sense of belonging seems to be critical in doctoral students' persistence, with several authors arguing that students' integration into the communities of their departments is important for their persistence (Herzig, 2002; Lovitts, 2001). Leder (1995) suggests that female higher degree students are more likely than their male counterparts to feel overlooked, neglected, and unsupported by staff—particularly in more informal settings.

Yap and Orazem (2001) reported that women in the U.S. who hold doctoral degrees are less likely to be married and to have children than similarly educated men. This suggests that for women, home life may be less compatible with jobs requiring higher levels of education. Women desiring marriage and family life may not opt to pursue a doctorate. Yap and Orazem (2001) concluded that marriage and children do not seem to interfere as much with the careers of males with doctorates as they do for females. Maple (1994) and Stage and Maple (1996) explored the background, educational experiences, and career expectations of female doctoral students in mathematics and science. Using the narratives of seven women who obtained a bachelor's degree in mathematics, they identified the reasons why most of these women left mathematics in order to pursue a doctorate in education including negative experiences in the mathematics program, "the nature of the mathematician did not match their early perceptions," and "a perceived conflict between mathematics as a profession and other roles—such as, parent, community member, and significant other."

The reasons why women return to pursue doctoral studies at a mature stage in their lives is less studied, and this perhaps reflects a continuing conception of the deficit model of women's lives and abilities. Yet, some factors have been identified. The feminist movement's contribution to women's lives came through its message of equal opportunity and programs such as affirmative action support. Such systems have, in turn, provided previously "excluded" women with the opportunity and determination to reclaim their desire and passion to participate in high-level mathematics (Hyde &

Jaffee, 1998). Anderson and Swazey (1998) found that the most common motives given for pursuing a doctorate revolved around the students' interest in and enjoyment of the subject leading to a desire to learn more and to improve their career prospects. Becher et al. (1994) reported similar responses, but with the distinguishing note that mature students enrolled for doctoral studies for personal development. Women receive far less income compared to their male counterparts (Reyes & Stanic, 1988) and are more likely to still be paying off their student debts later in life while men are more likely to have finished paying their debts in their early 30s. It is unsurprising that women tend to participate in PhDs later in life when they have, at least theoretically, a bit more financial stability and are more likely to have adult children.

The literature is far richer in supplying reasons why women leave mathematics than why they return to their studies. The narrations that follow expand on this aspect.

SEVEN EXCEPTIONAL WOMEN'S STORIES

The women who relate their experiences here come from several countries and all have had very different career paths. The answer to the question of how these women were selected is a surprisingly simple one. The women almost spontaneously came together with very little effort from the authors' side. The first two authors are active researchers in mathematics education, both with a solid post graduate mathematics background, one holding a PhD in mathematics, the other in mathematics education. Being acquainted with the mathematics community and having observed the investigated phenomenon with keen interest, the authors were familiar with a number of women embarking on doctorates at a mature age. When approaching a few possible participants, the authors found the response to be overwhelmingly positive and were surprised at others requesting to join the project on hearing about it.

We asked the participants to write about five pages with the following points to assist with the structure of their stories.

- Description of life story, career path
- Motivation for doing a PhD at a late stage
- Problems experienced—personal and work
- Influence of doing a late PhD on her career path
- Personal growth experienced
- Personal perspective on phenomenon of females doing late PhDs

The brief to the participating women was to narrate their life stories, to give their motivation for doing a late PhD, to describe the influence of doing a late PhD on their career paths and personal lives and to relate problems / highlights experienced while doing a PhD. All women shared their life and career paths with astounding eloquence and honesty, and some of them described the process of writing as quite an emotional experience. Their writings are touching, full of pathos, narrating about full lives through which the quest of doing a doctorate is a thread of triumph.

The stories told are those of family commitment and of putting others before self. These are stories of devotion to children, husbands and parents while trying to scaffold a career and of disappointment, perseverance and determination despite difficult circumstances. The stories tell of hope and self-fulfillment, of maturity and a positive self-image. Above all, these stories are of compassion for people and a passion for beliefs. (Note: Real names have been used in these stories.)

Lucia Falzon: From Receptionist to Researcher

> *My journey to get to this point has been a long one—it probably began when I first started school in 1959 and was immediately attracted to the brightly colored counting frames and arithmetic workbooks. It has been a complex journey and yet it somehow all fits together into a nice sequence.*

Lucia was born in Malta in 1955 and educated in a private girls' school that specialized in languages, literature, and history. Upon finishing secondary school she realized that her poor science background prevented her from fulfilling her real ambition of entering university to study mathematics. She started working as a hotel receptionist. She immigrated to Australia as a newlywed at the age of nineteen and had to reluctantly accept work as a sales assistant.

> Three years later pregnancy released me from the drudgery of working in a shop and I became a full-time mother. I had a second baby eighteen months later and became far too busy to contemplate a career outside home for a while.

Lucia enrolled for science and mathematics at school level, finding time to study and do assignments while her husband worked shifts and the pre-schoolers were in bed. She found this arrangement perfect for unwinding and stimulating her brain. She passed with flying colors and at the age of 29 enrolled as a part-time bachelor of science student at Flinders University in Adelaide.

First-year mathematics was a real treat for me. Most of my fellow students were about ten years younger than me and did not appear to have much appreciation for the excellent lectures I thought I was getting. Despite the difficulties involved in fitting the new demands of attending lectures and managing a home with virtually no support from a husband whose views of marriage were rather traditional, I felt truly energized.

At the beginning of her third year at Flinders she started specializing in mathematics. It was also the time that she separated from her husband and became a single mother with full financial and caring responsibility for her two children. She survived on a tertiary expense allowance, a supporting parent benefit, and part-time tutoring of high school students in mathematics at home.

Given the demands on my time and my lack of finances, I did not get to enjoy much of a social life at university. I had made some good friends though and felt accepted by my fellow students despite being so much older and a mother.

When making enquiries about doing honors in pure mathematics, she did not get much encouragement from the head of school. "I was not sure whether this was because it was felt I did not have an extensive background in functional analysis or whether my age, gender, or motherhood status might have played a role." She decided to get a teaching diploma instead, majoring in mathematics education, an interest stimulated by the young students she had by then started coaching.

Teaching jobs in an urban area for a newly qualified teacher were hard to come by, and she was forced to register as a relief teacher. Her first assignment was at a "tough" high school where she felt "no amount of theory or enthusiasm" could help her. She really wanted to advance her knowledge of mathematics and returned to Flinders to become a tutor for first-year students while also doing computer science subjects in a bid to improve her chances of getting more interesting work.

In 1990, at the age of 35, a new stage in her mathematics career started when she obtained a position as "professional officer" at the Defence Science and Technology Organisation (DSTO), a government research facility that employs over 2000 scientists. Her job was to advise on mathematical techniques for analyzing the performance of digital communications systems.

I was in the minority on many accounts: the only female in my section, the only mathematician among technical officers and engineers, the only single parent, and one of very few non-Anglo Saxons. I suspect I also differed in my political, religious, ideological, and world views as well.

She started enjoying the job, felt encouraged, and was invited to do a two-year, part-time Graduate Certificate in Telecommunications at the University of Adelaide, run jointly by Applied Mathematics and Electronic Engineering and aimed at teaching teletraffic analysis.

> The course gave me a good grounding in stochastic modeling, particularly queuing theory and network analysis. I was able to apply my newly acquired knowledge in the workplace and became involved in teletraffic research and participated in seminars and workshops. I started learning about techniques for conducting research and for the first time started contemplating doing a PhD.

She was accepted to do a master's degree by research at the University of Adelaide with the option to convert to a PhD after one year. By aligning the PhD topic with DSTO's research requirements, she was able to study part-time on a full-time wage.

> This meant an increased workload that I had to balance with my family responsibilities. I enjoyed my whole doctoral experience despite the stressful workload.

Lucia graduated five years later at the age of 42. She was ready for new challenges and ventured into a new research direction; more recently she established a multi-disciplinary team to carry out research on modeling and analysis of social systems in support of intelligence analysis.

> My current responsibilities include co-leadership of this interdisciplinary team of social scientists and mathematical modelers who work together to develop new techniques and software tools for social modeling. I also lead a DSTO collaborative research program on social network analysis with the University of Melbourne, where I hold an honorary senior fellowship.

Sibba Gudlaugsdottir: One Step at a Time and Having Fun All the Way

> *If for some reason or another, females have not been able to do it and then the opportunity comes, why not?*

Sibba was born in Iceland in 1962. She grew up on a farm and attended boarding school until the age of 15, then completed her final year of compulsory schooling in a local school in a rural town. In Iceland, compulsory education stops at the age of sixteen. Adolescents who wish to continue with their education then have the option of going to a technical college to acquire a trade, such as becoming a hairdresser or builder, or going to college to study a further four years, which qualifies them for entrance into university. Such extended secondary education means that the average age of someone entering university was 20 years. At the time Sibba was growing up, women in Iceland were often in relationships in their late teens and having children in their early 20s. This meant that many females didn't go straight into university, and delayed their tertiary education until their children had grown up.

> My story is a little different. I was ill at the time when I should have started college, so I started two years after the others. I had to support myself and did college part-time in the evening. During that time I met my husband; we got married and had our only child. By the time I was 21, I was still trying to do college work, unsuccessfully.

After the birth of her daughter, Sibba dropped out of college. However, at the age of 24, Sibba resumed her studies and finished college four years later. Soon after completing college, she and her husband decided to move to Australia.

> There were many reasons for such a bold move. We'd been to Australia before as tourists and had fallen in love with the people and the country. Also there were good job opportunities there for my husband, and I would have more choices in what I wanted to study at university.

Sibba started a part-time bachelor's degree at Macquarie University at the age of 30. She thought that her main interest would be mathematics or actuarial studies, but ended up finding that her real passion was statistics. It took her six years to complete her undergraduate degree while also taking care of her daughter. She then went straight into a master's degree in applied statistics, which she completed while working full-time for the university as an associate lecturer.

> The reason for the Master of Applied Statistics was that I intended to work outside the university environment; that was the intention. It was a lot of fun to do and I do not regret having done that degree. The reason that I am doing the PhD is that I am stuck on a very low level, so the only reason I am doing a PhD is for purposes of promotion because I have decided to stay within universities.

Sibba started her PhD at the age of 41. She is now six years into her PhD, and intends to finish within the next two years. Her research area involves using statistical analysis and statistical modeling techniques to throw some light on certain theories within evolutionary genetics.

When my supervisor and I wrote down our expectations, one of the first things I wrote down was "I want to have fun." The best way of doing it is to have fun on the way and not mind getting side-tracked.

When she initially told her husband that she wanted to go back to college, he was supportive, but he never realized that she would still be studying 22 years later!

Sibba feels that her studying all the time has influenced her daughter considerably. Her daughter looks at higher education as the norm, and there was never a question of her doing at least a master's degree.

I sometimes couldn't get a babysitter, so my daughter came with me to lectures and to the library when I was studying. She thought it was fun to come with me to the university after school and go to lectures on statistical theory.

The issue of females doing PhDs later in life is one that is close to Sibba's heart. She feels it is common for females to do PhDs later than men. This is particularly true in Iceland, where men have a better chance of getting good jobs. Females tend to support the male partner through the education system until he finds a job, and then it is the female's turn.

I know this was very much the pattern when I was supposed to go through the education system. Today things have changed a little because females are not getting involved in serious relationships as early on as my generation did. Also, when I finished school there was not much opportunity of doing a PhD in Iceland so people tended to go overseas when pursuing post-graduate degrees, which could be hard for females especially because you cannot uproot children at an early age.

I feel strongly about this whole thing of women doing their PhDs late. On occasions when talking to my female peers that have not yet obtained a PhD, they ask "Why should I be doing it? I am over 45 years of age and why should I start now to do a PhD?" I think this is the wrong attitude. They are going to be stuck at a low level unless they do something about it. As academics we stay in the system until we are 60—or over 60 even—45 is nothing. So I feel very strongly that they should start a PhD as soon as opportunity arises. Okay, yes, it takes time and it is additional stress factor in our lives. But they should go for it.

Sibba maintains that it is easier at the age of 45 to climb the ladder in an academic environment than in a corporate environment. However, in order to be able to start to climb that ladder one needs to have a PhD.

Barbara Edwards: Saying Goodbye to Home and Family

I felt guilty about leaving my family, especially my younger son, who decided he did not want to move to Pennsylvania. But an even greater issue was dealing with the

fact of being a student again after having lived in the world as a "responsible adult"
for so long.

Barbara was married the summer after receiving a bachelor's degree in English. She taught English in public schools, something she did not really enjoy, before their first child was born two years later. Over the next eight years they had two more children, and Barbara was a stay-at-home mom. When her youngest son started grade school she decided to go back to the school.

> I was determined to do *anything* but teach. Someone in my neighborhood dared me to take a calculus course; one thing led to another and four years later I had earned a master's degree in mathematics. At this point in my life I had no intention of having a "career"—I just wanted to dabble in work, to teach a few algebra or calculus courses at Portland State University just to supplement the family budget.

While studying she worked as a graduate teaching assistant in the mathematics department and realized that she actually did like teaching—as long as she was teaching "big" kids and teaching them mathematics, not English grammar.

A few years into this new routine she was separated from her husband and needed to have a job, so she started working full-time as an instructor. She taught a pre-calculus course at first but one summer was assigned to teach Introductory Real Analysis.

> It was during this year that two things happened that would challenge my notions of how mathematics is learned and the role of the instructor in the learning process. First, these were all mathematics majors; therefore, I reasoned, they would be "good" students who would eventually succeed if I could explain things clearly enough. However, for some students all my expert explaining of the ideas didn't seem to help much. How could they have so much trouble?

> Second, I decided to do a small survey of my pre-college algebra students and I was amazed to learn that some of these students had very good grade point averages—even some with almost all A's in their other courses. How was it possible that such good students would struggle with mathematics that was below the college level? These two experiences left me with many questions on how and why students learn mathematics.

April 1993 is a date for her to remember when her career changed dramatically. She drove to Seattle, Washington, with Jeanette, a colleague who was attending the annual meeting of the National Council of Teachers of Mathematics. She never intended to attend any of the NCTM sessions and simply wanted to have lunch with a good friend of hers who lived in Seattle.

However, while in Seattle, Jeanette introduced her to Kathleen Heid and Glenn Blume, mathematics education researchers from Pennsylvania State University.

> One thing led to another and by the end of that week I decided I should quit my teaching position and go back to graduate school at Penn State. That August, the day before my 50th birthday, I packed up my car, left my house to the care of my oldest son and with my youngest son started the four-day drive to State College, Pennsylvania.

She did not set out to get a PhD. She saw this as an opportunity to learn something about doing research to find answers to the questions she had about how and why students learned (or really, didn't learn) mathematics in her classes. It wasn't until she got into the program that she realized that if she wanted to do research as a career she needed to get the credentials—a PhD.

> There are problems when one gives up a job and leaves home to pursue a new path. I worried about money somewhat, but mainly I felt guilty about leaving my family, especially my younger son, who decided he did not want to move to Pennsylvania. But an even greater issue was dealing with the fact of being a student again after having lived in the world as a "responsible adult" for so long. I had been my own boss in a lot of ways in my teaching position. As a student I felt like I was at a somewhat lower position in life. My three children— my sons were 17 and 21 and my daughter, who had been recently married, was 25—and my parents were very supportive of my decision to return to school. My children were amused when they started getting letters addressed to "The Parents of Barbara Edwards" and even arranged at the end of the first semester for me to receive a "Penn State Finals Week Survival Kit."

She managed to complete everything, handed in her thesis in August 1997, and was surprised that even though she was 54 years old she had no trouble finding a position—and Oregon State was not her only offer. She enjoys her career now but finds it somewhat odd to have been in a career only ten years and already be at "retirement" age.

> My degrees in English and mathematics, my belief in life-long learning, the fact that I thoroughly enjoyed my earlier career raising three children, my interest in and observations of how people think and learn—all have contributed to my current position as a mathematics education researcher. In some ways maybe it was a good thing to launch into this career after 50 because I had lived long enough to have all these experiences. I certainly will never regret my decision to pursue my PhD.

Tobia Steyn: No Regrets

Although the balance between work and play can easily get disturbed, "survival" depends on one's ability to adapt temporarily and restore the balance.

Tobia graduated in 1969 majoring in mathematics and history. Forty years ago in a rural area of South Africa and being one of the first generation in her family to study at university, her parents' main concern was that she should get a degree.

Thinking back about my choice of what to be and what to study, I cannot remember even having considered many options. The choice may have been between becoming a teacher, a nurse or doctor and I went for the first.

She did a teaching diploma the year after graduating and also met her husband, Jasper. She then joined the Department of Mathematics at the University of Pretoria as a lecturer responsible for first-year courses. Her formal training in mathematics ended at honors level in 1972, and education became her focus.

After getting married, both of them continued with their studies, Jasper in engineering and Tobia did a post-graduate degree in orthodidactics with specialization in mathematics education. Then her children came.

To continue working was never even a consideration. Two reasons influenced my devotion to be at home with my children. The one was that I wanted to give them the attention I presumed only I could give. The other reason was in the labour law of the 1970s; my contract with the University stipulated that pregnancy would be regarded as termination of service! So in April 1976 I had to resign and left the University.

Staying at home as spouse and mother, looking after our two sons and daughter, implied a loss of my single identity. My identity became a "one plus four" [spouse plus three kids] identity. The implication hereof then was that whatever was planned and done, I had to take the "one plus four" into consideration.

Tobia spent about fifteen years caring for their three children and has no regrets. She feels that babies and toddlers (and older children) do not ask an hour of quality time in a single session—they are in need of quality time split up in units of minutes or seconds throughout a day. She did not lose the "one" identity totally, doing correspondence courses in psychology and in information technology.

Tobia returned to the job market and started working part-time in the early 1990s, just at the time when computers were surfacing as a new educational tool and also at the time in the history of South Africa that was marked by educational initiatives to establish programs to provide for students who had previously not shared in the opportunity of getting a tertiary education.

She started by coordinating a computer-based bridging course in mathematics and gained valuable experience that laid the foundation for her interest in using computer-based graphing technology to enhance students' understanding of fundamental mathematical concepts.

> In 1999 I did a certification course in the interpretation of the Herrman Brain Dominance Instrument in Lake Lure in the United States. I had the privilege to meet Ned Herrman, whose whole brain model has served as one of the conceptual models on which I base my teaching strategy. I also have the privilege to know Professor Richard Felder, who is well-known for his contribution to engineering education in the United States, and I use his learning style model as part of my teaching strategy. Recently I also had the opportunity to meet Tony Buzan, creator of the Mind Map concept. This was a bonus opportunity as I have been training students to use Buzan's mind mapping since 1997.

Tobia completed a master's degree in mathematics education in 1998 and a PhD in 2003, also in mathematics education. All three children were engaged in completing high school and university study at the time. Tobia remembers the late 1990s and early 2000s as a time of survival of the "one identity" alongside the "one plus four" identity.

> Having achieved this academic milestone, I have always thought that my parents, who had by then both passed away, would have been very pleased and proud to share in this achievement. My supervisor for both endeavours started as a first-year mathematics student when I left academia for motherhood. He had pursued a career as counselling psychologist, obtained PhDs in education and psychology and had become a dedicated researcher.

In 1999 she took up a position as senior lecturer in the School of Engineering responsible for a course in the academic development program for first-year engineering students and has gone from strength to strength. The learning facilitation strategy that she has developed in the Professional Orientation modules at the University of Pretoria has proven to provide the required preparation to students who have shortcomings in their educational background. In 2006 she received the Laureates Award for Education Innovation from the University of Pretoria in recognition of this contribution.

> The physical and emotional wellness of my family has always been a priority. Although the balance between work and play can easily get disturbed, "survival" depends on one's ability to adapt temporarily and restore the balance. A resume of my life will be incomplete without acknowledging the significance

of my interaction with all the people who have crossed my path through life. I believe that any experience in life is a learning opportunity. I also believe that the grace of God is the golden thread that pilots the destiny of my life.

Belinda Huntley: Pursuing Possibilities of Promotion

A PhD has given me the licence to be taken seriously in the academic environment.

Belinda has always considered teaching as a calling, rather than as a career. She began her career as a secondary school teacher of mathematics and physical science after graduating from the University of the Witwatersrand (WITS) in Johannesburg, South Africa, and completed an honors degree part-time.

> I particularly enjoy the challenge of helping students with weak mathematical abilities. One of the greatest rewards in teaching for me is when you see the negative attitudes of your students towards mathematics being transformed into a more confident, positive attitude.

She married Brian, her best friend since school days, and completed a master's degree in education in 1985, the same year her first son, Byron, was born. She continued teaching while cherishing the dream in the back of her mind to complete a PhD, after raising her children.

Four years later a change in her career came when she left teaching to supervise prospective mathematics teachers at WITS. Shortly afterwards she became a full-time mathematics tutor in the mathematics department. This post involved tutoring first-year students registered for mathematics courses, as well as supervising postgraduate students completing their secondary school teaching diplomas. Within this post, she was promoted to Senior Tutor and eventually to the position of Principal Tutor.

Her second son, Christopher, was born ten years after the first and she remembers:

> This was such a joyous event in my life as we had been struggling to have a second child. My responsibilities grew both at home and at work. I was appointed to co-ordinate the first-year mathematics major course, involving over 500 students. I also served on various committees. With a heavy teaching, marking, and administrative load, I had very little time to even consider registering for a PhD. The later I was leaving it, the more difficult it was going to get back into studying towards a further degree.

Belinda started experiencing pressure to do a PhD—externally from the head of department and internally from a wish to fulfill a lifelong dream.

In addition, she had reached the top of her salary scale, and there was no possibility of promotion in her career path without a PhD.

> Finally, in 2004 at the age of 46, I took the brave step to register for my PhD. Due to the nature of my research, I was referred to a supervisor at another tertiary institution, sixty kilometers away. I actually found it quite exciting to spread my wings and "enlarge my territory," so to speak, yet it also meant traveling the busy road weekly to see my supervisors.

Belinda did not find it easy to cope with studies, work commitments and running a household.

> To exacerbate the situation, my husband was offered a three-year contract to work in the United States of America. This was an opportunity of a lifetime for him, and I didn't want my studies to stand in his way. So for three years, during my PhD studies, I became a "single" mother. I often felt guilty when I was spending time with my children and should have been working on my PhD, and vice versa. I felt torn between my studies and my family commitments. I started working later and later every day, and often went to work only having had a few hours sleep at night. Consequently, my health suffered and I developed a sleeping disorder.

Belinda says that writing did not come easy to her. As all her previous studies had been in the scientific field, she found it difficult to express herself other than just stating the facts. Even writing up the students' course notes did not prepare her sufficiently for the necessary writing skills required for a thesis.

> One highlight of my period of research was when I was granted a full year's sabbatical leave. This year turned out to be an incredibly productive year for me. Another highlight during my sabbatical was that my husband and I adopted a teenage daughter. Suddenly, my family had grown even more and my responsibilities with it.

> My "low" was when my data analyst realized that some contradictory results were being produced using the model that I had developed in my study. I felt like giving up. With constant encouragement from my husband and parents, I continued with my studies. With sheer determination and tenacity, I met the deadline of the submission date of my thesis. I had fulfilled the first requirement of my PhD.

The next goal was to have a paper accepted for an international conference. Belinda found the experience exciting and almost overwhelming. She also found it quite amazing to attend presentations by mathematics education researchers whose work she greatly admired and had read.

For the first time, at the conference, I started feeling like a respected member of the PhD community. In my career, I had always been recognized for my excellence in teaching and my contribution to mathematics education. My ultimate reward was when I was awarded the Department of Science and Technology 'Women in Science' Fellowship Award for my research in an area where participation by women is traditionally low. I accomplished a lifelong dream and received national recognition for my research. A new chapter in my life had begun.

Barbara Miller-Reilly: Just in Time for Retirement

Completion of my PhD was now NOT going to influence my career path. But then I decided that I wanted to finish it for my own satisfaction.

The year 1966 saw Barbara Miller graduate in Wellington, New Zealand with a bachelor of science in mathematics. She started in the teaching profession but soon ventured into computer programming. By then she was married to Ivan Reilly; their first child, a son, had been born; and they were living in the United States temporarily. In the next few years she did a master's degree in Computer Science in the U.S. Returning to New Zealand, she worked as assistant lecturer in mathematics until the birth of two daughters. Then she combined parenting with a bit of part-time teaching and programming.

> When my youngest child started school in 1981, I contacted another wife of a mathematician, in more or less the same position as myself and asked whether she would be interested in sharing a job with me. I was then in my thirties and did not want a temporary job. She was interested and we were jointly appointed in a programmer/analyst position at the University of Auckland.

Barbara's career path started developing in several ways in the mid-eighties. She initiated the EQUALS project in Auckland, a movement that formed at the University of California Berkeley in the U.S. with the aim of increasing the access of girls and women in all areas of study and work in mathematics, science, and computing. This led to the initiation of a careers seminar on mathematics for high school girls, a yearly event that has been running ever since.

> I also started my first research project in the area of gender and mathematics and it was important to know how my EQUALS colleagues would value this work. I joined the university's fledgling Student Learning Centre and in my teaching of mathematics to mathematics-avoidant students incorporated

many ideas from EQUALS. Their support gave me confidence as my teaching practice was atypical within the university.

One thing led to another. In 1986 Barbara initiated the planning of a successful conference on Women in Mathematics, Science, Engineering, and Technology. She then went on research and study leave on a Winston Churchill Fellowship to the U.S. to study programs for encouraging women in computing, mathematics, and science. A Fullbright grant saw three EQUALS staff visit Auckland and help build a national network. Barbara was elated when a new position was established at the University of Auckland—that of Liaison Officer for Women in Engineering and Science in 1988.

> I felt all my efforts, and those of my colleagues, were being acknowledged and felt a sense of achievement that our work would carry on with this funded position. For me this was a start in growing confidence that led to presenting findings at Australia and New Zealand over the next few years.

In 1992 she was called to supervise student-teachers' research projects, which expanded her knowledge of the link between teaching and research. An invitation to attend the 1993 ICMI conference on Gender and Mathematics Education in Sweden and the award of a Claude McCarthy Fellowship inspired her to start doctoral studies.

> Starting a PhD suddenly seemed like an interesting challenge, possibly extending some of the results of my previous research. I later decided to change the topic, and I don't think I realized exactly what a huge challenge this would turn out to be.

A decade followed that was full of family commitments with children getting married, involvement with aging parents, her main supervisor dying of cancer, and her daughter being seriously ill for a period. It was not easy to do research.

Barbara says that during the 1990s the Mathematics Education Unit slowly became a well-established unit within the Mathematics Department, and she always tutored part-time within the unit as well as in the Student Learning Centre. She finally moved out of the Student Learning Centre and fully into the Mathematics Education Unit in 2000.

Barbara took a year without teaching (and pay) in 2005, in order to give herself time to finish. In order to do this, she needed to technically retire and then was offered a contract for two more paid years of teaching before actual retirement came into effect.

> I debated during the first half of that year (without pay) whether to finish or to, instead, write journal papers. Completion of my PhD was now NOT

going to influence my career path. But then I decided that I wanted to finish it for my own satisfaction. I put my head really down for those last three months, and the family was wonderfully cooperative. I was lucky that everyone kept well and there were no unexpected events. The floor of the house was covered with my notes in several rooms! I handed my PhD in a week before we went overseas for three months on conference leave. What a relief. That elation lasted months!

Despite being of retirement age, Barbara is still doing research as an Honorary Research Associate and finds that doing a late PhD has positively influenced her retirement.

Jillian Knowles: Unabated Passion for Social Justice

There were people I could turn to when my committee proved toxic for me. And I and my family would have been unbearably disappointed in me if I had quit . . . we had all made really big sacrifices.

Jillian was born in 1948 in Brisbane, Australia, into an upper middle-class Australian family. Her mother, a trained physiotherapist, was a stay-at-home-mom due to her father's resistance, who was a chartered accountant.

We all went to private single-sex high schools. From early grades I was "top of the class," which meant that I was literally seated in the back left hand corner. While my brothers were above average, neither sat in the back row, so I was considered the "bright" one. They had to do well in sport: cricket, rugby union . . . I didn't do well in sport. I was Dux of my school in 12th grade and earned a full Commonwealth Scholarship to go to the University of Queensland in whatever faculty I chose. I chose mathematics because I could and I wanted to show everyone that I could.

At honors level Jillian found the abstract theoretical approach in mathematics daunting, and she was more interested in being involved in civil rights activism, especially on behalf of Aboriginal Australians who were not then citizens of Australia, and trying to integrate that with her evangelical Christian faith that she had embraced as an early teen. When she had the opportunity to consider pursuing a doctorate in mathematics right after earning her undergraduate degree in mathematics, she was diverted to the much more interesting and challenging (to her) field of mathematics education . . . a field that had a real civil rights component in it.

My passion for social justice for indigenous people and my Christian faith led me to join the Summer Institute of Linguistics and qualified as a field linguist.

She worked doing field linguistics with two different groups of Aboriginal people and also with university and college students urging them to consider field linguistics. It was in 1979 while doing field linguistics that she met her "Yankee" future husband. He was a 36-year-old carpenter who was volunteering in Australia, and she was 31 years old. After they married they drew straws to decide where to settle. The U.S. won the draw, so they settled in West Newbury, Massachusetts, where he had been born. He was self-employed and she worked part-time while taking care of their three daughters. She did a variety of part-time jobs such as home tutoring students with learning disabilities or mathematics issues, and tutoring and teaching mathematics courses at a local community college.

> Looking back, living poor in America with three children and a self-employed husband with no health insurance (or retirement savings or insurance of any kind) did not feel safe, and this was an important part of my decision to begin my PhD studies. I was 46 years old. I realized that with my present qualifications I could not get a permanent full-time position in a tertiary institution.

Jillian decided on a self-designed PhD in Educational Studies at Lesley College in Cambridge, Massachusetts. It was the closest to an Australian/European doctorate. She feels that her doctoral studies provided answers to her professional questions.

> I had considered a PhD program in mathematics, but that would not have enabled me to answer my question. I needed to combine study of advanced mathematics, mathematics education theory and practices, and psychological counseling theory for that.

Her studies and discoveries enhanced her practice, and, in turn, her practice inspired her deep questions and research. "It took me 10 years to complete my doctorate and I spent at least $30,000 net (after teaching fellowships and grants, etc.) on it that we did not have."

Jillian describes this period of ten years as often extremely stressful. There were family and financial hardships. They had to take out a second mortgage partly to pay for the family expenses and also to help pay for Jillian's doctoral program. In addition, the PhD studies proved to be a rough ride.

> Mine was the first dissertation for which my senior reader was reader. Looking back, I realize she had a picture of what a dissertation should look like, but she had no idea of how to guide me in that direction. In my first attempt, I wrote 1,000 pages, some of it brilliant, a lot of it very mediocre, much of it not needed, and she became increasingly frustrated without being of much help.

Jillian could not change readers, so she had to seek writing help outside her committee, which had only heard of the struggle from the senior reader's viewpoint. Jillian's writing-self was healed, the committee was satisfied, and she graduated at the age of 56, nearly four years after she first began writing the dissertation, ten years from when she began her doctoral program.

Why and how did I persist through this and complete my PhD? Fewer than 20 percent of the people who begin doctoral programs in the U.S. complete them. My program was small and not rigid with respect to timetable. There were people I could turn to when my committee proved toxic for me. And I and my family would have been unbearably disappointed in me if I had quit... we had all made really big sacrifices. I could have kept my assistant director position, but I would have deprived my profession of what I believe to be very important findings.

When I graduated, almost immediately, I heard of a faculty position at a small private comprehensive college closer to my home. I applied and was hired, almost certainly because of the PhD. When I sought a higher beginning salary based on my experience and qualifications, I was rebuffed. Subsequently, young male PhD in mathematics hires have automatically been offered more as starting salary. Was it my gender or the fact that my PhD was in mathematics education and not in mathematics? I almost immediately regretted my decision—mostly because my passion and my focus is on individuals' mathematics self-healing and development; classrooms of students challenged and frustrated me. I have turned this around, improving my practice and incorporating my doctoral findings into my work with my students, which fuels my ongoing classroom-based research and writing, but I would like to go back to serve in the only educational setting in America that redresses the stark inequities of opportunity in elementary and high school education in this country—the community college. My passion for social justice remains unabated and I realize I want to apply my doctoral findings and subsequent discoveries in that setting—helping rehabilitate students who have been impoverished by the system. I am excited.

Since I completed my doctorate I have been desperate to share my findings and my post-doctoral research findings with my colleagues. But it is very difficult since my college is not a research institution and requires me to teach four undergraduate mathematics courses a semester. So getting my paper published this year in *ZDM: The International Journal of Mathematics Education* felt almost like giving birth—I said to myself "I can die now. My findings are no longer buried in a dissertation most people won't access."

NARRATIVE ANALYSIS: COMMONALITIES

The narrative of each participant was circulated to the three authors who read, empathized and, in some cases, wept, with the joy and commitment of the stories. We used a thematic analysis to draw out the commonalities.

Family Commitments

The predominant impression left after reading the life stories of these women is how interwoven their professional lives and family commitments are. In all the cases, family life played a major role in determining their career paths, and in turn, their careers decidedly influenced their family's lives. In almost all cases having a career was secondary to embarking on family life.

Barbara Miller-Reilly maintains that women themselves were not ambitious enough in pursuing careers: "Many of us, as women in my generation, did not expect our career ambitions to be taken seriously. *We* did not take them seriously! We were educated/socialized to take the main parental responsibilities without question." She links parental responsibility to doing a PhD later in life:

> I think this is why many women in their 50s, 60s 70s or even 80s take on the challenge and interest of academic study which may continue as far as a PhD. They are finally free of the major responsibilities of parenting, although other family responsibilities do continue.

The importance of family life also emerges in other stories:

> At this point in my life I had no intention of having a "career"—I just wanted to dabble in work, to teach a few algebra or calculus courses at Portland State just to supplement the family budget. (Barbara Edwards on doing a masters degree while her children were still small)

> My dream was to eventually get a PhD, but only after I had raised my children. (Belinda)

> I think the reason for females going later is the fact that they support their male partners in getting their education first and then it is their time to go through.... (Sibba)

We submit that women still have the role of nurturer and that parenting responsibilities are synonymous to prioritizing time and money for family (particularly children) and sometimes even partners. Higher education, such as doctorates, come secondary to these and is often an afterthought in the self-actualization phase of life. This view supports that of Mura (1987) and Yap and Orazem (2001).

Staying at Home or Combining Family Life and Research

The women who related their stories here dealt with their young families in different ways, a number taking the deliberate decision to stay at home for a period of time in order to raise a family.

Lucia had two babies in quick succession and simply became "far too busy to contemplate a career outside the home for a while." For Tobia it was a matter of principle to be available for her children in their early years. When she eventually resumed her professional career and started studying, she dealt with the dilemma of juggling a career and family.

> While I was doing the research and writing up of the MEd and PhD, all three of the children were engaged in completing high school and university study. Thinking back, I can now not imagine how I had managed and I have no intention to try something similar again soon!

The choice of staying home during the initial few years of children's lives irrevocably delays the ascent of the professional ladder and is probably the biggest difference in career paths of women and men.

Women who tried to combine a career/studies and family life found that this was not all plain sailing.

> My family, especially my children, found it difficult to adjust to the fact that I was no longer available at any time they desired. I also struggled to cope with a full teaching load, doing research for my PhD, being a mother, wife and running the household. (Belinda)

> It was an extremely stressful time. (Jillian on doing a PhD while being fully committed to family life)

> Although I was somewhat worried about the money I mainly felt guilty about leaving my family. (Barbara Edwards on doing a PhD away from home)

One of the lucky ones to have experienced full support from her husband is Sibba, who in turn took a deliberate step to ensure his continued support: "He has always supported me, but we made a deal, you see, that this wouldn't affect our time together. So when you set out like that you have to make sure that your family doesn't suffer, and that is what I tried to do."

We propose that combining family life and a career/studies while caring for a family is extremely stressful and requires commitment and the ability to juggle activities.

Aging Parents

Doing a PhD later in life does not only mean that women have to deal with teenagers/young adults while coping with the rigorous and intense routine required by doing research—but it also means, in most cases, dealing simultaneously with aging parents.

Barbara Miller-Reilly says of coping with elderly parents:

> I don't think I realized exactly what a huge challenge this would turn out to be! What I thought would be a quiet decade on the family front turned out to be extremely eventful. Some events were great highs (although time consuming)—the wedding of each of our three children and the birth of our first three grandchildren. I also was very involved supporting my mother during the last five very difficult years of her life. I had to put the PhD away frequently. Ivan's parents also died during these years. Then my main supervisor who was also my fellow researcher and friend died with cancer over almost a year.

Jillian, who took ten years to complete her PhD, describes how her studies were influenced by personal circumstances involving, among other things, the passing away of her parents: "My mother died one week before I began (my PhD studies), my parents-in-law died four months and fourteen months later, and my father two years before I finished."

We submit that added to young children as a factor for delaying PhD studies is the care and concern that aging parents require, often more so from the female partner in a relationship.

Credibility

A common awareness in all the women's journeys is that having a doctorate gives you credibility and an academic standing. The importance of doing a PhD for professional progress is also mentioned by Belinda as an incentive:

> I cannot really pinpoint one specific incident that motivated me to register for my PhD. I think it was a combination of external pressure from the head of my department and intrinsic motivation to fulfill a lifelong dream. In addition, there was no possibility of promotion in my career path without a PhD.

Others agree:

> It wasn't until I got into the program that I realized it would take more than one year to satisfy my curiosity and I also slowly began to realize that if I wanted to do research as a career I needed to get the credentials—a PhD. (Barbara Edwards)

> You see, you haven't had a chance up until now, you are going to be stuck on the level you are at unless you do something about it and get a PhD. (Sibba)

We propose that females who follow an academic career without having a PhD experience a "glass ceiling" in their promotion possibilities and that this realization often serves as an incentive for doing a PhD, no matter how late in life.

Intellectual Excitement

However, the credibility that a PhD offers is not the only motivation for enrolling for the degree. A stronger motivation seems to be curiosity and conviction.

Although Jillian starts off by saying that she started on a PhD because she did not feel financially safe, she continues to give the more significant reason for furthering her studies:

> But I REALLY wanted to find out an answer to my burning professional question—that was my MAIN reason for wanting to do a PhD—I wanted to know *why* some apparently capable students seemed to sabotage themselves while taking math courses in college while others who seemed less likely to succeed did.

Others agree:

> My PhD research had served a need in an area in South Africa where expertise was needed—assessment in mathematics education. I had produced research that could benefit the mathematics education community. (Belinda on the importance to her personally of contributing to a knowledge base)

> I could have kept my assistant director position but I would have deprived my profession of what I believe to be very important findings. (Jillian on how her convictions dominate the desire to climb the promotion ladder)

We submit that promotion as an incentive for women for doing a late PhD is secondary to realizing their dreams and their ability to contribute, supported by the findings of Anderson and Swazey (1998) and Becher et al. (1994).

Impact on Career Path

Doing a late PhD necessarily means that your career path, in the academic world, will deviate from that of someone who does a PhD early in his or her career. Promotion suddenly becomes a possibility, a career previously stationary could be kick-started into action, and there could even be retirement benefits!

A positive turn in her career path as a result of doing a PhD is described by Jillian:

> When I graduated, almost immediately, I heard of a faculty position. It was somewhat better pay for a 9 month rather than 12 month position. I applied and was hired, almost certainly because of the PhD.

Others agree:

> The PhD had given me the licence to be taken seriously in the academic environment. (Belinda on completion of her PhD)

Even completing a PhD at a very late stage can bring pleasant surprises.

> Normally I would not have expected, since I do not have a high ranked academic position, to have been given any space (i.e., a desk) in the department but now, as an Honorary Research Associate, I have a desk and the entire infrastructure to continue to do research. I am very, very pleased. (Barbara Miller-Reilly on retiring shortly after completing a PhD)

> It has been a little strange to be a "junior" faculty member at this age and this has had an impact upon my career path. I do not have the drive that I might have had twenty-five years ago to become "famous." I want to pursue my research interests and I want to communicate what I discover. (Barbara Edwards on obtaining a permanent position at the age of 54)

We conclude that no matter how late a PhD is obtained, the qualification has a direct and beneficial influence on women's career paths.

Personal Growth

A major benefit of doing a PhD late in life is the personal growth experienced. Quoting Barbara Edwards: "In terms of my own personal growth, getting my PhD and starting a new career at a later age seems to have been a culmination of all my previous experiences, both in life and in my education."

Jillian feels that the maturity of thought that age brings is a benefit for doing a PhD:

> Well, I have developed a strong sense—I think deepened by my experience of doing the doctorate—that it is generally wiser not to go straight into a doctoral program following undergraduate completion. That is how I advise my daughters and advisees at college... it is better to go into your field and work... if developing in your field or changing to a more compatible field requires further study, do it then when you know what your questions are.

We propose that maturity of thoughts is a benefit stemming from doing a PhD late in life and that personal growth is the profit.

CONCLUSION

We first make two general observations:

1. Although all the women started out doing a mathematics degree, only Lucia completed her PhD in mathematics, and Sibba has almost finished hers, whereas all the others turned to research in mathematics education.
2. Also, it seems that the career paths of women doing late PhDs are intricate, with many twists and turns along the way, different from the often steady incline of careers of people doing early PhDs.

There is no doubt that the gender gap in mathematics exists. However, in the narrations of these women little mention was made of negative stereotyping during their lives that could have served as a reason for not doing an early PhD, as surmised by Carr and Jessup (1997).

A lack of confidence observed in women as compared to men (Forgasz, Leder, & Vale, 2000) also did not feature strongly either in the lives of these women as a deterrent for initially doing a PhD. What did emerge from the narrations was that confidence does play a role when returning to studies after a long period of time. Also, lack of support from fellows and peers in mathematical departments, mentioned by Etzkowitz et al. (2000), was experienced by few females as problematic. A sense of belonging proved to be critical in the persistence of doctoral studies and supports the findings of Herzig (2002). We further agree with Herzig (2004) that prospects of career and/or family commitment often proves to be the reason for women dropping out of the academic pipeline.

In conclusion, although the women's narrations tell of hardship, obstacles, and perseverance, there was never a trace of self-pity; needless to say these women are paragons of tenacity and, as such, exceptional women.

In all cases, the challenge of a doing a PhD later in life was turned into a story of triumph. The women often saw doing a PhD at a later stage in life as more beneficial to personal growth and maturity of thought, as well as to research.

The reasons for the trend of women completing PhDs later in life are an interaction of socio-cultural, political, intrapersonal, and other factors. These factors highlight intrinsic differences in women's and men's attitudes and dispositions. Women are the principle caretakers of the young family and as such often willingly sacrifice career prospects for pursuing this vocation. Whether this tendency will change is doubtful and should not be seen in a negative light. What should change is that women returning to the career environment and embarking on PhD studies at a later stage in life should be given a softer landing. They should be welcomed back into the academic environment, and action should be taken to ensure that confidence is built in a caring environment. Practical measures, such as scholarships targeting women, would assist in ensuring that their contribution to society through mathematics and mathematics education is enhanced.

REFERENCES

Anderson, M., & Swazey, J. (1998). Reflections on the graduate student experience: An overview. In M. S. Anderson (Ed.), *The experience of being in graduate school: An exploration. New directions for higher education, No. 101* (pp. 3–14). San Francisco: Jossey-Bass.

Becher, T., Henkel, M., & Kogan, M. (1994). *In pursuit of the PhD*. Princeton, NJ: Princeton University Press.

Carr, M., & Jessup, D. L. (1997). Gender differences in first grade mathematics strategy use: Social and metacognitive influences. *Journal of Educational Psychology, 98*(2), 318–328.

Etzkowitz, H., Kemelgor, C., Neuschatz, M., & Uzzi, B. (2000). Athena unbound: Barriers to women in academic science and engineering. *Science and Public Policy, 19*(3), 157–179.

Forgasz, H. J., Leder, G. C., & Vale, C. (2000). Gender and mathematics: Changing perspectives. In K. D. Owens & J. Mousley (Eds.), *Research in mathematics education in Australasia 1996-1999* (pp. 305–340). Sydney, Australia: MERGA.

Herzig, A. H. (2002). Where have all the students gone? Participation of doctoral students in authentic mathematical activity as a necessary condition for persistence toward the Ph.D. *Educational Studies in Mathematics, 50*(2), 177–212.

Herzig, A. H. (2004). "Slaughtering this beautiful math": Graduate women choosing and leaving mathematics. *Gender and Education, 16*(3), 379–395.

Hyde, J. S., & Jaffee, S. (1998). Perspectives from social and feminist psychology. *Educational Researcher, 27*(5), 14–16.

Koehler, M. S. (1990). Classrooms, teachers and gender differences in mathematics. In E. Fennema & G. Leder (Eds.), *Mathematics and gender* (pp. 128–148). New York: Teachers College Press.

Leder G. C. (1995). Higher degree research supervision: A question of balance. *The Australian Universities Review, 38*(2), 5–8.

Lovitts, B. E. (2001). *Leaving the ivory tower: The causes and consequences of departure from doctoral study.* Boston: Rowman & Littlefield.

Maple, S. A. (1994). *A way of life: Background and experiences of women doctoral students in mathematics and science.* Unpublished doctoral dissertation, Indiana University, Indianapolis.

Mura, R. (1987). Sex-related differences in expectations of success in undergraduate mathematics, *Journal for Research in Mathematics Education, 18*(1), 15–24.

National Opinion Research Center. (1999). *Doctorate recipients from United States universities: Summary report.* Retrieved June 8, 2009, from http://www.nsf.gov/statistics/srs00410/htmstart.htm

Reyes, L. H., & Stanic, G. M. A. (1988). Race, sex, socioeconomic status and mathematics. *Journal for Research in Mathematics Education, 19*(1), 26–43.

Stanley, J., & Benbow, C. (1980). Sex differences in mathematical ability: Fact or artifact. *Science, 210,* 1262–1264.

Stage, F. K., & Maple, S. A. (1996). Incompatible goals: Narratives of graduate women in the mathematics pipeline. *American Educational Research Journal, 33*(1), 23–51.

Thurgood, L., Golladay, M. J., & Hill, S. T. (2006). *U.S. doctorates in the 20th century.* Retrieved June 8, 2009 from http://www.nsf.gov/statistics/nsf06319/pdf/nsf06319.pdf

Yap Y., & Orazem P. (2001). Differences in earnings and sectoral choices of doctorates in the United States, by sex and race. Retrieved November 16, 2008 from http://209.85.173.132/search?q=cache:83ww7JRs2JAJ:www.nber.org/~sewp/events/2004.05.28/Orazem_employed_PhDs.pdf+Yap+Y.+%26+Orazem+P.+,+(2001)+Differences+in+Earnings+and+Sectoral+Choices+of+Doctorates+in+the+United+States,+by+Sex+and+Race.&hl=en&ct=clnk&cd=1&gl=au

CHAPTER 18

RECOGNIZING GENDER IN MATHEMATICS RELATIONSHIPS

A Relational Counseling Approach Helps Teachers and Students Overcome Damaging Perceptions

Jillian M. Knowles
Northern Essex Community College, Haverhill, MA

Learning assistance centers where students come for help with mathematics, writing, and content tutoring are ubiquitous in U.S. colleges[1] and universities. In order to earn an associate's (2-year) or bachelor's (4-year) degree, students are generally required to take at least one mathematics course. I have found that the reason that up to 25% of students withdraw, or fail on their first attempt, is not only because of poor high school preparation. Through my experience as mathematics director of a learning center at a small state university and teaching at other colleges in the Northeast, I observed students' puzzling behaviors that seemed to be sabotaging their

International Perspectives on Gender and Mathematics Education, pages 421–450
Copyright © 2010 by Information Age Publishing

efforts. These behaviors affected even students who seemed adequately prepared mathematically for their course. I hypothesized that understanding students' relationship with themselves, the instructor, and mathematics might provide insight into the origin and development of these behaviors. To investigate my hypothesis, I piloted a brief relational mathematics counseling approach with the ten students in an introductory statistics class who signed up for individual mathematics counseling (Knowles, 2004). In my analysis of this new counseling approach, past and present gender stereotyping emerged as a substantial influence on the participants, on our interactions, and on their progress. This called for further investigation.

Questions that fuelled the current analysis of gender effects on mathematics teaching and learning were: How did the interaction of a college student's gender and the gender of the student's mathematics teachers past and present affect the student's approaches to the current mathematics course being taken? And what, if anything, did a mathematics tutor-counselor, or instructor, need to be aware of with respect to gender in order to be able to help the student succeed?

In my study, six of the ten students from an introductory statistics (STAT104) class of 13 who participated in mathematics counseling were women and four were men. Data were collected on students' background and progress throughout the course. In addition to that gathered during counseling sessions with participants, data were collected through survey and observation of the class as a whole. When I analyzed these data, a consistent relationship emerged between students' level of mathematics preparation in relation to course demands and their mathematics self-esteem, discussed in detail below. I specifically examined gender patterns that emerged in the way these students approached the course and related to me in counseling. I found that gender stereotyping in past mathematics relationships had negative effects on the progress of the mathematically underprepared students, which in turn negatively influenced the student–mathematics counselor relationship (Knowles, 2008). When I failed to address these gender issues or addressed them too late, progress of tutor-counseling was sabotaged, resulting in student withdrawal from the course or continuing poor performance. On the other hand, when I recognized and addressed these gendered relational patterns, progress was made and students succeeded. The effect of gender stereotyping on the well-prepared and adequately prepared students was also pertinent, not only to their relationship with me in counseling and with Ann, the instructor in class, but also to these students' future mathematical aspirations.

What emerged in this study has the potential to bring about change and recovery. When I, as a tutor-counselor, noticed and confronted gender labels associated with mathematical learning, students began to shed spurious gender barriers and grow, less fettered, into their individual potentials.

The way I faced my own gender biases in relation to the tutor-counseling of the students became crucial in this process. What also emerged was the very real effect of conscious and unconscious gender biases, originating from families and schooling, on students' mathematics success in college. This has implications not only for academic support professionals and college academics (faculty), but also for teachers and parents whose unexamined gender biases inevitably affect the mathematical learning progress of individuals in their care, both male and female.

MATHEMATICS RELATIONALITY AND CHANGE

As a framework for my study I needed a holistic theory to situate both mathematics learning and mathematics affect. So, recognizing the need to address root causes of mathematics affective problems, I turned to theorists of mathematics affect such as McLeod (1992). Because I believed that effects of unconscious motivations rooted in past experience were likely key, I looked closely at their endorsement of classical Freudian-type analysis and counseling approaches, albeit for cases of extreme mathematics emotionality (see McLeod, 1992, citing Tahta, 1993). Weyl-Kailey (1985) found a psychoanalytic approach to be necessary in understanding unconscious motivation behind pathological mathematical behaviors of her clients. Because my interest was in the mathematics mental health (Tobias, 1993) of ordinary students, not just those with extreme difficulties, I had earlier rejected psychoanalytic theory. However, psychoanalysis in its attention to the unconscious and effects of the past on present behavior gave insights that other approaches did not. Thus, I concluded, it should be pertinent for all students. Supporting this, Buxton (1991) suggested that some struggles of ordinary students with mathematics might well be understood in terms of Freud's concept of the superego (McLeod, 1997). Therefore, I resolved to explore Freud's theory and the relational theories that evolved from it. Tony Brown (2008) and his colleagues apply relational psychoanalytic theories to teaching, learning, and researching in mathematics education, lending further support to this admittedly unconventional approach to mathematics learning support.

Relational psychotherapies rest on the premise that repetitive relationship patterns derive from the human tendency to preserve the continuity, connections, and familiarity of a personal interactional world.[2] They recognize that the task of understanding the person and helping the student to disembed from counterproductive interactional patterns may be more complex and indirect than cognitive therapy concedes. However, like cognitive therapists and unlike classical Freudian psychoanalysts, relational theorists regard people as able to consciously choose to change their own patterns of thinking and behavior (Mitchell & Black, 1995).[3]

Mitchell (1988, 2000) uses a form of relational conflict psychotherapy derived from Freudian psychoanalysis. His approach helps ordinary adults who have goals that are not being fulfilled because they are embedded in conflicting relational patterns with themselves and their significant others. Mitchell's (1988) theory integrates the three major relational strands of psychotherapy that emerged from Freud's classical psychoanalysis. Each strand emphasized one dimension of what Mitchell terms a person's "relationality" or current behavior that the outcome of the development of self, external and internalized objects, and interpersonal attachments (Mitchell, 2000). When these dimensions were considered in the context of a student's mathematics learning experience, I interpreted them as follows:

1. Mathematics self or selves (using Kohut, 1977)
2. Internalized mathematics presences or objects (using Fairbairn, 1952)
3. Interpersonal mathematics relationships or attachments (using Bowlby, 1973)

I hypothesized that understanding a student's mathematics relational dimensions and the way they interacted with one another to express the student's relationality might provide insight into the origin and development of my students' puzzling behaviors, affect, and conflicts in doing mathematics.

Mathematics Self

Self psychology (Kohut, 1977; Mitchell, 1988) reexamines early self development in light of adult relational difficulties in order to identify current adult self needs. This perspective provides a way of understanding the mathematics self of an adult student, that is, the core of a student's mathematics identity. The mathematics self may be seen as part of a person's academic self, which is situated in the person's nuclear self. According to Kohut (1977), to develop a healthy self, the child must experience mirroring: unqualified recognition, delight, and admiration from a parent or primary caretaker—or in classroom terms, the teacher's firm belief in the student's growing mathematics ability. The child also needs the opportunity and the invitation to idealize and incorporate into self a (mathematics teacher) parent image,[4] which becomes the mathematical ideal and values for the self (i.e., the superego; Kohut, 1977). Further development depends on the (teacher) parent providing tolerable reality to modify the child's early natural grandiosity,[5] that is, feelings of mathematical omnipotence. This means that students must be then pushed to build more realistic growing selves as the teacher allows manageable frustration by not meeting all demands, and not teaching all the procedures. Thus students develop their own competence in finding mathematics solutions and realistic confidence that they can do so.

A student's mathematics self may be underdeveloped because of teacher/parent early and consistent neglect or over-indulgence. Another's mathematics self may have been developing appropriately until it was undermined by inappropriate placement or a negative experience or relationship. These students have likely developed defenses to protect themselves from further damage. Although a student's conscious goal is success in a mathematics course, the student may act in ways that jeopardize that goal—typically with underconfidence (related to depression and anxiety) or overconfidence (a typical expression of unhealthy grandiosity). Self-esteem may be low or compromised, so there may be little belief in that success. The unconscious goal of protecting the mathematics self may be in conflict with the conscious goal of success in the course (Knowles, 2004; Mitchell, 1988).

Internalized Mathematical Presences

Object relations theory focuses on the way early interpersonal relationships are internalized and inner images of the self and the other are formed, shaping perceptions and ongoing relationships with real and internalized others (Fairbairn, 1952). From this perspective, a student's internal reality is influenced by teacher presences internalized in the past that affect the mathematics self and the way current external reality is perceived. This is the context in which the mathematics instructor and course are found. Past internalized presences may be so prominent that they take precedence over current reality; the student might experience the present teacher as the teacher in the earlier negative classroom. Conflicts may arise when this mismatch between a student's internal and external reality negatively affects progress in the current course. If these conflicts are not recognized and resolved satisfactorily, the student's conscious desire to succeed or even survive in the course may be thwarted.

Interpersonal Mathematics Relationships: Teacher and Mathematics Attachments

Attachment theory offers further insight into the development, significance, and challenges of a student's interpersonal dimension of relationality. In particular, attachment theory examines the ways in which a person forms ongoing relationships with significant others and work (Bowlby, 1982). A student's tendency towards a problematic dependent, detached, or ambivalent relationship, or towards a healthy, self-reliant relationship provides clues to the security of early relationships and subsequent experiences of loss or change in those relationships. The extent to which a college student seeks help when needed from the instructor or learning assistance personnel has been found to be an important factor in student success (Zimmerman & Martinez-Pons, 1990). The student's established attachments and relational patterns might determine whether and how contact is likely to be made with those who could help. If a student has developed attachment patterns to mathematics teachers or tutors characterized by investing either too little or too much reliance in the teacher or tutor, self growth in mathematics learning might be compromised. The student needs to develop a healthier balance between taking responsibility for working and getting help. The student therefore faces a likely conflict between maintaining familiar but counterproductive attachment patterns and the risk of trusting a relative stranger for help.

BRIEF RELATIONAL MATHEMATICS COUNSELING PILOT STUDY

To conduct the pilot study, I was a participant-observer in every class of the 10-week, 5 hour per week course. I also conducted relational mathematics counseling sessions with 10 volunteer participants (for an average of 5 sessions each). I adapted and developed instruments and approaches to explore students' mathematics learning; their history, beliefs, attitudes, and emotions about mathematics learning; and their relational patterns as they participated in course. Some of these instruments—for example, *My Mathematics Feelings* (Ferguson, 1986, 1998; Richardson & Suinn, 1972; Rounds & Hendel, 1980)—were administered pre- and post-course to the whole class. Other instruments—for example, the *College Learning Metaphor, JMK Mathematics Affect Scales* (Knowles, 2004)—were administered during counseling sessions. Each session was audio-taped, and two classes were videotaped.

Much of my data were subjectively experienced. After meeting several times with each participant, I met with Dr. P., a psychological counselor familiar with my study, to discuss the participants and the approach I had used with each one. This, along with the data from semi-structured interviews with the course instructor, Ann, provided triangulation for my data. I analyzed the transference–countertransference interactions between me and the participant, classroom observations, mathematical products, assessment and survey data, and the transcripts of sessions. I used integrated cognitive-relational analysis that included mathematics educational analysis (e.g., of class tests, counseling session mathematics work, and assessments that are described below) and relational episodes analysis (Luborsky & Luborsky, 1995). Using these approaches I developed a case study of the class as a whole, in-depth case studies of three of the ten participants, and shorter case analyses of the other seven participants and the two non-participants who remained in the group beyond the first class (Knowles, 2004).

MATHEMATICS SELVES: MATHEMATICS PREPARATION INTERACTING WITH MATHEMATICS SELF-ESTEEM

When I analyzed the results of my study, I found a predictable association between the students' level of mathematics preparation with respect to the current course demands (see Table 18.1) and the state of their mathematics self-esteem, yielding three major categories of mathematics self.

There were three principal sources of data that led me to classify students according to their levels of mathematics preparation for the class (see Table): (1) students' class products including exams, (2) the *Algebra Test,*[6] which yielded algebraic variable developmental levels 1 (low) through 4 (high), and (3) the *JMK Arithmetic for Statistics Assessment* (Knowles, 2004), which examines small and large number and operation sense and numerical problem solving. As can be seen from Table 18.1, the data demonstrated

TABLE 18.1 Criteria for Determining Level of Mathematical Preparedness of STAT104 Participants

	Well Prepared	Adequately Prepared	Underprepared
Course grades:			
Examination#1	B+ through A	D through A	F through C–
Final Course Grade	A– (B+) through A	B– through A–	F, AF, WP through B
Algebra:			
Algebra Test (Sokolowski, 1997) and class, exam and counseling session work	Level 4+	Level 4	Level 1 or 2
Arithmetic:			
JMK Arithmetic for Statistics Assessment and class, exam, counseling session work	Good ($\geq 85\%$)in all 8 categories	Adequate or above ($\geq 70\%$) in all but one or two number or operation sense categories; variable in other sections	Ranges from adequate or above ($\geq 70\%$) on at most 6 categories to inadequate ($<50\%$) or marginal ($50\% \leq$ x $\leq 59\%$) on three or more categories

that a student's grade in the first examination could not reliably be taken alone as indicative of the level of preparation for the course.

Evidence of students' sound, compromised, or low mathematics self-esteem came from both relational (e.g., Mathematics *Metaphor, JMK Mathematics Affect Scales,* transference–countertransference analysis) and affect survey assessments, and from observations. Kohut (1977) described a process whereby a person's internal self-structure was consolidated to "a storehouse of [realistic] self-confidence and basic self-esteem that sustains a person throughout life" (p. 188, footnote 8). Mathematics self-esteem is defined here as the mathematics-specific result of this process. Participants' expressed confidence levels did not generally correlate with their levels of self-esteem. For example, students with low self-esteem generally expressed either underconfidence or overconfidence, while those with sound self-esteem were realistically confident.

Students' mathematics self-esteem and their preparation level were associated as follows: Two female students, one a non-participant, had sound mathematics selves stemming from good preparation and sound mathematics self-esteem. Five students, three female and two male, had undermined mathematics selves: they were adequately prepared but their mathematics self-esteem had been compromised. Five students, two female and three male, had underdeveloped mathematics selves, characterized by underpreparation and low mathematics self-esteem (Knowles, 2004). In the first two groups, noteworthy gender-related effects emerged, but it was in the last group that I identified the most distinct gender-related patterns linked with past experiences that negatively influenced present mathematics learning (Knowles, 2008).

GENDER IN MATHEMATICS RELATIONALITY

My search for a theoretical framework for understanding gender in mathematics relationships led first to the work of educators like Sadker and Sadker (1994, 2002) and Boaler (1997), and social psychologist Dweck (2000, 2006). However, psychotherapist and sociologist Chodorow (1978, 1999) provided the relational psychotherapy framework for understanding gender in order to analyze gender interactions between student, instructor, and tutor-counselor in this study.

Chodorow (1999) noted, "Each person's sense of gender is an individual creation.... Each person's gender identity is ... virtually a fusion of personal and cultural meaning" (pp. 69–70). Gender stereotyping related to mathematics abilities continues to be ubiquitous in U.S. culture. However, the effect of this stereotyping on each individual is dependent on interactions of personal identity with family, school, and community. Thus, I had to reject the temptation to universalize or essentialize the effect of gender stereotyping on mathematics students (cf. Thorne, 2002). A difficulty I face here in discussing gender—mine, participants', and that of influential figures in their mathematical development—lies in how to discuss each individual's *unique* femininities or masculinities, while still identifying gender-related *patterns* and devising *generalizable* change strategies. Chodorow suggested a solution that she attributes to Fast (1984, p. 77): one must distinguish between "subjective and observed gender." This "enables us to sort out the relation between individual uniqueness on the one hand [subjective gender] and commonality, generalization, or universality on the other [observed gender] that may be reported as statistical trends" (Chodorow, 1999, p. 103).

Identifying gender differences in children's paths of self-development can contribute to understanding how the type and pathology of mathematics self-development and instructor attachments differed for boys and girls. In U.S. society in the 1970s, the majority of women stayed home to raise their children, while their husbands worked outside the home (Chodorow, 1978). Despite the fact that American men now spend twice as much time in childcare as they did in the 1960s (Sullivan & Coltrane, 2008, citing Fisher, Egerton, Gershuny, & Robinson, 2006), women remain the principal childcare providers of young children. Even within the U.S. classroom, children are still primarily socialized by female teachers.[7] Consequently, it is likely that current U.S. college students are most likely to have received their early "mothering" from women. Thus, it still remains that boys must differentiate themselves from their mothers or female teachers in order to achieve "maleness," while girls' achievement of a female identity allows them to continue to identify with their female parent or teacher figures. This process results in girls and women who "depend" and boys and men who "do" and who defend against their desire to depend on a female figure (Chodorow, 1978).

Although holding the stereotypical belief that "males are better at mathematics than females" is no longer socially acceptable in the U.S., the belief is pervasive (Leedy, Lalonde, & Runk, 2003), even among teachers who consciously repudiate such a claim (Sadker & Sadker, 1994). U.S. female elementary teachers continue to be socialized to believe that their mathematical ability is lower than males' and that their efforts are unlikely to improve their achievement levels. As girls' developing identities do not require the sharp differentiation from mothers or female teachers that do boys', they are likely to internalize this belief without protest as their teachers mirror low expectations of their mathematics ability. On the other hand, boys, whose behavior and lack of effort are often criticized by the teacher, are likely to be assumed to have "natural" mathematics ability. Thus boys who are striving to differentiate themselves from their female teachers may internalize a "capable but lazy" mathematics self as part of that effort. The concomitant belief that "people [men] who are smart at mathematics don't have to work at it" prevails in the U.S. This belief leads to male-identified boys learning not to study mathematics in order to maintain their "capable at mathematics" status, and to female-identified girls studying, while achieving poor or adequate or even good grades and maintaining their "less-capable" status.

Herbert and Stipek (2005) reported that from the beginning of the third grade, girls rated their mathematics abilities lower than did boys, even though there was no difference in achievement. Parents also rated boys' mathematics abilities higher than girls' in both the third and fifth grades. When students incorporate the beliefs of their parents and teachers that their mathematics ability was a fixed entity, inevitable because of their gender, they tend to avoid challenge, treat poor grades as judgment of their ability, and limit the possibility of growth (Dweck, 2006). It is not only those who believed they were "bad" at mathematics who were affected in this way; students who believe they were "good" at mathematics are also limited by these fixed mindsets.

COUNSELING AND THE TUTOR–STUDENT DYAD

It was only as the study proceeded that I began to understand the implications of a relational psychoanalytic approach. I recognized that it was not only about the students' need to see and change. It was probable that we both would have to change. My puzzling behaviors, affect, and conflicts also had to come under scrutiny. The tutor–student relationship, rather than the student, had to become a key focus. Thus, perhaps the most significant difference between my previous traditional tutoring practice and relational mathematics counseling involved my change in focus from the student to the tutor–student dyad. Since our dyad was in the context of the STAT104 class, the principle players and foci of our interactions included the course material, the classroom environment, and our individual relationships with Ann as the instructor.

Key to this new focus on the dyad was the need to be aware of the unconscious subtexts of our relationship—the transference of our past relationships into the present one. In the past, when I had tutored students, under pressure of time I took what the students told me about themselves at surface value, and attended more closely to the course content and the mathematics course products (e.g., quizzes, tests). Now, as a counselor, I began to listen not just to what the student said, but to what was behind the statements and behavior. I became attuned to the student's transference, the role the student cast me into, and to my countertransference, my impulse to act out or react against the role imposed on me by the participant's transference.

This new role of counselor alerted me of the effect of gender stereotyping on participants, on me, and on our interactions. I was forced to become aware of my own gendered preferences, reactions, and biases—previously largely unexamined—based in my family and schooling history and my current status. Through outside counseling, I was able to understand the effect of my own family and community dynamics on my beliefs about gender and mathematics. I realized that I had myself struggled under gender stereotyping. As a cute petite blond in Australia in 1960s, I pursued mathematics in an effort to prove to the boys that I was capable, to counteract pervasive stereotypes. As I considered this, I became aware that I reacted quickly and judgmentally to stereotypically "feminine" students and to "masculine" grandiosity. I realized that growth in self-awareness could be crucial to the development and success of the student–counselor relationship.

To investigate further the unconscious mindset at work behind each student's behaviors and statements toward mathematics, I asked each student to create a mathematics metaphor. Participants were asked to compose a list of metaphors such as weather, color, or fictional character describing their view of and relationship with mathematics. They were then required to expound upon one and explain how they felt the metaphor related to their mathematics performance. Interrogating metaphors together helped us better understand their underlying motivations.

THE STUDENTS: RELATIONAL COUNSELING AND GENDER

All students in the STAT104 class had completed at least a year in college prior to the course. This was Mulder's and Robina's first college mathematics course. Of those who had previously taken college mathematics courses, only Autumn, Catherine, and Lee had been successful; the rest had either failed or earned Ds. Of these Jamie, Rod, Karen, and Mitch were repeating STAT104. Space precludes detailed discussion of each participant in the study so I have chosen to focus on the one participant of the two students in the class with sound mathematics selves (Robina (f)) and only one of the five participants with undermined mathematics selves (Jamie (f)), so that I can discuss all four of the participants with underdeveloped mathematics selves, the ones for whom gender effects had most current toxic affect on their mathematics.

Students with Sound Mathematics Selves

The two students with sound mathematics selves had good mathematics preparation for the course (see Table 18.1) and sound self-esteem. Both were women in their forties. One of these, Robina, signed up for tutor-counseling.

Robina (f)

Robina was a registered nurse (RN) returning to college to complete her bachelor's degrees (BSN). In our experience, RN to BSN students were typically older women, with distant and poor mathematics backgrounds, who struggled with STAT104. Consequently, the course instructor, Ann, and I expected her to have trouble and probably do poorly. Robina's "feminine" classroom behaviors—her almost constant frown of puzzlement, flustered air, and practice of asking questions and checking her understanding whenever she was uncertain—reinforced our initial judgment.

Getting to know Robina in the counseling setting dramatically changed my initial perceptions. I observed Robina's competence and her realistic confidence, given the appropriate effort she expended (10 hours of homework per week), in approaching her first formal mathematics course in 25 years. Her metaphors for mathematics learning were positive: white, owl, sun, and Belle. The metaphor Robina most identified with was Belle from Walt Disney's animated movie *Beauty and the Beast* (1991). Robina described Belle as follows:

> **Ro:** She had a quality that, you know, for a female cartoon character, fictional character, you know, she was surprisingly grounded and bright and she took things seriously. She was always very clear about who she was and what she had no time for and what she was interested in. She is a character that, to me, relates to math.

Robina's mother had insisted that she was "just like" her grandmother, who was known in the family to be good at mathematics. Robina did find herself to be good at mathematics, yet had difficulty in the Roman Catholic elementary school she attended. She would often know the answer to the mathematics problem but had trouble explaining and writing out her steps in the solution, so the nuns had marked her down or wrong. Like Ann and I, her earlier teachers probably also experienced Robina's "feminine" flustered classroom persona in a way that fueled their judgment of her mathematical products as lacking. Despite this, Robina was resilient mathematically and developed sound mathematics achievement and self-esteem.

Interestingly, Ann and I initially expected Rod (see below), another older student in the same degree program as Robina, to do well, perhaps because he presented as a confident and relaxed male. It turned out that our initial perceptions of both were wrong. Robina went on to earn A–s while Rod struggled to earn Ds and C–s. Robina asked "on task" questions and answered correctly approximately twice as often as she questioned or answered incorrectly or was off task, just as Rod did. Yet Robina gave us the impression of "feminine" incompetence, while Rod gave off an air of competence. Robina seemed to be exhibiting the often-noted tendency of women, compared to men, to be considerably more tentative about what they know, particularly in a "masculine" domain such as a college mathematics classroom (Johnson, Funk, & Clay-Warner, 1998; Palomares, 2008).

As the class proceeded, Robina's good grades slightly modified Ann's initial perceptions of her as a struggling nursing student. However, her in-class persona prevailed. Ann was never convinced that Robina was mathematically sound. To the end, Ann remained convinced that Robina and Rod "didn't know what they were doing" in the problem-working classroom sessions. She also noted that Robina was always the last to finish (Interview 3, August 3).

Robina expressed uncertainty about her ability to do "harder" mathematics, the research statistics that she would need for her graduate degree. Through our counseling sessions, we examined together and detoxified the sources of that uncertainty. We observed her present appropriate effort and sound achievement, allowing her to become more realistically confident of her mathematical future.

Students with Undermined Mathematics Selves

Three women and two men, all participants, were adequately prepared mathematically for the class (see Table 18.1), but their mathematics self-esteem had been compromised. They had undermined mathematics selves.

Jamie (f)

Relationship patterns based on a student's internalized teacher presences of the past may pull or push the tutor into assuming the teacher role demanded. Jamie's shy personality and transference of past mathematics teachers onto Ann and me placed us as dangerous (c.f. "shark" in her metaphor list or "storm"). Any direct attention elicited extreme discomfort as evidenced by her heightened color and physical retraction. Ann, experiencing Jamie as "fragile," left her alone, as Jamie wished. But Jamie wanted to do better this time, since she had received a D in her previous attempt at STAT104.

Analyzing Jamie's metaphor for mathematics as a "storm" during which she would "hide," it became clear to me that Jamie was an adequately prepared mathematics student with an undermined mathematics self. History taking revealed that she had "bad" teacher internalized presences. There had been a fifth grade teacher who "yelled" and led shy Jamie to hide in class; Jamie put it this way: "You sit down and shut up so you don't bother her." There had also been a high school teacher whom Jamie perceived to be a good Algebra I teacher, yet this teacher had become unapproachable in Pre-calculus. These experiences damaged Jamie's teacher and mathematics attachments. Jamie's "bad" teacher internalized presences were evidenced in the way she avoided contact with Ann and me despite evidence that we were not dangerous to her. She "hid" in class, making no eye contact, succeeding to the point that Ann and I had trouble recalling if she had been present, even though the class met around a large table. She signed up for counseling, but she made no attempt to meet with me. I had to overcome my countertransference reaction to leave her alone in order to give her the help she both asked for and ran away from.

Although Jamie had experienced success doing mathematics (through 4th grade, and again in Algebra I), her confidence had been further undermined by her mother's belief that the women in the family were "not good" at mathematics. She reported, "My dad, he's very good at math. My mom always said that, unfortunately, me and my sister got our math genes from her, 'cause she's not good" (Session 3). I questioned her mother's theory and reminded Jamie that we were gathering evidence that refuted that claim.

I helped Jamie become conscious of her adequate mathematics competence (e.g., Level 4 on the *Algebra Test*) and the relational roots of her anxieties and avoidance behavior that followed her into the current classroom. I gave her relational assignments: Jamie was to ask Ann a question during class and she was to initiate the next appointment with me. She did both. I also tutored her in the mathematics. Through my outreach, she accessed help and reattached to mathematics and mathematics teachers. Her anxieties abated somewhat and she earned a B+ in the course. To her, the mathematics classroom was no longer a "storm," but "a partly sunny day . . . I can go out in it . . . if I have my umbrella."

Students with undermined mathematics selves were all either anxious or depressed or both, despite adequate mathematics preparation for the course. However, unlike the females, the two males in the group did not express this overtly. Both males had active but ineffective approaches to coping with their compromised self-esteem, and both resisted relational exploration and my strategic advice until it was almost too late. The three females showed no such resistance, although each of them had damaged relationships with female teachers that needed repair. For Mitch, depression seemed more prominent than his anxiety. His mathematics metaphor of himself fleeing Inspector Javért (Boublil & Schönberg, 1980) aptly captured his feeling that mathematics was hounding him through the years. He rigidly continued to do just what he had done in his last attempt at the course, regardless of the fact that circumstances and I indicated the need to adjust; he finally believed the evidence and "heard" my increasingly louder urgings after Exam #3. Pierre clung to doing 17 hours of homework a week, covering much more than was being examined, despite this approach yielding only D's and C–'s. When I suggested he focus, he ignored me. It was only when I forcefully confronted him with the inevitable consequences of continuing in this manner that he a changed his approach. Pierre successfully ended the class with a B–. Mitch continued to defend against my push to change his approach. However, now knowing that he had a sound sense of the algebraic variable (Level 4 on the *Algebra Test*—see Table 18.1), doing just enough work and taking my strategic advice just in time were sufficient for him to replace his previous F with a B.

Students with Underdeveloped Mathematics Selves

Two females and three males in the class were mathematically underprepared (see Table 18.1) and had low self-esteem. Their mathematics selves were underdeveloped. All but Frank signed up to participate in the study. Here are the stories of the participants with underdeveloped mathematics selves.

Karen (f)

Karen was a tall, blond 22-year-old woman, a psychology major and the first in her family to attend college. She was an assistant teacher of developmentally challenged children, a girls' basketball coach, and she drank beer (Class 5). She was not stereotypically "feminine." Karen's metaphors for mathematics were: "black," "stormy," "cloudy," and "bear." She highlighted "cloudy," "because there are some aspects of math that are more clear to me, but mostly math is my worst subject and has *always* been hard for me to understand." Ann's, and my initial, experience of Karen, as she defensively held us at arm's length, suggested that her demeanor may have become a factor inhibiting even well-intentioned teachers from reaching out to her. They, and we, expected her to continue to do poorly. Karen strongly disagreed with the statement "Mathematics is a subject men do better at than women."

However, she restricted herself to solely doing the work while not expecting to understand the mathematics, not because she was a non-mathematical girl, but more probably because she had been perceived as one. It seemed that she had been seen and treated by her teachers as a "typical girl" who could not understand mathematics even when she tried. The effect was poor achievement and underconfidence in mathematics, accompanied by negativity and depression because of low teacher expectations.

I began to suspect that Karen failed to develop a secure attachment to any mathematics teacher. According to Bowlby (1982), *secure* attachment is achieved when caregiving by the mother figure is characterized by being sufficiently available and responsive (Bowlby, 1982). The mother figure, and by analogy the teacher, becomes the *secure base* from which the child can move out and explore the world, but return for comfort and reassurance in times of distress. It seemed that no mathematics teacher had offered to be a secure base with whom Karen felt safe to connect. She had learned to care defensively for herself and expected little from the teacher. Such low expectations seemed to have made her angry with the neglectful female teacher/s who failed adequately to mirror her growing mathematics self. She became anxious and depressed because she knew she did not have what was necessary to do it on her own. Now, in her STAT104 course, she needed support, yet she was suspicious of her current female instructors. She had previously failed STAT104, and her need to succeed this time drove her to sign up for mathematics counseling. My task was to mirror Karen's real competencies and challenge her unrealistic negative thinking and the repetition of failing behaviors. I needed to challenge her attribution of failure to lack of ability, and help her to attach herself to mathematics and to me and Ann as secure base mathematics teachers.

During the session before Examination #3 we worked on practice problems in parallel, as usual, and I noticed her impressive grasp of the material. At first Karen demurred, citing external factors such as my help and "cheating from the book" as giving that appearance. But taking the examination, she was sure, would not be the same.

> **K:** See this is going to be my problem. This was already done you know what I mean? I cheated in the book. No, well I just looked at this [decision chart formula sheet she was creating that she could use during the exam] actually.
>
> **JK:** … you'll have this on the test [examination]. Right? And you've become aware of how you might be tempted to choose one [statistical test] rather than another! Right? So I think you probably won't [choose the wrong test], right?
>
> **K:** 'Cause I could probably get half way through the problem and realize that it wasn't right.
>
> **JK:** I would think so!

Karen went on. She realized she had now internalized the material and ways of troubleshooting on the examination if she got into trouble. She was going into Examination #3 a very different person from the one going into Examination #1, for which she earned a D-. She was strategically prepared, adequately supported by her formula sheet, and now newly aware of her own grasp of the material and ability to monitor and troubleshoot if things went wrong. Karen earned an 85%, losing only two points on the computational section.

Mulder (m)

Mulder was a short, athletic (three sports a year in high school, Session 1), strongly masculine-identified biology major. Both of his parents had attended college, although his mother did not complete a degree. Ann thought Mulder seemed "smart" and "on the ball [with respect to] his research experience"—but not likely to put in the effort needed to succeed in the class. Mulder's metaphors gave clues to his central conflict that I did not pick up at first: "For me math is like Mulder searching for aliens. I am searching why I make math so difficult for myself." He was referring to Fox Mulder from *The X-Files* (Carter, 1993). This science fiction television series featured FBI paranormal detective Fox Mulder (and his partner Scully) who was searching for the aliens he believed had abducted his sister.

Mulder's identification with the male mathematics stereotype was explicit: "You know my uncle's getting his masters; he's a great student. My dad's really smart so it's kind of like a thing I know I have. It's just a matter of applying it and taking advantage of it" (Session 3). But Mulder did not seem to have learned how to apply it or take advantage of it, probably because he wasn't expected to—he was just expected to "get" it because he was male. Like Karen, his mathematics competence had not developed appropriately (both Level 2 on the *Algebra Test*—see Table 18.1) but, as a male, he was not free to admit it. The effect was understandable: poor achievement, resistance, and covert effort, covered by overconfident grandiose insistence that if he (overtly) tried he would be fine. Mulder's attachment style was defensive—defense of his masculine self against the shame of not being capable, and against dependence on me, his female tutor. Our relationship was often combative. He sought help but deflected it with overconfidence and resistance.

I, like Ann, had expected this "smart but lazy" male to do better than a D– on Examination#1. But when he did not, part of me wanted to put down his inflated male ego to prove my female superiority. I was not attending to the very real mathematical limitations revealed by the D–, that were being defended by his masculine grandiosity. He needed me to change before he could; we were unlikely to resolve his conflict unless I did. He was certainly not interested in addressing his issues seriously while I was trying hard to change him.

Mulder did poorly on the conceptual part of Examination #2 and again on Examination #3. When I met with Dr. P, my psychological counseling consultant, to review progress with my participants, he confirmed that my direct "trying to change Mulder" approach was not only unsuccessful, it was likely contributing to his resistance. Dr. P suggested a strategy developed by Adler (Mosak, 1995) that had been named *paradoxical intention* by Frankl (1963). The theory suggested that "the symptomatic patient unwittingly reinforces symptoms by fighting them . . . to halt this fight, the patient is instructed to intend and even increase that which he or she is fighting against" (Mosak, 1995, p. 74). At the next session (Session 5), I suggested that Mulder's own search for "why I make mathematics so difficult for myself" might entail exploring his underlying and seemingly growing resistance to mastering the conceptual part of the course. I proposed that Mulder take a mini-test of conceptual multiple choice questions (typical of the conceptual portion of the examination questions that he was failing) from the textbook's study guide. While taking the test, I suggested that he should strongly resist answering these conceptual multiple choice questions while talking aloud about his resistance. He seemed intrigued. As he worked through the test, I commented on his process. Mulder's narration alternated between the illogical "eeny-meeny-miny-mo" answer choice, expletives, and logical thinking. At one point I commented:

> **JK:** . . . every now and then you stand up against your own resistance . . . first of all, you say, "I don't want to do this, I don't want to come back here", "She's [JK? or Ann?] a pain in the neck." . . . but then [you see] what happens when you try them [the questions].

After this interchange I felt a distinct change in Mulder's demeanor, a turning point. He began to ask "why" questions. Mulder became actively engaged in the discussion of my responses. I commented again:

> **JK:** There's a part of you that is really listening and engaging in it and there's another part that's like "uugh" [pushing away with my hand]. It's almost like you've got this little battle going on [Mulder chuckles]. You don't think that's happening?"

I suggested, "It's like a cooperative part that does this [engaging with the problems] and a resistant part [that resists engaging at all]." He seemed intrigued with the concept of his two conflicting parts doing battle. The focus of his grumbling began to change. It became less about her [Ann] making him do the conceptual work, and more about its cognitive demands, which, he discovered, he could master when he put the effort in.

At the end of our session, Mulder announced to his finite mathematics instructor who had just walked in, "I have a resistant side of me." To her dismissive, "Don't we all?" he insisted, "I lost 18 points on the multiple-choice." This was the first time that Mulder verbalized the theory of his two battling sides and owned it. It felt good to pull myself out of the fight and let Mulder battle himself. He did this successfully. That evening he earned a B+ on the "conceptual" part of Examination #4 and went on to earn a B in the course.

Kelly (f)

Kelly was a 19-year-old feminine-identified woman who had a history of poor mathematics achievement. Her classroom presence was not unlike Robina's. She needed to do well in STAT104 in order to pursue a social work degree. She had deficits in number and operation sense, and understanding the algebraic variable (see Table 18.1). She expressed high levels of anxiety on all scales, and had a learned helpless orientation to mathematics learning. She and her mother called both Ann and me on several occasions to discuss whether Kelly should stay in class, and to posit a learning disability as a reason for Kelly's poor performance. Kelly was blaming her mathematics difficulties on something she felt was out of her control (a learning disability). Her "sudden storm" metaphor for mathematics and her stereotypically "feminine" helpless dependence on Ann and me, along with her mother's inappropriately active role in her adult daughter's mathematics course, all pointed to Kelly seeing her problems with mathematics and their solution to be external to herself.

> **Ke:** ... a stormy day best describes my view of mathematics. Um, a stormy day comes upon us suddenly. We feel as though we understand the weather, and then suddenly, it becomes all stormy. In math I always feel confident in the beginning of the problem, of a class, and then suddenly I'm lost, kind of like the storm just arrives.

I was not able to help Kelly avoid another failing experience. She, like Karen, had earned a D– on Examination #1, but her main difficulty seemed to be more a lack of strategic practice and failure of nerve than an inability to do the mathematics required. I should have mirrored her competence by reference to where she had done well in Examination #1 in order to help her reattach to her own ability to do the work, but instead I re-taught and coached her on course content practice problems. The storm was indeed more internal than I realized. I needed to help Kelly see that storms were inevitable and help her devise ways to take back control and weather them rather than remain lost and helpless. She had frantically externalized responsibility for her outcomes to me and Ann. However, what she needed from me was validation of the storm experience and also mirroring of an increasingly competent mathematics self. With this self-understanding, strategic preparation, and a strategic formula sheet, she *could* have handled the storm herself. Kelly dropped the course just before the second examination and earned an AF (see Table 18.1).

Rod (m)

Rod was a tall, athletic nurse in his forties returning to study, as noted above, to gain a bachelor's degree. Rod seemed confident and competent in the lecture portion of the class, but in the problem working portion of the class, he struggled with substituting into the formulas and with order of operations. I bristled at his we're-the-adults-here way of relating to Ann and me in class. Rod had tried STAT104 once and failed it, a fact that he inadvertently let slip in study group. He had, in contrast, written on his initial survey that he expected an A in this class and had earned an A in his last mathematics class, Algebra. Although Rod's grandiosity was more extreme and unrealistic than Mulder's, it seemed to have stemmed from a similar source—his underdeveloped mathematics self. It brought forth a similar but more extreme countertransference reaction in me. He wrote on his metaphor survey:

> **R:** Sherlock Holmes reminds me of how I feel about math.
> Both have a problem to be solved with several steps in-
> volved... with an answer, very rigid, with many rules. There's
> usually only one correct answer. Math is time-consuming with
> many rules. If you allow enough time to learn, one can do well.

Yet Rod seemed ambivalent about doing that himself, or at least to admitting to trying it. He was surrounded by women at work, and had a woman as his superior. He reported that his motivation for doing this class was to get a degree that would allow him to change to a more male connected position such as nurse, anesthesiologist, or physician's assistant, where he said he would be more likely to be working with male surgeons or doctors. He seemed to have an underlying fear of not being capable of doing the mathematics that he, as a male, "should be able to do," as well as a desperate need to be able to do it to escape the female-dominated world he was in. I listened to what I heard as his shifting stories and grandiose attempts to cover up and excuse past mathematics struggles in high school:

> **R:** Yeah. Yeah. Let's see, it was—basically, I didn't put
> enough—I didn't do well in math a long time—I did—
> I took—Geometry was the last math I took... Algebra I,
> Algebra II and Geometry... But, um. And I got 'B's in all of
> them, but I never spent any time.

And in college in his first attempt at STAT104:

> **R:** A lot of the stuff on the thing [formula sheet] was use-
> less. I got like a 35 [%] on the test... I just dropped it [the
> course]. Yeah, I was like, I'm not interested... I don't need
> it... to be honest with you, if I want a C, then I don't need
> to be here.

Rod was underprepared mathematically for the class with low mathematics self-esteem and vulnerable masculinity. He was taking a risk enrolling again, and in signing up with me. Instead of recognizing this, I reacted to his "lies," blew his cover, and scolded and pushed him as I sensed his former female teachers had done. His transference of past female mathematics teachers placed me as threatening to his vulnerable mathematics and masculine self, so he defensively tried to keep me from seeing and tackling his mathematical difficulties with him. From the outset, I saw and reacted to this unrealistic grandiosity and overconfidence. In my countertransference I was indeed threatening to him. Unlike Mulder, Rod did not stand up to me but oscillated between avoidance, mild protest, and non-strategic effort in a way that did not achieve any more than marginal results.

Before our last session, I met with Dr. P. and confessed my strong negative reaction to Rod's "lying" grandiosity. Dr. P. suggested a way to repair the relationship. In Session 4, I apologized, and repair began as Rod continued to defend himself: "I could do well . . . if I practiced each problem two or three times . . . but it is summer . . . I am not willing to give it the time" and "I don't need this"— even though STAT104 was a required course for his degree. Nevertheless, he tried more as I backed off and stopped scolding. He did better on the exam the next night (improving from a 59% to a 72%), but not the big improvement he expected. A family crisis intervened, and he withdrew from the class without penalty (WP—see Table 18.1). I was unable to extricate myself in time from my defensive need to put this man in his place in order to help him.

Of all the students in the class, the mathematically underprepared group had the clearest gender-related patterns that negatively affected how they had developed mathematically and how they currently related mathematically. The men expressed grandiosity and overconfidence with respect to their mathematics ability, and they resisted female help. The women were depressed or anxious, and both were underconfident. They were either detached from or overly dependent on female help. It was with this group that my own gender biases most challenged my ability to help, and I was forced to become aware of my own gender and its effects on my relationships with these students, particularly the men.

STUDENTS' ATTACHMENT TO MATHEMATICS

It is not only student–teacher attachments that are affected by the way the teacher manages the learning environment. Student–mathematics attachments are also affected. U.S. elementary teachers are likely to lack a secure base[8] in the arithmetic they teach (Ma, 1999), and those with insecure attachments are less able to provide secure attachments to those in their care (Ainsworth, 1989; Bowlby, 1980). Karen's arithmetical difficulties and her poor number and operation sense indicated that she had not developed secure attachment to arithmetic—neither had Kelly or Rod. Karen and Rod's shame in not knowing "children's mathematics" made it almost impossible for them to consciously admit it and to accept my offer of extra tutoring. Mulder's relatively sound arithmetic indicated that he may have developed a relatively secure attachment to mathematics in elementary school that seemed to have been disrupted when his transition to algebra faltered. Both Karen and Mulder had negative reactions to Ann's expectations that they work on problems in class without seeing them previously demonstrated; they instead expressed approval of teachers who had demonstrated problems in class before students tried them, even though neither of them had succeeded under this teaching approach. It seemed that they were accustomed to the traditional U.S. high school approach of teacher lecture and demonstration of procedures, followed by student practice. They had thus developed insecure attachments to high school mathematics (level 2 on the *Algebra Test*) and a resulting unhealthy dependence on the teacher's mathematical authority (Skemp, 1987). I helped them see that Ann's approach, though uncomfortable for them, forced them to make mathematical decisions on their own and grow mathematically. To help reattach these students to mathematics, I suggested we analyze together their exams, looking for error patterns and evidence of sound understanding. I noticed that all the males in the study tended to resist and defend against this analysis, possibly because it meant that I, a woman, would clearly see their mathematics frailties. On the other hand, the women showed no hesitation to our doing exam error analysis together. However, they were generally too accepting of their errors as defining them. I had to work hard at pointing out growing evidence of their reattachment to mathematics and I had to challenge out loud any continued negativity.

CONSEQUENCES OF UNEXAMINED PRACTICE: FAMILIES AND SCHOOLS

This study calls attention to the effect of gender biases of families, students, and teachers on mathematics learning success. We can see implications not only for academic support professionals and faculty in colleges, but also for families and early teachers whose unexamined gender biases affect the mathematical learning progress of people in their care.

Families

One student from each category in the study revealed an explicit family gender-related math ability theory. Robina's family theory had proved positive for her. She was like her Grammie, who was "always doing math," and "if my mother said it, it must be true . . . well, I think I WAS good at math." It was not clear to me whether her family's assessment that Robina was like her grandmother was the result of early displayed mathematical abilities or because she was like her grandmother in other ways, looks perhaps, or mannerisms, and was thus assumed to be good at math by association. Whichever was the case, this family math theory seemed to have helped insulate her considerably from the negative effects of assaults from teachers who disapproved of her way of doing mathematics. From early on, the societal "males are better at math than females" did not apply to Robina, nor did the "I'm not a math person" classification. My visceral reaction was to accept and approve of this positive math story with its positive effects. However, there remains the disquieting knowledge that no family math gene theory, even a positive one, is, in itself, a realistic basis for the growth of a sound mathematics self.

Jamie, a student with an undermined mathematics self, expressed an explicit negative family mathematics ability theory. Jamie was told she had inherited her mother's female "not good" at mathematics gene. My response was immediate and strongly reactive:

> **JK:** Tell her she's wrong! . . . That's not true. It's nothing to do with your math genes. Tell her it's not that. I'll have to get her in here!

Jamie responded with an appreciative, "Yeah!" as we both laughed. But what I should have done instead was ask, "Jamie what do you believe? What is your understanding?" It would have been better for me to give her the opportunity to realize her own growing sense that maybe she was not "not good" at mathematics after all. The theory, with its implied acceptance and expectation of Jamie's poor math achievement, may have perpetuated her feeling of math class being beyond her control, that "the storm that will come back" even if she performs well on an exam or a course. In counseling she was beginning to discover that she could take control of outcomes and that she had what was needed mathematically to do well—being female (something she had no control over) was not a factor of any consequence in this.

Mulder's positive family theory did not yield positive results. Instead, believing that he would be good at mathematics like his father and uncle, the males in the family, continued to perpetuate his unsuccessful resistant attitude. There were several essential differences between Robina's and Mulder's positive family-related mathematics ability theories. Robina's had been explicit and known to her from childhood; Mulder's seemed to have been implicit until he voiced it to me. Robina experienced her family mirroring her mathematical potential and she grew into that, doing mathematics and succeeding at it. Mulder had observed male relatives to be "smart" but he reported no family recognition of his actual mathematical prowess, so his remained hypothetical.

Schooling

The struggle to adequately develop a cohesive sense of self and relationship is more likely to affect people "for whom stereotypical gender experience plays a part in the defens[es] . . . they adopt" (Carper, 2001, p.12). Cheng (2005) reported that gender-stereotyped Chinese (Hong Kong) university students faced with a stressful situation were less able to cope. These students chose a stereotypically gendered response regardless of whether the strategy was appropriate to the situation. The stereotypical "masculine" choice was direct action, whereas the "feminine" choice was acceptance and social support. Non-gender-stereotyped students were able to choose a strategy that was appropriate to the situation, be it a "feminine" or "masculine" response, regardless of their gender. It seems likely that stereotypically feminine or masculine elementary students are more likely to experience teacher stereotyping in the mathematics classroom and to choose the stereotypical responses. They are consequently more likely to have underdeveloped mathematics selves as adults than non-stereotypical students.

If that had indeed happened for students in my study, I had expected that mathematically underprepared males would probably be strongly masculine-identified and females, strongly feminine-identified. While this was true for the males, it was true for only one of the underprepared females, Kelly. With each, the evidence I have presented above does indeed indicate that all but Karen had seen themselves to be and all including Karen had been treated as stereotypical boys or girls in elementary (and later) mathematics classrooms, with the negative consequences of underdevelopment of their mathematics and low mathematics self esteem. My struggles to recognize and overcome my own gender biases in order to help them illustrate the severity of their plight. The experience of Robina illustrated that even for the mathematically well-prepared, realistically confident female, stereotypically feminine behaviors may jeopardize recognition of competence, and thus confidence to continue mathematics education that was otherwise clearly within reach.

Karen, who persisted and did well with counseling support, was the only one in the underprepared category who was not stereotypically gendered. Nevertheless, it seemed that her being female had played a destructive role in how she was perceived and treated by her early mathematics teachers and in the underdevelopment of her mathematics self. As Chodorow noted above (1978), girls predominantly mothered by women have more latitude in the expression of their gender than boys who were predominantly mothered by women. Thus, Karen's more androgynous style may have been a factor in her persistence through the course this time, albeit with healing and strategic help. She, unlike Kelly, did not resign to a passive, dependent feminine approach but rather, appropriately chose "direct action," and that was crucial to her success in the class.

While mathematically underprepared students were the most identifiably impacted by gender stereotypes, those with undermined mathematics selves were also negatively influenced, despite adequate mathematical preparation. While all had developed more resiliency than the mathematically underprepared students—they had experienced achieving mathematically and being recognized in that achievement—their success in this class was uncertain. All of these students at one point or another experienced poor advising, improper math placement, or inadequate strategic support. The influence of my and my students' gender on our interactions indicated that their female teacher attachments had been affected by masculine and feminine stereotyping of mathematics. While not needing the defense of grandiosity that the underprepared males used, the males nevertheless found it difficult to see value in, and receive help from me, perhaps in part because I am female. The women needed a new mathematically competent female teacher-tutor who believed in and mirrored their growing mathematical competence to replace past female teacher-mothers who had not mirrored their mathematical competence.

THE STRUGGLE

Walkerdine (1998), speaking in particular of accepted analyses of the plight of girls in mathematics, notes:

> Such concepts as stereotype, role, and conditioning suggest that girls [and boys] are molded by mothers and teachers. They are oppressively women-blaming and do not account for the positioning of the women themselves. We prefer to see femininity [and masculinity] as a site of struggle, where socialization does not easily work and where abnormality is always breaking through. (p. 163)

It is easy to blame mothers and former teachers for my participants' troubles. However, as a mathematics tutor, teacher, or counselor, even I with my new insight, found that it is hard not to believe Karen when she claimed she was hopeless at mathematics and not worth spending time with. It is difficult not to give Jamie a wide berth so as not to damage her further. It is hard to accept Mulder's and Rod's vulnerability and not react to their grandiosity, and almost impossible not to believe that Robina's and Kelly's "feminine" ditsiness does not mean that they are mathematically hopeless. Yet the teachers' (and parents') struggles remain: to recognize and accept students with their well-constructed defenses; to believe unequivocally in their ability to understand the mathematics and to find ways to help them let down their defenses. Then they can find and develop the mathematics prowess in themselves.

A crucial subtext to the struggle is the power differential between men and women in mathematics and academia. The men in my study, all of whom were needy, were in the uncomfortable position of relying on a female mathematics instructor and a female mathematics tutor-counselor in order to succeed. This possibly contributed to their resistance to the assistance offered. Mulder's changing his response to the statement "Mathematics is a subject men do better at than women" from neutral to strongly disagree signaled that he no longer needed the masculine defense for his vulnerable mathematics self, since he had discovered that he was capable of doing mathematics after all. Mitch made the same shift in opinion. All others in the class who had disagreed initially changed to strongly disagree, except for Karen who strongly disagreed from the beginning. What they experienced in class and in counseling seemed to have contributed to a change in their belief in the male–female mathematics hierarchy—for them, at least, perhaps there was no hierarchy after all.

THROUGH BLAME TO HEALING AND GROWTH

Gender played a part in the development of the particular mathematics self of each participant in this study and, in turn, contributed to the defenses we each developed to protect our particular mathematics selves. The new counseling approach of attending to transference and countertransference immediately gave me conscious access to a fund of analyzable and usable data, previously largely ignored in mathematics tutoring practice. I sought to make conscious our unconscious gendered motivations—mine and theirs—so that we could assess and adjust them in light of present constraints. In some cases I lived my countertransference—perpetuating previous harm done as in the case of "feminine" dependent Kelly and grandiose "masculine," yet vulnerable, Rod. But with the others my new attention to my own biases caught me. Robina's ditsy "feminine" classroom persona irritated and embarrassed me, but in counseling, I discovered and respected her competent, intelligent female "Belle" persona. My own gender identity was very prominent in these reactions, but relational psychotherapeutic insights provided the guidance and restraint I needed to accept students where they were and believe in, affirm, and mirror their growing mathematical prowess.

In typical tutoring situations, transference usually remains implicit as both student and tutor continue to act out old patterns of interaction without the conscious reflection that my new approach encourages. As with Jamie, the resolution of each student's central relational conflict required mathematical and relational repair.

So, for us—teachers, tutors, and parents—it is not too late for those in our charge. We must recognize and shed fixed mindsets about their and our mathematics abilities based on gender, and replace them with our beliefs in the continuing growth and development of the mathematics ability of all (Dweck, 2006), regardless of gender. The struggle to move from unexamined practice and blame to healing and growth is hard, and "success" is not guaranteed, but my study reveals that we can address our mathematical gender biases that have endangered our students' learning and our ability to help them succeed.

NOTES

1. College is a generic term in the United States for post-secondary educational institutions. It includes two-year junior and community colleges and four-year universities and colleges. The term does not apply to secondary educational institutions.

2. "[P]sychopathology, in its infinite variations, reflects our unconscious commitment to stasis, to embeddings in and deep loyalty to the familiar...we experience our lives as directional and linear, but like Penelope...we unconsciously counterbalance our efforts, complicate our intended goals; seek out and construct the very restraints and obstacles we struggle against" (Mitchell, 1988, p. 273).

3. Freud's view was that a person's choices were largely determined by unconscious instinctual drives and forces outside personal conscious control. Recognizing that individuals, in contrast, are responsible for their own choices and actions, implies that helping people to become conscious of their own hidden motives should provide both more insight into puzzling behaviors and also the possibility of modifying hidden motives in light of conscious goals. The consistent relationship between academic/mathematics achievement and locus of control (see McLeod, 1992; Nolting, 1990) was pertinent here. Students who failed to see and take responsibility that was realistically theirs in achieving success in a course, holding others or external factors responsible instead, consistently achieved less well than those who owned and exercised that responsibility (internal locus of control).

4. Usually seen as the father, although the role rather than the gender is the central factor.

5. A mathematics student's natural early grandiosity that has not been appropriately modified into a realistic assessment of self becomes unhealthy grandiosity, usually masking a sense of worthlessness, rooted in this developmental fail-

ure. A grandiose person seeks an audience to admire one's self-proclaimed inflated sense of importance, competence, potential, or success (Kohut, 1977; Stern, 1989).

6. Sokolowski's (1997) adaptation of Brown, Hart, and Kuchemann's (1985) Chelsea Diagnostic Algebra Test

7. According to the United Nations Statistics Division, in 2005, 89% of elementary teachers in the U.S. were women. http://unstats.un.org/unsd/demographic/products/indwm/tab4e.htm. Retrieved March 8, 2008.

8. What Ma (1999) calls "a profound understanding of fundamental mathematics" (p. 120).

REFERENCES

Ainsworth, M. D. (1989). Attachments beyond infancy. *American Psychologist, 44*(4), 709–716.

Boaler, J. (1997). *Experiencing school mathematics: Teaching styles, sex and setting.* Buckingham, UK: Open University Press.

Boublil, A. & Schönberg, C-M. (1980). *Les misérables, the musical.* Lyrics by Kretzmer, H. Directed by R. Hossein. Paris. Based on Hugo, V. (1862) *Les misérables.* Belgium: Lacroix and Verboeckhoven.

Bowlby, J. (1973). *Separation: Anxiety and anger* (2nd ed. Vol. 2: Attachment and loss). New York: Basic Books.

Bowlby, J. (1980). *Loss: Sadness and depression* (2nd ed. Vol. 3: Attachment and loss). New York: Basic Books.

Bowlby, J. (1982). *Attachment* (2nd ed. Vol. 1: Attachment and loss). New York: Basic Books.

Brown, T. (Ed.) (2008). *The psychology of mathematics education: A psychoanalytic displacement.* Rotterdam: Sense Publishers.

Brown, M., Hart, K., & Kuchemann, D. (1985). *Chelsea Diagnostic Mathematics Tests and teacher's guide.* Windsor, UK: NFER-NELSON Publishing Company Ltd.

Buxton, L. (1991). *Math panic.* Portsmouth, NH: Heinemann. (Original work published in 1981)

Carper, A. (2001). *Hysteria and narcissism: Two concrete representations of the experience of gender.* Unpublished Doctoral Dissertation, Adelphi University, Garden City, NY.

Carter, C. (1993). *The X-Files (USA).* 20th Century Fox Television.

Cheng, C. (2005) Processes underlying gender-role flexibility: Do androgynous individuals know more or know how to cope? *Journal of Personality 77*(3), 644–673.

Chodorow, N. (1978). *The reproduction of mothering: Psychoanalysis and the sociology of gender.* Berkeley: University of California Press.

Chodorow, N. (1999). *The power of feelings: Personal meaning in psychoanalysis, gender, and culture.* New Haven, CT: Yale University Press.

Dweck, C. S. (2000). *Self-theories: Their role in motivation, personality, and development.* Philadelphia: Psychology Press,

Dweck, C. S. (2006). *Mindset: The new psychology of success.* New York: Random House.

Fairbairn, W. R. D. (1952). *An object-relations theory of the personality.* New York: Basic Books.

Fast, I. (1984). *Gender identity: A differentiation model.* Mahwah, NJ: Lawrence Erlbaum.

Frankl, V. (1963). *Man's search for meaning, an introduction to logotherapy* (I. Lasch, Trans., rev. ed.). New York: Washington Square Press.

Ferguson, R. D. (1986). Abstraction anxiety: A factor of mathematical anxiety. *Journal for Research in Mathematics Education, 17*(2), 145–150.

Ferguson, R.D. (1998). *Mathematics anxiety: The Phobus Mathematics Anxiety Inventory.* Retrieved March 1, 2004 from http://www.accd.edu/sac/math/FACULTY/RFerguso/MathAnx.htm

Fisher, K., Egerton, M., Gershuny, J. I., & Robinson, J.P. (2006). Gender convergence in the American heritage time use study (AHTUS). *Social Indicators Research.* DOI 10. 1007/s11205-006-9017-y

Herbert, J. & Stipek, B. (2005). The emergence of gender differences in children's perceptions of their academic competence. *Journal of Applied Developmental Psychology, 26*(3), 276–295.

Johnson, C., Funk, S. J., & Clay-Warner, J. (1998). Organizational contexts and conversation patterns. *Social Psychology Quarterly, 16*(4), 361–371.

Knowles, J. M. (2004). Brief relational mathematics counseling as an approach to mathematics academic support of students taking introductory courses. Unpublished doctoral dissertation, Lesley University, Cambridge, MA.

Knowles, J. M. (2008). Gender in mathematics relationality: Counseling underprepared college students. *ZDM: The International Journal on Mathematics Education, 40*(4), 673–692.

Kohut, H. (1977). *The restoration of the self.* New York: International Universities Inc.

Leedy, M. G., LaLonde, D., & Runk, K. (2003). Gender equity in mathematics: Beliefs of students, parents, and teachers. *School Science and Mathematics, 103*(6), 285–292.

Luborsky, L., & Luborsky, E. (1995). The era of measures of transference: The CCRT and other measures. In T. Shapiro & R. N. Emde (Eds.), *Research in psychoanalysis: Process, development, outcome.* Madison, CT: International Universities Press, Inc.

Ma, L. (1999). *Knowing and teaching elementary mathematics.* Mahwah, NJ: Lawrence Erlbaum.

McLeod, D. B. (1992). Research on affect in mathematics education: A reconceptualization. In D. A.Grouws (Ed.), *Handbook of research on mathematics teaching and learning* (pp. 575–596). New York: Macmillan.

McLeod, D. B. (1997). Research on affect and mathematics learning in the JRME: 1970 to the present. *Journal for Research in Mathematics Education, 25*(6), 637–647.

Mitchell, S. A. (1988). *Relational concepts in psychoanalysis: An integration.* Cambridge, MA: Harvard University Press.

Mitchell, S. A. (2000). *Relationality: From attachment to intersubjectivity* (Vol. 20). Hillsdale, NJ: The Analytic Press.

Mitchell, S. A., & Black, M. J. (1995). *Freud and beyond: A history of modern psychoanalytic thought.* New York: Basic Books.

Mosak, H. H. (1995). Adlerian psychotherapy. In R. J. Corsini & D. Wedding (Eds.), *Current psychotherapies* (5th ed., pp. 51–94). Itasca, IL: F. E. Peacock Publishers.

Nolting, P. (1990). *The effects of counseling and study skills training on mathematics academic achievement.* Pompano Beach, FL: Academic Success Press.

Palomares, N. A. (2008, May) *Women are sort of more tentative than men, aren't they? How men and women use tentative language differently, similarly, and counter-stereotypically as a function of gender salience.* Paper presented at the International Communication Association Annual Conference, Montreal, Canada. Retrieved June 23, 2009 from http://web.ebscohost.com/ehost/pdf?vid=8&hid=103&sid=eb3a020c-ab12-4e5d-8b25-4135f41a9f90%40sessionmgr3

Richardson, F. C., & Suinn, R. M. (1972). The Mathematics Anxiety Rating Scale: Psychometric data. *Journal of Counseling Psychology, 19*, 551–554.

Rounds, J. B. J., & Hendel, D. D. (1980). Measurement and dimensionality of mathematics anxiety. *Journal of Counseling Psychology, 27*(2), 138–149.

Sadker, M., & Sadker, D. (1994). *Failing at fairness: How America's schools cheat girls.* New York: Charles Scribner's Sons.

Sadker, M., & Sadker, D. (2002). The miseducation of boys. In S. M. Bailey (Ed.), *Jossey-Bass reader on gender in education* (pp. 182–203). San Francisco: Jossey-Bass Publishers.

Skemp, R. R. (1987). *The psychology of learning mathematics* (Expanded American ed.). Hillsdale, NJ: Lawrence Erlbaum Associates, Inc.

Sokolowski, C. P. (1997). *An investigation of college students' conceptions of variable in linear inequality.* Unpublished Doctoral Dissertation, Boston University, Boston, MA.

Stern, E. M. (1989). *Psychotherapy and the grandiose patient.* New York: Hawthorn Press.

Sullivan, O., & Coltrane, S. (2008, April). *Men's changing contribution to housework and child care.* Paper presented at 11th Annual Conference of the Council on Contemporary Families. University of Illinois, Chicago. Retrieved March 8, 2008 from http://www.contemporaryfamilies.org/subtemplate.ph[p?t=briefingPapers&ext=menshousew.

Tahta, D. (1993). Victoire sur les maths. *For the Learning of Mathematics, 13*(1), 47–48.

Thorne, B. (2002). Do boys and girls have different cultures? In S. M. Bailey (Ed.), *Jossey-Bass reader on gender in education* (pp. 125–150). San Francisco: Jossey-Bass Publishers.

Tobias, S. (1993). *Overcoming math anxiety* (Rev. ed.). New York: W. W. Norton & Company.

Walkerdine, V. (1998). *Counting girls out: Girls and mathematics* (rev. ed.). London: Falmer Press.

Weyl-Kailey, L. (1985). *Victoire sur les maths.* Paris: Robert Laffont.

Zimmerman, B. J., & Martinez-Pons, M. (1990). Student differences in self-regulated learning: Relating grade, sex, and giftedness to self-efficacy and strategy use. *Journal of Educational Psychology, 82*, 51–59.

CONTRIBUTORS

Joanne Rossi Becker is a Professor of Mathematics at San José State University in the U.S. Her research interests include gender and mathematics, teacher professional development, and generalization in algebra. She currently teaches content and methods courses for prospective teachers and directs several professional development projects.
E-mail: becker@math.sjsu.edu

Sarah Burke Berenson is the Yopp Distinguished Professor of Mathematics Education at the University of North Carolina at Greensboro. Her research focuses on the development of high-achieving young women in mathematics as well as the teaching and learning of algebra. The majority of her teaching is done at the doctoral level.
E-mail: sarah_berenson@uncg.edu

Sigrid Blömeke is Professor for Instructional Research at Humboldt University of Berlin. Her research interests include teaching and teacher education, international comparative studies, and teaching and learning with ICT. She is especially interested in the measurement of teacher competencies.
E-mail: sigrid.blomeke@staff.hu-berlin.de

Susan Bracken is an Assistant Professor in Adult and Higher Education. Her research is primarily qualitative and her interests fall under two umbrellas: adult learning in organizational contexts and feminist/critical theory studies.
E-mail: susan_bracken@ncsu.edu

International Perspectives on Gender and Mathematics Education, pages 451–457
Copyright © 2010 by Information Age Publishing
451

Kimberly Dadisman is the Associate Director of Intervention Research at the Center for Developmental Science at the University of North Carolina–Chapel Hill. She also is a Researcher with Chapin Hall at the University of Chicago. She received her PhD from the University of Wisconsin–Madison in Educational Psychology. Her research includes the identification of community and school-based afterschool and extracurricular activity programs and the evaluation of their impacts on indicators of rural and urban youth development including school engagement and educational attainment.
E-mail: dadisman@email.unc.edu

Helen Forgasz is an Associate Professor and Associate Dean in the Faculty of Education at the Clayton campus of Monash University. Helen's research and teaching interests include the interactions of equity, attitudes and beliefs, mathematics learning with technology, and grouping practices. She has an extensive publication record in these fields.
E-mail: Helen.Forgasz@education.monash.edu.au

Ansie Harding is an Associate Professor in the Department of Mathematics and Applied Mathematics at the University of Pretoria. Her research focuses on mathematics education at tertiary level and finds pleasure in thinking creatively and innovatively about teaching and applying mathematics.
E-mail: aharding@up.ac.za

Jennifer Hall is a doctoral candidate and part-time Professor in Mathematics Education at the University of Ottawa, Canada. Her research interests include gender issues in mathematics and the impact of societal views of mathematics. Her teaching interests focus on the education of teacher-candidates from primary to secondary mathematics.
E-mail: jhall061@uottawa.ca

Gabriele Kaiser is Professor for Mathematics Education at the University of Hamburg. Her research interests include the teaching and learning of mathematical modeling and applications in school, international comparative studies, gender and cultural aspects in mathematics education and empirical research on teacher education. She works at the Faculty of Education at the University of Hamburg.
E-mail: gabriele.kaiser@uni-hamburg.de

Jillian Knowles, as Assistant Professor of Mathematics at Northern Essex Community College, teaches and also consults with their new upper-level math support lab. She has taught and tutored undergraduate mathematics and mathematics education a number of colleges and Universities in the Northeast United States. Her research focuses on development and analysis of on-

going assessments of affective, relational, and cognitive components of students' mathematics learning for effective tutoring and classroom teaching.
E-mail: jknowles@necc.mass.edu

Amanda Lambertus is a PhD candidate in mathematics education at North Carolina State University. Her research interests are in graduate mathematics departments that support students in inviting and positive environments as well as the development of high-achieving young women.
E-mail: ajlamber@ncsu.edu.

Thuy T. La, PhD, is Project Manager of Formative Assessment in a Networked Classroom in the Curriculum Research & Development Group, University of Hawai'i, Honolulu, Hawai'i. Research interests include effects of gender on mathematical learning and career motivation, classroom formative assessment, and educational policy, especially related to the Asia Pacific region.
E-mail: thuyla@hawaii.edu.

Eun-Jung Lee works as a researcher in the WISE centre, Korea National University of Education. Her research and teaching interests include gender differences in mathematical achievement, parental support, and social support.
E-mail: dkrlvn@lycos.co.kr

HeiSook Lee is the Dean of College of Natural Science, Ewha Woman's University. She works as professor at the Department of Mathematics and director of WISE National Center. Her research and teaching interests embrace mathematics education, and particularly promoting female students to pursue their careers in mathematics and science fields.
E-mail: hsllee@ewha.ac.kr

Kyeong-Hwa Lee is Associate Professor, Seoul National University. Her research and teaching interests embrace mathematics education, and particularly gender issues, gifted education, and analogical reasoning.
E-mail: khmath@snu.ac.kr

Gilah Leder is Professor Emerita, La Trobe University and Adjunct Professor, Monash University. Her research and teaching interests embrace mathematics education, gender issues, affect, exceptionality—particularly high achievement—and the interaction between learning and assessment. She has published widely in each of these areas.
E-mail: Gilah.leder@education.monash.edu.au

Maria-Dolores Lozano works as a teacher of mathematics in the Mathematics Department at Instituto Tecnológico Autónomo de México, a private university in Mexico City (ITAM). Her research interests focus on the teaching and learning of mathematics at different school levels using an embodied cognition perspective. She is currently teaching linear algebra.
E-mail: lolis.lozano@gmail.com

Michelle Muchatuta is a Senior Research Assistant at Macquarie University, Australia. Michelle's research interest is in Equity and Social Inclusion in Higher Education. As a solicitor, her teaching interests are Corporations, Business Law and Jurisprudence. Michelle will begin a PhD in the near future.
E-mail: michelle.muchatuta@vc.mq.edu.au

Claire H. Okazaki, MEd, is a Junior Specialist in the Curriculum Research & Development Group at the University of Hawai'i, Honolulu, HI, United States. She explores the role gender plays in students' understanding of mathematics, investigates mathematical ideas of children, and is a teacher researcher/developer of Measure Up, a mathematics curriculum for grades 1–5.
E-mail: cokazaki@hawaii.edu

Judith Olson, EdD, is a Mathematics Researcher in the Curriculum Research & Development Group, University of Hawai'i, Honolulu, HI, United States. Her interests include the interaction of technology related to gender issues, the influence of gender on mathematics learning, the role of technology in classroom formative assessment, and addressing these issues in professional development.
E-mail: jkolson@hawaii.edu

Melfried Olson, EdD, is Assistant Professor of Mathematics Education in the Curriculum Research & Development Group at the University of Hawai'i, Honolulu, HI, USA. His interests include the role gender plays in mathematical learning and understanding, problem solving and reasoning of children, classroom formative assessment, and addressing these issues in professional development.
E-mail: melfried@hawaii.edu

Carolina Rubí Real Ortega works as secondary school mathematics teacher in a public school in Mexico City. She collaborates as external supporting researcher in the Department of Mathematics Education. She has collaborated in studies concerning students' attitudes towards mathematics and mathematics teachers' professional development. Her research interests are gender and algebraic thinking, difficulties in mathematics' learning

and teaching at elementary level, and relations and properties of whole numbers. Her teaching interests include secondary school mathematics and teachers' training.
E-mail: crealortega1983@yahoo.com.mx

Pamela Paek, PhD, is a Senior Associate with the National Center for the Improvement of Educational Assessment, working with educational agencies in the United States to design and implement effective assessment and accountability systems. Her interests are in secondary mathematics, assessment policy, design, and implementation issues.
E-mail: ppaek@nciea.org

Seoung-Hey Paik is Professor, Korea National University of Education. Her research and teaching interests include science curriculum development and development potential of female students in science through out-of-school activity.
E-mail: shpaik@knue.ac.kr

Teri Perl is a PhD in Mathematics Education from Stanford University. She is a Co-founder of *The Learning Company*, prominent early educational software company and a founding member, recent President, *Expanding Your Horizons Network* devoted to encouraging young women toward math-related careers. She is also an author of three books involving women and mathematics: *Math Equals* (1978), *Women & Numbers* (1993), and *Notable Women in Mathematics*, co-edited with Charlene Morrow (1998).
E-mail: teriped@aol.com

Martha Patricia Ramírez works as teacher trainer and primary school teacher in a public school in Mexico City. Presently she is developing her PhD research at the Department of Mathematics Education at the Center for Research and Advanced Studies of the National Polytechnic Institute (Cinvestav-IPN). Her research interests are gender and mathematics, classroom interactions in the mathematics class, and classroom intervention to promote equity. Her teaching interests include primary school mathematics and teachers' training.
E-mail: pramirez222@yahoo.com.

Dylan Robertson is a Research Analyst for the Chicago Public School where he is investigating the long-term impacts of early educational interventions on children's developmental outcomes, examining the impact of various school policies on children's educational attainment.
E-mail: dlrobertson2@cps.k12.il.us

Claudia Rodríguez works as primary school teacher in a public school in Mexico City. Presently she is developing her PhD research at the Department of Mathematics Education at the Center for Research and Advanced Studies of the National Polytechnic Institute (Cinvestav-IPN). Her research interests are gender and mathematics, social representations and mathematics, and the use of technology to support mathematics teaching and learning. Her teaching interests include primary school mathematics supported by technology.
E-mail: claurom65@yahoo.com

Kristjana Steinthorsdottir is a teacher at Krikaskoli in Mosfellsbaer, Iceland. Her interests include the role of the teachers in creating mathematically rich learning environment in the playschool to develop young children's, age 3 to 6 year old, understanding of numbers.
E-mail: Kristjanas@simnet.is

Olof Bjorg Steinthorsdottir is an Assistant Professor at University of North Carolina in Chapel Hill. Olof's scholarly interests include equity, gender and mathematics, and the teaching and learning of mathematics among students in pre-kindergarten through 5th grade, specifically how teachers can use students understanding to make instructional decisions.
E-mail: steintho@email.unc.edu

María Trigueros works as Professor of Mathematics at the Mathematics Department at Instituto Tecnológico Autónomo de México (ITAM), a private university in Mexico City. Her research interests are teaching and learning of abstract mathematics concepts at university level, learning of elementary algebra through the concept of variable, the use of technology in the classroom, and gender and mathematics education. Her teaching interests include, dynamical systems, calculus, and linear algebra. She uses technology and modeling to support her teaching.
E-mail: trigue@itam.mx

Sonia Ursini works as researcher at the Department of Mathematics Education at the Center for Research and Advanced Studies of the National Polytechnic Institute (Cinvestav-IPN). Her research interests include difficulties in the teaching and learning of elementary algebra at different school levels, in particular, concerning the concept of variable; the use of technology to support mathematics teaching and learning; gender and mathematics; and social representations and mathematics. Her teaching interests include problems concerning the teaching and learning of elementary algebra; the use of technology for mathematics teaching; historical development of mathematical thinking—in particular, non-euclidean geometry.
E-mail: soniaul2002@yahoo.com.mx

Colleen Vale is an Associate Professor in the School of Education at Victoria University, Australia. Her research interests include digital technologies and equity and social justice in mathematics education. She has publications in a diverse range of fields concerning higher education including pre-service teacher education and professional learning of mathematics teachers.
E-mail: colleen.vale@vu.edu.au

Fiona Walls is a Senior Lecturer in mathematics education at James Cook University, Australia. Her research focuses on the social and political processes of education. Her teaching aims to build pedagogical approaches founded on mathematics as culturally, socially and historically constructed, and which embrace the principles of social justice.
E-mail: Fiona.Walls@jcu.edu.au

Lynda R. Wiest is a professor of mathematics education at the University of Nevada, Reno. She attained a PhD. at Indiana University in Bloomington, Indiana, in 1996 after teaching elementary and middle school students for eleven years in Pennsylvania. Her main scholarly interests are mathematics education, educational equity, and teacher education. She is Director of the Northern Nevada Girls Math & Technology Program, which she founded in 1998, and she is a member of the advisory board for the U.S. organization Women and Mathematics Education.
E-mail: wiest@unr.edu.

Leigh Wood is the Associate Dean, Learning and Teaching in the Faculty of Business and Economics at Macquarie University, Australia. Her research areas include curriculum design, assessment, and graduate capabilities. As Chair of the Education Standing Committee of the Australian Mathematical Society, she is implementing an Australia-wide professional development program for mathematics teachers at university level.
E-mail: leigh.wood@mq.edu.au

Xin Ma is Professor of Mathematics Education and Education Statistics at the University of Kentucky in the United States. He is a Fellow of the (U.S.) National Academy of Education. His main areas of research and teaching include mathematics education and education statistics.
E-mail: xin.ma@uky.edu

Lightning Source UK Ltd.
Milton Keynes UK
UKOW04f1117300714

236039UK00001B/32/P